COMPREHENSIVE BIOCHEMISTRY

COMPREHENSIVE BIOCHEMISTRY

SECTION I (VOLUMES 1-4)
PHYSICO-CHEMICAL AND ORGANIC ASPECTS OF BIOCHEMISTRY

SECTION II (VOLUMES 5-11)
CHEMISTRY OF BIOLOGICAL COMPOUNDS

SECTION III (VOLUMES 12-16)
BIOCHEMICAL REACTION MECHANISMS

SECTION IV (VOLUMES 17-21)
METABOLISM

SECTION V (VOLUMES 22-29)
CHEMICAL BIOLOGY

SECTION VI (VOLUMES 30-39)
A HISTORY OF BIOCHEMISTRY

COMPREHENSIVE BIOCHEMISTRY

Series Editors:
ALBERT NEUBERGER
*Chairman of Governing Body, The Lister Institute
of Preventive Medicine, University of London, London (U.K.)*

LAURENS L.M. VAN DEENEN
*Professor of Biochemistry, Biochemical Laboratory,
Utrecht (The Netherlands)*

VOLUME 38

SELECTED TOPICS IN THE HISTORY
OF BIOCHEMISTRY
PERSONAL RECOLLECTIONS. IV.

Volume Editors:
E.C. SLATER
*Department of Biochemistry, University of Southampton, Southampton
(United Kingdom)*

RAINER JAENICKE
*Institut für Biophysik und Physikalische Biochemie,
Universität Regensburg, D-93040 Regensburg (Germany)*

GIORGIO SEMENZA
Laboratorium für Biochemie, ETH-Zentrum, CH-8092 Zürich (Switzerland)

ELSEVIER
AMSTERDAM·LAUSANNE·NEW YORK·OXFORD·SHANNON·TOKYO
1995

Elsevier Science B.V.
P.O. Box 211
1000 AE Amsterdam
The Netherlands

ISBN 0 444 81942-8 (Volume)
ISBN 0 444 80151-0 (Series)

This book is printed on acid-free paper

PREFACE TO VOLUME 38

As in previous volumes (Vols. 35–37) in the Comprehensive Biochemistry series encompassing Selected Topics in the History of Biochemistry, the chapters in this volume complement The History of Biochemistry in Vols. 30–33 by M. Florkin and Vol. 34A by P. Laszlo.

The aim of the editors was to invite selected authors who had participated in or observed the explosive development of biochemistry and molecular biology particularly in the second half of this century to record their personal recollections of the times and circumstances in which they did their work. The authors were given a completely free rein with respect to both content and style and the editors have made no attempt to impose any sort of uniformity in the chapters. Each reflects the flavour of the personality of the author.

This series of personal recollections was started some time ago by one of us (G.S.) who was struck by the fact that the explosive development of the biochemistry and molecular biology and related sciences had led to the almost unique situation that these fields had become of age at a time when their founding fathers, or their scientific children, were alive and well. In the intervening years, time has taken its toll and sadly many leading figures have died in the last few years. In this volume, two posthumous accounts are included, the first by Feodor Lynen written a few years before his death in 1979, the second by Claude Rimington, who died shortly after his 90th birthday and after delivering the manuscript for this volume.

The contributors to this volume encompass a wide variety of experiences in many different countries and in very different fields of biochemistry. Some have worked close to the laboratory bench throughout their scientific life and are continuing to do so. Others have been closely engaged in organisational matters, both nationally and internationally. All mention incidents in their own career or have observed those in others that will be of interest to future historians who will record and assess the period in which our contributors have lived and worked. It was an extremely exciting time for the life sciences. It was also a period of major and often tragic historical events that deeply affected the life and work of the generation to which our contributors belong.

The editors wish to express their gratitude to all those who made this series possible, especially the authors.

University of Southampton
Southampton, 1994   *E.C. Slater*

University of Regensburg
Regensburg, 1994   *R. Jaenicke*

Swiss Institute of Technology
Zürich, 1994   *G. Semenza*

CONTRIBUTORS TO THIS VOLUME

A.A. BAYEV
*Engelhardt Institute of Molecular Biology, Russian Academy of Sciences,
32, Vavilov Street, Moscow 117312 (Russia)*

H. BEINERT
*Department of Biochemistry and Biophysics Research Institute,
Medical College of Wisconsin, 8701 Watertown Plank Road, Milwaukee,
WI 53226 (U.S.A.)*

P.N. CAMPBELL
*Department of Biochemistry and Molecular Biology, University College
London, Gower Street, London WC1E 6BT (United Kingdom)*

F. GIBSON
*The John Curtin School of Medical Research, Australian National
University, G.P.O. Box 334, Canberra City, A.C.T. 2601 (Australia)*

E.J.M. HELMREICH
*Medical Clinic, Division of Clinical Biochemistry and Pathobiochemistry,
The University of Würzburg School of Medicine, Versbacherstrasse 5,
Würzburg (Germany)*

F. LYNEN[†]
Max-Planck-Institut für Biochemie, Munich (Germany)
[†]*deceased, August, 1979*

C. RIMINGTON[†]
The Norwegian Radium Hospital, Oslo (Norway)
[†]*deceased, August 8, 1993*

R.N. ROBERTSON
P.O. Box 9, Binalong, NSW 2584 (Australia)

T. WIELAND
*Max-Planck-Institut für medizinische Forschung,
Jahnstrasse 29, 69115-126 Heidelberg (Germany)*

B. WITKOP
*Laboratory of Cell Biology and Genetics, NIDDK, National Institutes of
Health, Building 8, Room 403, Bethesda, MD 20892 (U.S.A.)*

Laurens van Deenen †

In Memoriam

To our great sorrow we have to report the death of Professor Laurens van Deenen, on September 4, 1994.

Laurens van Deenen was an outstanding scientist who made great contributions to our knowledge of the structure and function of lipids of biological importance, and their interaction with other components of the cell. He also increased our understanding of the nature of membranes in biological systems and the relevant fields of enzymology. He was a pioneer full of original ideas, and carried out his work with the best available methods, always being cautious in the interpretation of his results.

Laurens van Deenen was the creator of a school which was internationally recognized and as a person was generous to his co-workers. He had wide interests outside his own discipline.

He joined the Editorial Board of *Biochimica et Biophysica Acta,* and was one of the Managing Editors from 1964–1993 and Chairman during the years 1983–1989.

From 1977 onwards, Laurens van Deenen was also involved in the editorship of the original *Comprehensive Biochemistry* series, and was one of the initiators in developing a second series, *New Comprehensive Biochemistry*, being responsible for the realization of this venture. Here his wide knowledge of biochemistry, stretching far beyond his own fields, his wisdom and his judgement were great assets.

Throughout his life, Laurens van Deenen set high standards for his own work; he was a stimulating colleague and a good friend.

London, October 1994   *Albert Neuberger*

CONTENTS

VOLUME 38

A HISTORY OF BIOCHEMISTRY

Selected Topics in the History of Biochemistry
Personal Recollections. IV.

Chapter 4. Recollections: Vacillation of a Classical Biochemist
by E.J.M. HELMREICH (Ed.: R.J.) 163

Chapter 5. These are the Moments when we Live! From Thunberg Tubes and Manometry to Phone, Fax and Fedex
by H. BEINERT (Ed.: E.C.S.) 193

xiv

Chapter 6. Chorismic Acid and Beyond
by F. GIBSON (Ed.: E.C.S.) 259

Chapter 7. Charge Separation: A Personal Involvement in a Fundamental Biological Process
by R.N. ROBERTSON (Ed.: E.C.S.) 303

E.C. Slater, R. Jaenicke and G. Semenza (Eds.)
Selected Topics in the History of Biochemistry: Personal Recollections, IV
(Comprehensive Biochemistry Vol. 38) © 1995 Elsevier Science B.V.

Chapter 1

Life, Luck and Logic in Biochemical Research*

FEODOR LYNEN †

Max-Planck Institut für Zellchemie, Munich (Germany)

The invitation to present the first Mason Lecture, sponsored by the Howard Hughes Medical Institute, is a signal honor, and I should like to express my deep gratitude for the opportunity. A Mason Lecture offers both an occasion and a stimulus to look back on one's own career and to place one's experimental work into the framework of modern biochemistry. Doing this, I realized the importance of the surroundings in which I grew up and which eventually led me into science. I also realized that frequently, just by accident, personal life and scientific work are directed. This is very true in my case.

My father's and mother's families lived for many generations in the Rhine area, not too far away from Holland and Belgium. Our family name, Lynen, very unusual in Germany, probably originated in these countries. I would certainly have become a Rhinelander if my father had not chosen engineering as a profession and accepted an offer from the Technische

---

† Prof. Lynen suddenly passed away in August 1979, a short time after having accepted the invitation to write an autobiographic chapter for this series. We are grateful to his widow, Eva Lynen, and to the Chicago University Press for allowing us to reproduce Prof. Lynen's Mason Lecture, which was published in 'Perspectives in Biology and Medicine' Vol. 12, 1969, pp. 204–218.

Hochschule in Munich. I was born there in Schwabing, which is, with its artistic atmosphere, the Greenwich Village of Munich. I was the sixth child, and two more were to be born after me. The fact that I grew up in Munich and in Bavaria with its mountains nearby and its pleasant social life is responsible for the steadfastness with which I stayed in this place.

As a little boy, I got in touch with chemistry through my brother Walter, twelve years older than I, who wanted to become a chemist and for that reason had installed a small chemical laboratory in the attic of our house. He chose me as his assistant, and I, of course, was very proud to be allowed around him in this advanturous enterprise. Walter was killed in World War I. Many years later, when I had chemistry as a subject in high school, I was attracted by the experiments performed by the teacher, Dr. Wolf. I tried to repeat them in the laboratory still existing in our house, but was not always very successful. One explosion ruined my best suit and hurt my face, and thus extinguished my passion for chemistry as well as my mother's sympathetic feeling toward my activity.

When I finished high school in 1930 with the German 'Abitur,' I had no definite ideas about my future profession. It could have been either medicine or chemistry. Only the shorter line at the registration desk of the Department of Chemistry determined me to choose the chemical subject. The Munich Chemical Institute was famous for its tradition. Following Justus von Liebig, its high reputation was built up by Adolf von Baeyer, Richard Willstätter, and Heinrich Wieland, all of them Nobel Prize winners. In contrast to many other chemical institutes in Germany, all of these famous organic chemists were already interested in biochemistry. The lectures of Heinrich Wieland, who was head of the Institute at my time, had already introduced the students to biochemical problems. Thus the ground was prepared.

Two events which were to be of utmost importance to my future life occurred within one year. On January 1933, Hitler came to power in Germany, and I think it is unnecessary to elaborate on the consequences of this. On the same day exactly one year before, I was ski racing in Kitzbühel, Austria, and hoped to win the 'Silberne Gams' (silver mountain goat). I did not win it, but I broke my leg and had to stay in the hospital for nearly one year. Although this accident was extremely sad for me at the time, because it stopped my career as a ski racer and made me lose one year in my studies, it turned out to be quite fortunate later. It saved me from any service in the Nazi organizations like the SA, escape from which was nearly impossible for a student at that time. Besides that, Theodor Wieland, the son of our 'Geheimrat,' whom I had supervised at the beginning of my studies, now became my colleague and very close friend. Through him and the Oktoberfest I got to know his sister Eva, to whom I was married in 1937. But before that I had to finish my studies for a Ph.D.

The problem I worked on was the toxic principle of the mushroom *Amanita phalloides*, commonly known as death cap, which is often mistaken for an edible mushroom and causes fatal accidents every year. I embarked on the purification of the toxin and was able to achieve a significant purification. Studying the chemical properties of the purified material, I realized its peptic nature. After acid hydrolysis, which led to complete loss of the toxicity, amino acids could be detected. I mention this work in some detail because in the course of these investigations I learned the enormous value of persistence in scientific endeavors. In the purification of a biologically active natural product, the goal for the chemist is always to get it in a crystalline state. Only in this way can he be sure that the product obtained is pure and that the properties found are not due to some minor contaminants of the main component. After two years of hard work on the purification of the toxic principle without reaching its crystallization, I got the feeling that there are organic molecules which cannot crystallize, however hard one tries. I was convinced that the toxins of *A. phalloides* would fall into this category and tried to persuade my teacher to the same idea. As I later found out, he did not share my pessimism. Although he let me finish my experimental thesis work, he continued to scratch, with glass rods, the walls of test tubes containing solutions of the purified toxin. Ironically, four weeks after I had finished my experimental work on this problem, Ulrich Wieland, a nephew of my teacher, continuing my thesis work, succeeded in crystalizing phalloidin, the first toxic compound ever crystallized from *A. phalloides*. When I heard this exciting news, I hurried back into the laboratory, dissolved my purified fractions – which I had kept – in a little solvent and when I scratched the wall of the test tube, you may believe me, nice crystals appeared. Nevertheless, I had missed one good chance because I was only second. By this early experience I learned persistence in research, which in my later scientific career was to become a very dominating factor. Students working with me have frequently suffered from this attitude, but when I look back now, I realize that it was mostly to their profit.

The oral examination connected with my Ph.D. thesis occurred in February, 1937. Three months later I married Eva Wieland and thus I became the son-in-law of my teacher. This very close relationship with Heinrich Wieland was to become another important determinant in my life. He was a man of very great personality in many respects, highly esteemed by all of his students. Besides his great scientific accomplishments, it was his broad scholarship and his brave humanity that impressed the young generation who came in close personal contact with him. It must be admitted that many German university professors in the years from 1933 to 1945 did not oppose Hitler's doctrines; on the contrary, they even supported him and his aggressive politics. But it must also be stated that some of them opposed Hitler's regime, and they served as shields against nazism for the young students working with them. Their institutes were real islands in which humanity prevailed and clear thinking was never suppressed. Wieland's institute fell into this category, and it was certainly to a great extent owing to his merit that many of his students survived this middle-age period without being harmed. Like many of his other students, I owe very much to Heinrich Wieland's influence. This I learned after the war in 1945, when Bavaria was occupied by American troops and the military government dismissed many scientists from our university. I was able to keep my position as lecturer.

In the meantime I had started work on a new biochemical problem. After I had received my Ph.D. degree, my father-in-law asked me to collaborate with him in cancer research. In collaboration with the professor of pathology in Munich, Max Borst, we embarked on studies of the oxidation reactions in tumor cells. It was known from the pioneering experiments of Otto Warburg that tumor cells differ from normal cells in their capacity to degrade glucose by glycolysis even under anaerobic conditions. But tumor cells also respire, and the question was how CO_2 is produced from carbohydrates. At that time it was already generally assumed that glucose was transformed into pyruvate by the glycolytic enzymes under aerobic conditions as well. Wieland concentrated on the idea that pyruvate is oxidized to form CO_2 and acetate, the latter then being further metabolized. In experiments with yeast cells Wieland and Sonderhoff had found that succinic acid and citric acid were formed from acetate. In order to explain their formation, it was assumed that two molecules of acetate (I) are dehydrogenated to form succinate (II) which, after being oxidized to oxaloacetate (III) by enzymes already known, can either lose CO_2 and form pyruvate (IV) or condense with a third molecule of acetate for form citrate (V) (Fig. 1).

Sonderhoff, who was then still in the Munich laboratory, fed deuterium-labeled acetate to yeast cells and could demonstrate that the citrate formed actually contained deuterium in the amount expected by this scheme. However, the deuterium analysis of the succinate gave unpredicted results. It was found that only 50 per cent of the carbon-bound hydrogen atoms were replaced by deuterium, instead of the 100 per

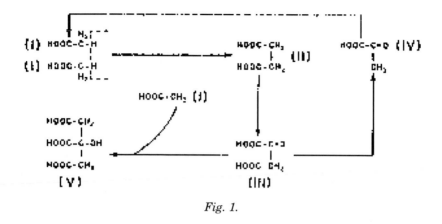

Fig. 1.

cent expected. This observation remained a puzzle for a few years.

I started work on the oxidative metabolism of tumor cells, using rat Jensen sarcoma, degraded in the meat chopper, and in agreement with our ideas could demonstrate the formation of acetate in small amounts.

But what was its fate? Fortunately, when I began to work on this problem, three fundamental discoveries were made. Albert Szent-Györgyi, working in Hungary, found the catalytic effect of C_4-dicarboxylic acids like succinate, fumarate, or malate on the respiration of homogenates of pigeon-breast muscle, but he interpreted these effects wrongly, as it later turned out. At about the same time Carl Martius in Tübingen discovered the new enzyme aconitase, and thus finally clarified the mechanism of citrate oxidation which had puzzled the biochemists for many years. Martius also demonstrated that, when a mixture of pyruvate and oxalacetate was oxidized with hydrogen peroxide in the test tube, citrate was formed. In a brilliant series of experiments Hans Krebs then integrated these various observations into the new concept of the citric acid cycle. Enzymatic citrate synthesis from pyruvate and oxaloacetate catalyzed by pigeon-breast muscle could be shown and was postulated to occur by an oxidative condensation, analogous to the one studied by Martius in his chemical experiments.

Krebs published the work on the citric acid cycle in 1937, the same year that I began to wonder about the mechanism of the biological oxidation of acetate. It occurred to me that the results of Wieland and Sonderhoff in their studies on yeast cells could be reinterpreted on the basis of this new metabolic cycle. When the sequence of events leading to the formation of succinic acid and citric acid was reversed, citric acid being synthesized from acetate before succinic acid, then the lower deuterium content of succinate in Sonderhoff's isotope experiments could easily be explained. Therefore, I was absolutely convinced that acetate, not pyruvate, was the partner of oxaloacetate in the enzymatic reaction of citrate synthesis, and I immediately embarked upon experiments with the two-fold purpose of (1) demonstrating the functioning of the citric acid cycle in yeast cells oxidizing acetate and (2) achieving the synthesis of citric acid in a cell-free system.

The first problem was solved within a few months. But all attempts to achieve the enzymatic synthesis of citric acid with cell-free yeast preparations were unsuccessful.

In one of the crucial experiments, acetate and oxaloacetate together were incubated with yeast cells, frozen in liquid air, and then thawed, in order to disrupt the plasma membrane otherwise impermeable to the highly charged dicarboxylic acid anion. I had learned to use this simple technique in experiments related to studies on alcoholic fermentation. These studies were very instructive for me because in the course of them I learned to read all the papers of Meyerhof, Embden, and Parnas, which have laid the foundation of the biochemical understanding of glycolysis. Furthermore, I became acquainted with the central role ATP plays in cellular metabolism, as an energy donor and also as a compound involved in regulatory processes. I am still proud of two papers that I published in 1941 and 1942. One contained a theory on the Pasteur effect; the other publication was a review article in *Naturwissenschaften* in which the role of phosphorylation reactions in cellular metabolism was discussed along lines similar to those of Lipmann in his classic paper in *Advances in Enzymology,* Volume I.

My studies of the Pasteur effect were to become very important for me in another respect. Through them I came in scientific and personal contact with Otto Warburg, who worked in the Kaiser Wilhelm Institut für Zellphysiologie in Berlin-Dahlem. This seemed rather surprising in view of the violent controversy ranging between Warburg and Wieland on the mechanism of biological oxidation, but I have always admired Otto Warburg as one of the greatest biochemists, who established one of the fundamentals of modern biochemistry. In that respect I regard the years between 1930 and 1945 as the most exciting for me as a young biochemist. In practically every volume of the *Biochemische Zeitschrift* a new sensational discovery was reported from the Dahlem Institute. Flavoproteins and pyridine nucleotides were found, their function in hydrogen transport was elucidated, and with it the puzzling problem of how vitamins act in metabolism solved. I still remember how excited I was on reading all these publications, and how I also wished that I might some day discover a coenzyme and elucidate its mode of action. I could not foresee that these wishes would eventually be fulfilled in our studies of biotin. Hugo Theorell, working with Warburg, crystallized the yellow enzyme from yeast and achieved for the first time in history the reversible splitting of an enzyme into protein and its functional group. In the following years, Warburg and his collaborators announced the crystallization of many glycolytic enzymes, among them glyceraldehyde-phosphate dehydrogenase, which was to have a special bearing on my own investigations, still centered in the mechanism of acetate oxidation. The crystallization of this dehydrogenase provided the first insight into another puzzling problem, that of the chemical mechanism of the coupling between oxidation and phosphorylation, that is, into the production of ATP. The still-missing link was the glyceric acid 1,3-diphosphate, found in Warburg's laboratory, which is produced in the dehydrogenation reaction and is used for the phosphorylation of ADP to form ATP.

But what was the connection with the acetate problem? Several observations, including the unsuccessful experiments on citrate synthesis with frozen yeast cells already mentioned, had led me to the idea that, instead of free acetate, some unknown derivative, called an 'active acetate,' could be involved in this synthesis. As it turned out, Fritz Lipmann had come to a similar conclusion. He had studied the oxidation of pyruvate by lactic acid bacteria and found that phosphate was required and acetyl phosphate was formed. This oxidation seemed to be analogous to the dehydrogenation of glyceraldehyde phosphate, and the acetyl phosphate that was formed seemed to be the 'active acetate' we were seeking.

In order to prove this hypothesis, I synthesized the crystalline silver salt of acetyl phosphate by a new, rather complicated method, but, again, the experiment to achieve the synthesis of citrate, using acetyl phosphate instead of acetate with our yeast system, was without success. Looking back at these experiments, I realize how unfortunate it was that I used yeast preparations and not preparations of *Escherichia coli*. It was found later that *E. coli* contains an enzyme system which can transform acetyl phosphate into the real 'active acetate.'

At this stage, which was reached in 1941, my work remained stagnant for many years. Hitler had attacked Russia, and the bomb attacks on our country were so intensified that research in our institute became more and more difficult. In December, 1944, the Chemical Institute in Munich was completely destroyed. Fortunately, a few months before, I had moved my laboratory equipment to a place outside Munich and found a temporary room in the Botanical Institute. It happened that at that time we again got access to foreign literature after many years of isolation. Only then did I become informed about the progress made in the meantime. 'Active acetate' was now recognized in many laboratories as occupying a central position, not only in carbohydrate catabolism but also in lipid metabolism. Lipmann had discovered coenzyme A, and there was experimental evidence that 'active acetate' was a derivative of this coenzyme. I was very eager to embark immediately on experiments with coenzyme A, but pigeon livers were required for the experiments, and at this time no pigeons could be found because they were being eaten and most of them were selling for high prices on the black market. Only after the revaluation of the German currency could I start on these experiments with the assistance of Ernestine Reichert. This led in 1950 to the identification of 'active acetate' with the thioester between acetate and coenzyme A.

I remember in every detail how this discovery came about. My brother-in-law, Theodor Wieland, was on holiday in his parents' house next to ours. He had worked in Richard Kuhn's laboratory on pantothenic acid, the vitamin discovered by Lipmann to be a constituent of coenzyme A. We spent a whole night in shoptalk about the possible link between acetate and pantothenic acid but could not come to any conclusion. On my short way back to our garden it suddenly came to my mind that the acetyl residue might be bound, not to pantothenic acid at all, but to sulfur. I recalled that in Lipmann's last paper on the composition of the purified coenzyme A preparations he had mentioned the presence of sulfur, but he did not pay much attention to it because his preparations were not yet pure. In addition, it was known that all enzymatic reactions studied in which coenzyme A was involved required the addition of glutathione or cysteine, presumably as agents for the binding of inhibitory heavy metals. Third, as a chemist I knew that sulfhydryl groups are more acidic than hydroxyl groups, which means that acetic acid bound to sulfur must have the properties of an acid anhydride and this should have the capacity to acetylate amines or alcohols. The crucial step for me was that I put the three things together. I became very excited, hurried into my study, and looked into *Beilstein*. Soon I found that thioacetic acid was known to react with aniline to form acetanilide. I thus became completely convinced that 'active acetate' must be a thioester. Ernestine Reichert was very surprised when I walked into the laboratory next morning and told her that we would now embark on the isolation of acetyl CoA from yeast Kochsaft. In two months' time this goal was reached, and by comparison with a synthetic thioester which was given to me by an organic chemist of the Munich laboratory, who happened to be working with such compounds, it could be proved that my prediction was correct.

This whole story was most exciting, but it was to become even more dramatic when my short publication was sent to *Angewandte Chemie* to be published in the December issue. Everything now seemed so simple to me that I could hardly believe that nobody else in the meantime could have had the same idea. Every day I expected a publication of this kind, and I was relieved only when I heard by letter from Otto Meyerhof and Carl Neuberg that my paper had come to the biochemists in the United States like a bombshell. I realized only later that most of the biochemists were prejudiced by the idea that active acetate could only be some kind of phosphate derivative. With the discovery of the thioester bond in acetyl CoA, the energy-rich phosphate bond had lost its unique role in metabolic reactions, and it was to be expected that still other energy-rich bonds would be discovered.

I mention this story in such detail because it illustrates how discoveries can happen. I had worked or thought continuously for thirteen years on the acetate problem and was relieved only when it was solved. As already pointed out, I have the philosophy that persistence is an essential element in science, which, however, should not exclude tackling more than one problem at a time. Also, a scientist wants to be happy, and if he does not progress with one problem, he might be more successful with others. In this respect I would like to mention that in the years during which I was concentrating on the acetate problem and achieving only little progress, my students and I, to our satisfaction, were more successful with other studies. During that period we also investigated the formation of ox-alate by *Aspergillus niger*, and we prepared crystalline oxalo-succinic acid, the dehydrogenation product of isocitric acid, and studied its enzymic decarboxylation to form α-ketoglu-taric acid. In our experiments on the Pasteur effect we became aware of the great importance of cellular compartmentation for metabolic regulations, and we developed an experimental technique to measure the efficiency of oxidative phosphoryla-tion in living cells.

The elucidation of the structure of acetyl CoA resulted in an invitation to attend a Gordon Conference in 1951. Everything was arranged, but, unfortunately, shortly before the confer-ence I again broke my leg in skiing and again had to spend several weeks in a cast. This provided me with the time neces-sary to think about new research projects. It was not at all difficult to find them. Once the structure of acetyl CoA was worked out, a detailed chemical scheme of fatty-acid oxidation could be written. I presented it in my first papers and called it the 'fatty acid cycle.' The next step then was to isolate and identify the four enzymatic components participating in this process. Here again my training as a chemist was a big help. The intermediates in the oxidative reaction are α, β-unsatu-rated-, β-hydroxy-, and β-ketocarboxylic acids bound to coen-zyme A. In order to learn something about their chemical properties, Luise Wessely and Werner Seubert synthesized simple models in which these carboxylic acids were bound, not to coenzyme A, but to N-acetyl cysteamine. It turned out that some of these thioesters possess characteristic ultraviolet-ab-sorption bands, and it seemed possible to use these properties in specific enzyme assays. When we started work on the fatty-acid oxidation enzymes, only acetyl CoA was available, and it had to be isolated from yeast Kochsaft. Thus the only reaction step which we could study in the beginning was the reversible condensation of two acetyl CoA's to form acetoacetyl CoA.

Luise Wessely tried very hard to demonstrate this reaction in experiments with liver extracts, using the absorption band of acetoacetyl CoA at 300 mμ. This band appears only at an alkaline pH because it is related to the formation of the enolate ion. For that reason she worked at pH 8. But even under these conditions, and though we spent much of our very valuable acetyl CoA from yeast, the absorption band did not appear. We were all very depressed, especially since I was supposed to go to the International Biochemistry Congress in Paris and would have liked to present new results. When I at last left the laboratory to go to Paris, in some kind of despair I turned to my other brother-in-law, Otto Wieland, who at that time worked with me, and he asked me what he could do during my absence. I said, 'I am sure it will not work, but why don't you see whether our model acetoacetyl-N-acetyl cysteamine can react with DPNH or TPNH in presence of liver extracts?' How great was my surprise and excitement when I heard after my return from Paris that this experiment had worked! By the use of this simple assay, the purification of β-hydroxyacyl CoA dehydrogenase, and soon also the demonstration of the enzyme thiolase, became possible. Once we knew that thiolase was rather active in liver extracts, Luise Wessely repeated the previous experiments, working now at pH 9 instead of pH 8. To our great astonishment the absorption band at 300 mμ now appeared, because at the more alkaline pH the thiolase reaction is shifted to the side of condensation. Thus I learned a new lesson for a scientist – be naïve and try an experiment, even if the chances for its success are very small. I have the feeling that many of my colleagues in biochemistry – and the same may be true also for other fields of science – often spend more time in discussing the pros and cons of an experiment than it would take to perform it. I think that a good many opportunities for new discoveries are missed that way.

I can give you another example of how naïveté and luck can help to solve a problem. It was during our studies on the enzymatic carboxylation of β-methyl crotonyl CoA to form β-methyl glutaconyl CoA, a reaction which, being reversible, was for many years believed to be involved in the biosynthesis of cholesterol and many other natural products. This carboxylation reaction required ATP as an energy source, and in that respect was similar to the enzymatic carboxylation of propionyl CoA studied at the same time in Ochoa's laboratory. We all realized that some kind of 'active CO_2' should be involved, and compounds like carboxyphosphate or carboxyadenylate were proposed. The situation reminded me of the pre-coenzyme A period in the search for the 'active acetate,' where all thinking also centered in some kind of phosphate derivative.

As I told you before, I had learned my lesson. Fortunately, I came upon another track, this one related to a review article on lipid metabolism I had written in 1954 for *Annual Reviews of Biochemistry*. During the preparation of this manuscript and reading all kinds of literature, I learned of the important research by Henry Lardy and his group showing that the vitamin biotin must be involved in the metabolism of propionate and isovalerate. This brought me to the idea that perhaps biotin is in some way involved in the formation of 'active CO_2.' This idea could be tested, and to our great satisfaction it turned out to be right. By a number of experiments it could be demonstrated that our methylcrotonyl CoA carboxylase, isolated from micro-organisms, contained protein-bound biotin. In the process of carboxylation, this biotin enzyme first takes up CO_2 and then delivers it to β-methyl crotonyl CoA, as could be proved by exchange experiments. But the important question remained, how is the carbonic acid bound to the biotin molecule? Having had good experience with substrate models in our studies of fatty-acid oxidation, I suggested to my collaborator, Joachim Knappe, that he should add free biotin to the enzyme to see whether it was carboxylated. He did not value this idea too highly and maybe felt that I was a bit crazy. Also, he was just at that time heavily engaged in the exchange experiments mentioned above. Luckily he ran out of radioactive phosphate and had to wait for another shipment. To fill the time, he threw some biotin into the reaction mixture used to assay the enzyme and found to his great surprise that it was carboxylated. By studying the chemistry of the CO_2-biotin thus formed, the mechanism of biotin action in enzymatic carboxylation reactions could be elucidated. I was really excited that my wish had finally come true. It later turned out that we had nearly unbelievable luck when we selected β-methyl crotonyl CoA carboxylase for these studies. In the meantime a number of other carboxylases and transcarboxylases with biotin as the functional group were isolated and studied, but not one of them could carboxylate free biotin.

What was our reason for especially selecting β-methyl crotonyl CoA carboxylase for these studies? Again my interest had originated in our studies on acetyl CoA. The distribution of carbon atoms of acetate distributed in the complex structure of cholesterol, so brilliantly determined by Konrad Bloch, suggested a branched-chain carboxylic acid as a precursor. When the thioester grouping in acetyl CoA was found, β-hydroxy β-methyl glutaryl CoA was a reasonable guess. Another guess was that β-hydroxy β-methyl glutaryl CoA is transformed into β-methyl crotonyl CoA, which contains the branched-chain carbon skeleton analogous to isoprene. These ideas were wrong, and only when mevalonic acid, the reduction product of hydroxymethyl glutaryl CoA, was discovered could the exact pathway of cholesterol synthesis be elucidated. It should be emphasized that nobody would have believed that the reduction of the thioester grouping would precede the decarboxylation. But Nature is always unpredictable, and the only way to attack biochemical problems is to make experiments.

I am frequently asked by young students how one should proceed in order to make significant discoveries. My feeling is that the nature of the problem that one studies is not the important question. What is important is the depth to which one probes the problem. To give you an example from my own experience: When Clark Bublitz came from Kennedy's laboratory in Chicago to work as a postdoctoral student with me, I was somewhat short of good ideas to work on, but I suggested he should try to isolate from liver the acetoacetyl CoA deacylase that had been postulated by various people, including myself, to be responsible for the formation of acetoacetate in ketosis. The basic idea proved to be quite wrong; nevertheless, Bublitz's persistence in every aspect of the idea finally led to the discovery of the hydroxymethyl glutaryl CoA cycle.

Another logical consequence of the work on acetyl CoA was our studies on fatty-acid synthesis. Again, we at first followed a wrong track when we believed that fatty-acid synthesis occurs by reversal of fatty acid oxidation. It was not until Salik Wakil had demonstrated the requirement of ATP and bicarbonate for the synthetic process that the right pathway could be seen. As in our studies on terpene synthesis, we expected to find free intermediates, and using radioactive precursors we explored this idea exhaustively until we were finally forced to the conclusion that we were dealing with a concerted process in which the several enzymes performed the entire synthesis without releasing the intermediates. Once we had accepted this idea, it was clear that we had to search for a complex of enzymes, the multi-enzyme complex. This idea proved to be correct, and we constructed a model to explain the way in which the sequence of reactions takes place. The particle we isolated from yeast which performed all these reactions seemed to be a physical entity because of the organized structure that we were able to see when it was examined under the electron microscope. What could not have been expected was that the organization of the particle is so beautifully precise that the individual complexes can align themselves so as to crystallize. Even remembering my early experiences with *Amanita phalloides*, where I could not believe that such complicated structures could crystallize, I considered it rather farfetched to expect that the fatty-acid synthetase multi-enzyme complex would crystallize. I may say that we were now dealing with a particle whose molecular weight was over 2,000 times greater than the original phalloidin. Nevertheless, Dieter Oesterhelt embarked upon this experiment and persisted for fifteen months until the day came when, by shaking the suspension of the sediment that had been formed, one could see the silky appearance which is the first recognition signal of protein crystallization. Then it was simply a matter of rushing to the microscope and holding one's breath until it was sure that not simply ammonium sulfate had been crystallized.

I could easily go on with these stories for another hour. If one starts delving into one's life history, many events come up that are worthwhile talking about, and many persons appear who played a decisive role in one's career. But I think it is better now to stop and mention only briefly that it was through my personal acquaintance with Otto Warburg and with Otto Hahn, then President of the Max-Planck Society, that in 1954 a new research institute was founded for me in Munich which provided most favorable conditions for research. Without that, and without a number of outstanding students, full of enthusiasm and fully devoted to science, working with me, we could not have achieved so much progress in our enterprise.

Looking back, I can say that I have been extremely happy in my chosen profession. I have never allowed it to become an obsession, and I have always found the time to enjoy life. Science has always progressed by one person's standing on the shoulders of others, and I am very happy that I entered at a time when one person could still hope to encompass all biochemical knowledge and even, perhaps, all organic chemical knowledge. These times have passed, and the heap has grown explosively. But man's capacity to absorb knowledge has not kept pace with the increase in knowledge. Instead of a beehive, we now have an anthill, and this means that we can now hope to have only a superficial knowledge of the whole of biochemistry, and such detailed knowledge as we have hardly extends beyond the boundaries of our own specialization. If I had to start again, I wonder whether I would choose science. I have sympathy with every young man or woman who is trying to choose a profession and sees this anthill before him. But I would advise that young person not to be too depressed, because even though he faces an anthill, there is still the possibility of developing his own personality.

REPORT FROM AMERICAN SCIENTISTS IN EUROPE:
SPRING-SUMMER, 1968

It was the season of assassinations.
Stricken, they fluttered like defective birds;
their wingbeats drooped and flickered in the words
of newssheets snapping at the windy crossings.
As specialists we knew that the danc-
ing blood's arrested, ignorant enzymes at
their quiet stations work as if the heart
still beat: The body never dies at once.
So we returned to work along old quays
whose worn stones no disorder could surprise
and talked of bonds and angles, while our eyes
sought the green river and the tidy trees.
We made intricate models to explain
the reasons enzymes work the way they can.

DONN KUSHNER

E.C. Slater, R. Jaenicke and G. Semenza (Eds.)
Selected Topics in the History of Biochemistry: Personal Recollections, IV
(Comprehensive Biochemistry Vol. 38) © 1995 Elsevier Science B.V.

Chapter 2

Memories of Heidelberg - and of Other Places

THEODOR WIELAND

Max-Planck-Institut für medizinische Forschung, Heidelberg (Germany)

I have always thought of myself not as a biochemist, but an organic chemist who sees chemistry as connected with living nature, according to Berzelius' definition of 'chemistry of the products of organisms'. The chemistry of natural products once played an eminent role in Germany and, in the course of time, gradually penetrated biology. Bio-organic chemistry, biochemistry, as well as molecular biology, all need the help of the chemist.

Early life in Munich

I was born in Munich on June 5, 1913, the second son of the chemist and University Professor, Heinrich Wieland, and his wife Josephine, née Bartmann. My father was the descendant of a Swabian family of clergymen. My grandfather, Theodor, was born in 1846 in the little village of Schlat, near Esslingen. He was one of the children of Heinrich Wieland, a clergyman. Theodor freed himself from this milieu and studied chemistry in Tübingen. He did his Ph.D. thesis under the guidance of Adolf Strecker, known for the synthesis of alanine, among others. Theodor had to serve in the war against France in

1870/71. He then founded a gold and silver refinery in Pforzheim. The firm still exists today and, until recently, carried his name, Dr. Theodor Wieland. By chemical means, such as nitric acid, potassium cyanide, and *aqua regia* (= nitric acid and hydrochloric acid), his laboratory recovered in pure form the components, mostly gold and silver, from the noble metal scraps of the jewelery industry which had just been established in Pforzheim.

His first son, my father Heinrich Wieland, was born in 1877 and attended the classical high school (Reuchlin Gymnasium) in Pforzheim. He studied chemistry at Stuttgart, Berlin, and then at Munich where he did his doctoral thesis under Johannes Thiele in the Institute of Adolf von Baeyer. There, he later on also established himself as an academic lecturer (Dr. habil.). My mother's family came from the upper Palatinate and lower Bavaria. Her father settled in Munich as a building contractor.

My childhood was not very eventful. In Munich we were hardly touched by the First World War. In 1921 we moved to Freiburg im Breisgau, where my father was appointed successor to Ludwig Gattermann. As his successor, he also took over the editorship of *'Practice of the Organic Chemist'* which he continued into the fifties, when I was asked to take on this task.

The pleasant time in the comfortable town with its beautiful, mountainous surroundings came to an end in 1926, when our family moved to Munich where my father was appointed Director of the Chemical Laboratory of the State of Bavaria as Willstätter's successor. There, together with my brother Wolfgang, I attended a classical high school (Ludwigsgymnasium), not eagerly, but not unwillingly either.

Before high school graduation ('Abitur') in the spring of 1931, my unremarkable school years were spent in a rather large bourgeois household together with my older brother, Wolfgang, my younger sister, Eva, and my brother Otto, seven years younger. We gathered regularly for meals. Sometimes

my father invited a colleague, collaborator, or foreign visitor
for lunch. My good mother was always present, but in the
background. To the best of her abilities, she guided the duties
and the well-being of her children. Of course, she could not be
everywhere, which guaranteed us a relatively large amount of
freedom.

Besides chemistry and a good cigar, my father loved nature
and hiking. It was not always our greatest joy to board a local
train on Sundays to explore trails in the surroundings of
Freiburg and, later on, Munich. Still, we learned to appreciate
the beauty of the landscape of Upper Bavaria, and, to the present day, the house in Starnberg, built in 1937, enjoys the
greatest of popularity in the family.

I remember these excursions very well. In a manner not at
all schoolmasterly, our father drew our attention to interesting plants and formations. He stimulated us to observe. On
one of the walks through the woods, he pointed out the green
mushroom, *Amanita phalloides*, called the death cap, with the
remark that its deadly effect ought to be analyzed by the
chemist. Not much later, in 1931, Hans A. Raab started these
investigations in the Munich Institute.

Opposite our official residence was the Glass Palace, a pom-
pous edifice of cast iron and glass, modelled after the Crystal
Palace in London. There, the annual exhibitions of various
Munich art associations took place. Unfortunately, in 1931 the
building fell victim to a big fire. At the time it contained an
exhibition of pictures of the Romantics, among them many by
Caspar David Friedrich. The cause of the devastating fire has
never been determined.

Years of Studies

To study chemistry was an almost inevitable choice for me,
since, after moving from Freiburg to Munich, we lived in a
house next to the Chemical Institute, and were connected to it

by a rain-proof passage. As a consequence of this close proximity, often noticeable through the nose, the schoolboy frequently visited the chemical laboratory with schoolfriends, and early on became familiar with unusual substances. The odors fascinated me most of all. Even a layman can differentiate with the nose between lilac, vinegar, gasoline, nail polish, spot remover, hydrogen sulfide (rotten eggs), or alcohol. In addition, for the chemist, there are thousands of other distinctive impressions, such as acetone, benzene, phenol, indole, bromine, chlorine, ozone, and many more. Organoleptics played a great role in former times. Richard Kuhn, my teacher later on, used to taste small samples of many substances. From to-day's perspective, with its fear of traces of chemicals, this might be considered quite daring.

I had enrolled for the study of chemistry at the University of Munich. The first term was dedicated to inorganic chemistry. The professor was Otto Hönigschmid from Bohemia, quite a character, popular and well respected. He was a specialist in the determination of atomic weights which was accomplished by the very precise weighing of silver chloride precipitated from the chlorides of the element concerned. At the time he was occupied with radium obtained from pitch-blende, found in his native country (Joachimsthal). Proudly, he showed a few decigrams of analytically pure radium bromide which glowed mysteriously in the darkened lecture hall. In those days no thought was given to radiation protection. The advance of mass spectroscopy, the method of F.W. Aston, the British Nobel laureate, which allowed a precise determination of the mass of atoms and their isotopes, was frustrating to Hönigschmid, since it successfully competed with his chemical method.

Hönigschmid sometimes visited the practice laboratory of the beginners. Unfortunately, at my work station he was hit by several drops of nitric acid which I was boiling down in a test tube, held in my clumsy fingers. This became quite notice-

able a short while later when yellow spots (xanthoprotein reaction) appeared on his beautyful white hair.

Together with friendly fellow students, I attended lectures,
some of them at more distant locations, e.g. at the Physical
Institute of Walter Gerlach, or I did the assigned experiments
in the Chemical Institute. This free life, yet regulated by duties, was agreeable and undisturbed by external events.

During the first semesters I could not get interested in organic chemistry. I enjoyed inorganic experiments. Colorful
sulfide precipitates were formed, solutions changed color, depending on the state of oxidation or acid concentration, and
gas bubbled up in the Kipp apparatus. These many impressions fortified my decision for chemistry, but did not point in a
particular direction.

A teaching assistant, Feodor Lynen, had the task of handing
out samples of the inorganic mixtures in which we had to analyze the components qualitatively. One day, this graduate student who always enjoyed jokes, gave me a mixture in a beaker
with the request to analyze it in a hurry, before the composition might change. The liquid was bubbling and emitting
greenish-brown vapors. It turned out to be a mixture of concentrated hydrochloric, sulfuric, and nitric acids.

'Fitzi' Lynen was an accomplished skier. In 1932, while
training for an international (FIS) downhill race, he unfortunately broke his left knee. It was a very complicated injury,
and he had to spend one year in the hospital. After a difficult
operation he regained a satisfactory ability to walk. He resumed skiing, better than any untalented skier. During the
course of my studies, the likeable young man soon became a
good friend, and, in 1937, my brother-in-law, when he married
my sister, Eva.

After two semesters in Munich, I went for a further year to
Freiburg im Breisgau, which I knew well from my early youth.
There, Dr. Fritz Schloffer, an assistant, reigned in the inorganic, analytical practice laboratory. He was a funny Austrian
with a cynical wit. He liked my cousin, Ulrich Wieland and me,

as evidenced by the many evenings playing skat or chess together, and also by joint skiing excursions. I passed the first
comprehensive examination ('Verbandsexamen', today: Vor-
diplomexamen). I was tested by Eduard Zintl, a very talented
inorganic chemist, who, unfortunately, died much too young,
and by the equally important Franz Gottwalt Fischer, as well
as by Gustav Mie, a physicist. I regretted that already in the
spring of 1933, I had to say good-bye to beautiful Freiburg.
However, my first flight in a tiny, one-engine, 5-passenger
Junckers plane to Munich Oberwiesenfeld via Böblingen was
a great experience.

In Munich it was already noticeable that the National So-
cialists had come to power. 'Now the others are the others',
satirized the comedian Karl Valentin. After a brief enrollment
in a student organization, I resigned. This also was a natural
consequence of the political standpoint of my parents. From
the beginning, my father was a declared opponent of the
Hitler-regime; being a Nobel laureate combined with his fear-
less attitude gave him enough authority to protect, throughout the times of the 'Thousand Year Reich', a number of scientific collaborators who could not be touched by the devastating
ideology of those in power. He provided refuge, protection, and
survival to many persecuted individuals. The exemplary conduct of Heinrich Wieland during the Hitler-years is highly esteemed and admired by those who know about it, but, generally, has not been sufficiently acknowledged by the post-war
German establishment.

My studies proceeded well. I soon completed the required
organic-chemical experiments (according to Gattermann-Wie-
land). Among the fellow students, assistants, and docents, I
found creative, interesting and stimulating personalities;
with some I have kept up contact to the present. We have
already met Feodor Lynen. After his serious sport injury, he
continued his studies with the best of success. For his doctoral
thesis, in collaboration with my cousin, Ulrich Wieland, he
isolated phalloidin, the first crystalline poison of the green

mushroom, *Amanita phalloides* in 1937. This was to occupy
my later scientific career to quite an extent.

Having obtained his Ph.D., F. Lynen turned to biochemical
problems. In the Institute of his father-in-law, Heinrich Wieland, the many topics of investigation were not only concerned
with synthetic-organic and natural product chemistry, but
also with biochemistry. In those days, for the biochemists,
(beer) yeast played an important role as a forbearing object of
study, perhaps, as an heirloom of the famous Louis Pasteur.
On his own initiative, Lynen undertook an examination of the
participation of phosphate in the sugar metabolism of yeast
(Pasteur effect). He also pursued the question of the mechanism of oxidative degradation of acetic acid by this micro-or-
ganism, a continuing subject of study in the Munich Institute.
Already in 1912, Heinrich Wieland had observed that hydro-
gen-containing compounds could be oxidized by noble metals
without participation of oxygen, but rather of hydrogen acceptors, such as quinone or methylene blue. For instance, hydra-
zobenzene was thus converted to the (colored) azobenzene.
This led Wieland to the formulation of his 'dehydrogenation
theory'. Through a long string of observations, it was proven
that many biological oxidations, such as ethanol to acetic acid
by acetic acid bacteria, consisted of the enzymatically catalyzed removal of hydrogen, which finally combines with oxygen to form water. This led to a dispute between Wieland and
Otto Warburg, quite embittered on Warburg's part. He
wanted to explain biological oxidation as an exclusive function
of the activation of oxygen by iron-enzymes, and, ironically, it
was he who isolated the hydrogen (electron) transferring coenzymes NAD^+ and the flavin-containing 'yellow enzyme'.

In the beginning of the thirties, H. Wieland turned his attention to the paradoxical behaviour of acetic acid which, in
the chemical laboratory, is resistant to the strongest oxidation
agents, but in the organism is smoothly 'burned' to CO_2 and
H_2O. Beer yeast was used to study this biological oxidation. As
was ascertained in 1932 with Robert Sonderhoff, it oxidizes

acetate quite rapidly under self-acceleration. The authors isolated succinic and citric acids from batches which were not completely oxidized. Much earlier, Torsten Thunberg had suggested an oxidative head-to-head union of the acetates to explain the biochemical formation of succinic acid which Pasteur had already found in fermenting yeast. H. Wieland initially agreed with this hypothesis. In the postulated degradation pathway, acetate was oxidized to CO_2 and H_2O via four C_4-dicarboxylic acids and pyruvic acid. Citric acid, which Sonderhoff had isolated as a degradation product of acetate, was considered a side product. But even then H. Wieland [1] contemplated for its formation: 'We consider it possible that oxaloacetic acid ......can condense with acetic acid via an aldol reaction'.

Later experiments by R. Sonderhoff and H. Thomas to oxidize tri-deuteroacetate, demonstrated in 1937 that the isolated succinic acid contained only about 2 deuterium atoms instead of 4, as was to be expected by the hypothesis of Thunberg. At the same time, in the Institute of F. Knoop in Tübingen, Carl Martius [2], former student of H. Wieland, had elucidated in beef liver the enzymatic degradation of citric acid via cis-aconitic acid, isocitric acid, and α-ketoglutaric acid (the steps responsible for the above deuterium proportion) followed by the known dehydrogenation and decarboxylation steps, leading ultimately to pyruvic acid via oxaloacetic acid.

In the same year, H.A. Krebs [3] verified in ground pigeon muscle the most important link in the sequence of aerobic sugar degradation, namely the enzymatic condensation of pyruvic acid with oxaloacetic acid to citric acid. The cycle was closed. Krebs integrated the Martius degradation into the Krebs cycle for which he received the Nobel prize in 1953. A brilliant conclusion of the 'house' theme of the Munich Institute was achieved by the discovery of coenzyme A (CoA) by F. Lipmann and the identification by Feodor Lynen of 'active acetate' as S-acetyl-CoA which, in the presence of 'condensing enzyme', effects the aldol reaction with oxaloacetic acid to cit-

30 THEODOR WIELAND

rate. This brought Lynen the Nobel prize in 1964. It also constituted a climax of the research started about seventy years ago, which can be called 'molecular metabolism dynamics'. My interest in S-acyl compounds and coenzyme A derivatives can be ascribed to an involved observation of this exciting organic-biochemical epoch.

I worked on my doctoral thesis at the Munich Institute during the time before the war. It was concerned with the elucidation of the structure of bufothionine, a component of the skin glands of the South American toad *Bufo arenarum*. In my second year, I was joined by Hsin Chi-yi, a young chemist from Peking who had been sent to Munich for further studies. Attempting the synthesis of a compound of the erroneous suspected structure of bufothionine, we obtained a different indole derivative. A literature review by Hsin gave strong evidence that this compound might be related to gramine, which had just been investigated by Hans von Euler in Stockholm, and the structure of which was still unknown. Our synthesis, applied to the basic indole, indeed, yielded gramine, the simple structure of which was herewith proven (Fig. 1) [4].

For our experiments, we needed non-aqueous hydrogen cyanide, known as a very poisonous liquid with a boiling point of 26°C. Hydrogen cyanide was used in great quantities in the Organic Chemistry Institute of the Polytechnical School not far from our laboratory, which was under the direction of Professor Hans Fischer, the famous porphyrin chemist. Each of his students had to prepare 0.5 liters of the dangerous substance. Hsin and I walked there to obtain quite a large quantity which we carried in a loosely stoppered round bottom

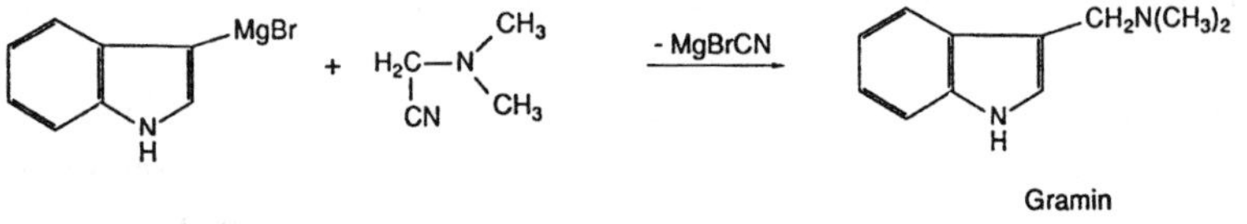

Fig. 1. First synthesis of gramine.

flask. This would be unthinkable under the present-day, super-cautious regulations. In those days, one was of the opinion that a chemist early on had to get used handling chemicals, even dangerous ones. In any case, after the conclusion of the experiment I passed my doctoral examination in January 1937 with very good results. Hsin Chi-yi became Professor at the Peking University, where he effectively participated in the famous insulin synthesis (see ref. 7). I met him 40 years later during a visit to China.

After my doctorate I undertook a spring skiing tour through Upper Bavaria with a few friends from the University and female cousins from Pforzheim, the birthplace of my father. A schoolfriend of my cousins, Irmgard Porcher, also came along. She was to play a particular role in my life. We skied from alpine hut to alpine hut, from Bayrischzell to Lengries.

Then came my decision for the future. I chose a position as assistant of Richard Kuhn, an organic-biochemical scientist of great potential, only 36 years old, and already Director at the Kaiser-Wilhelm-Institute for Medical Research in Heidelberg.

The first years in Heidelberg

My introductory visit to Heidelberg in March 1937 left me with a strong impression. The Institute, recently (1930) built by the architect Freese, was very much different from the old building in Munich, dating from the times of J. von Liebig and A. von Baeyer. It was erected in a timeless, prosaic clinker brick stile, according to the newest technical specifications. An odor of violets, stemming from ionone, wafted through the building, since, among other subjects, there was work in progress on the synthesis of carotinoids, derived from ionone. Professor Kuhn was friendly, but reserved and not very warm. One could hardly believe the tale that only a few years earlier he ran one night through the main street of Heidelberg with his collaborators who were of the same age as he, kicking an empty tin can. Obviously, he now observed more distance.

I remember many colleagues, young assistants, from these still care-free times. They made their reputation in science later on, among them also a few who had left the Institute before my arrival, such as Edgar Lederer [5], Theodor Wagner-Jauregg, and Hans Brockmann. Leonhard Birkofer from Erlangen (R. Pummerer) was concerned with carbohydrates. Friedrich Weygand, coming not long before me from Sir Robert (Robinson) in Oxford, successfully worked in the field of flavines. Kurt Wallenfels, with a Ph.D. from Graz, added to the flavor of the Institute with cinnamaldehyde which he needed for his synthesis of phenylpolyenes. Otto Westphal after his doctorate with Professor K. Freudenberg, joined the Institute half a year after me. Over the years I formed a close friendship with both of them. With Kurt I had many conversations, critical of the regime, while I was still a bachelor and we lived in two adjoining apartments in the domicile for assistants. Otto's family and mine were especially close later on.

We worked, lived and partied together with foreign 'postdocs', such as Pierre Desnuelle [6] from Marseille, C.J.O.R. Morris from London, and Wang Yu [7] who had just received his Ph.D. in Munich. Almost three decades later he became the focal point of the great Chinese team which accomplished the total synthesis of insulin. Besides other colleagues like Dietrich Jerchel, Hans-Joachim Bielig (later, with R. Kuhn, editor of *Liebig's Annalen der Chemie*), Otto Dann, Klaus Schwarz is worth mentioning; in the Heidelberg Institute during the war, he recognized the symptoms of selenium deficiency in laboratory rats. After the war, in Long Beach, California, he discovered further essential trace elements (tin, silicon, vanadium, fluorine). Helmut Beinert after the war soon went to the U.S.A. where he became a successful biochemistry professor at Madison, working on sulfur-iron enzymes. Also, I should not forget to mention Heinrich Trischmann, who started on his professional path in the Institute shortly after its opening

(1930). He became a masterly chemo-technician and the personal assistant of the 'boss'. Richard Kuhn had many of his ideas tried out in his private laboratory, to develop them further or to reject them. When the Nobel prize for chemistry was conferred on him in 1938, there was great jubilation in the Institute, which, however, was dampened by the order of the Nazi regime prohibiting him from accepting it. In Stockholm after the war, Kuhn was presented with the scroll and the medal, but not with the prize money which according to the statutes had to be forfeited, unless accepted within a year of the bestowal.

Kuhn's choice of subject like his teacher's Willstätter, was organic chemistry, in the truest sense of the word, namely the chemistry of organism, a domain reaching from physiology to chemical synthesis. At the time of my entry, vitamins were the focal point of interest. By taking up again the chromatographic separation procedure of M. Tswett (1903), Kuhn between 1931 and 1933, with Edgar Lederer [5], had separated the long-known carotene, the color principle of the carrot, into 3 components, α-, β- and γ-carotene, and had elucidated the constitution of the isomers.

By chromatography it was also possible to isolate beneath lycopine a pretty red azodye from commercial italian tomatoe pulp. Hans Brockmann further developed the technique in convincing fashion, e.g. by describing a scheme to reproducibly standardize the adsorption agents.

In the liver β-carotene is oxidatively split into 2 molecules of vitamin A which had been obtained by Paul Karrer in Zurich from shark liver oil at the same time. The pro-vitamin A effect of β-carotene had raised the interest in this new field of vitamins. In collaboration with the pediatrician, P. György, Th. Wagner-Jauregg in 1933 crystallized vitamin B_2, 'lactoflavin' (later 'riboflavin') from whey. Again, simultaneously with Karrer, it was synthesized by Kuhn and F. Weygand in Heidelberg. Later on followed the synthesis of lactoflavin-5'-phosphate (with H. Rudy) which had been recognized by H. Theo-

rell and O.Warburg as a constituent of the hydrogen transferring 'yellow enzyme' of yeast.

The finding of a water soluble co-factor as a component of an
enzyme system made the exploration of the vitamins even
more exciting. Already in 1932, Ernst Auhagen had found that
the antineuritic vitamin B_1 was a component of pyruvate decarboxylase in yeast. Now a further B-vitamin, B_6, called ader-
min by Kuhn and pyridoxine in the anglo-saxon laboratories,
was the focal point of very competitive research. While the
determination of its structure was successful (also in other
laboratories), and the synthesis was being worked on, a further vitamin, the so-called 'filtrate factor', needed exploring.
Lack of this factor in the food of experimental animals led to
inhibition of growth and to loss of hair.

Fortunately, it had been recognized by E.E. Snell and others
that the B-vitamins also are essential for certain micro-organisms. Making use of the rate of multiplication of micro-organisms, such as *Streptobacterium plantarum*, as an assay system, the concentration, that is the enrichment in a solution of
vitamin B_2 and B_6, of p-aminobenzoic acid and also of the 'filtrate factor' could be determined. After a rather unsatisfying
start-up period with a component of the complicated blood
clotting system Kuhn steered me to the isolation of a growth
factor for lactic acid bacteria which soon turned out to be the
unknown essential supply. In intense competition with the
American team of the Merck Company, enrichment and purification of pantothenic acid was undertaken. This was monitored with a bacterial assay by E.F. Möller. In the mean time,
the war had started. Fortunately, this scientific endeavor fitted into a vitamin research program, considered 'essential for
the war effort'. In addition, the production of hard cheese from
cow's milk as a provision for the army had become increasingly
difficult. This probably was a consequence of watering of milk.
Therefore, the exploration of growth conditions of many varie-
ties of lactic acid bacteria had priority.

In spite of the unpleasant and increasingly unbearable political system, the time before the Second World War was quite carefree. One stayed away from politics. Counter demonstrations in the present sense would have landed the participants immediately in a prison or concentration camp. In principle, the head of the Institute, an Austrian, was opposed to the Nazi government. In spite of his prestigious position (Nobel prize in 1938, president of the Society of German Chemists), Kuhn could afford resistance only in a very veiled manner, even in the most intimate circle of friends. This acquiescence was sometimes misunderstood, but it saved nearly all his collaborators from serving in the war, and probably many a young scientist from certain death.

At the time of the annexation of Austria (1938) and the 'Protectorate Bohemia and Moravia', naively, we could not seriously conceive that war was imminent. We led an almost carefree and modest life. Irmgard Porcher, a friend of my cousins in Pforzheim and the companion on the ski tour after my doctoral examination, enrolled in Heidelberg to study medicine. Together with friends from the Institute, we played tennis, went on excursions to the Odenwald, and swam in the then still clean Neckar. In winter, we went skiing in the Black Forest. At the Institute during breaks, we also had sport activities in which R. Kuhn liked to participate, such as table tennis or fist ball, a precursor of volleyball.

In the fall of 1939, German troops marched into Poland; the Second World War had begun. Ration cards to curtail food consumption had already been prepared. The black-out of all windows and streets was ordered. All this was received with a certain resignation; there was no trace of patriotic enthusiasm which supposedly prevailed at the start of the First World War. Of course, the use of cars was no longer possible; we were lucky to get a permit to buy a bicycle tire or other necessary things, such as shoes, textiles, or cooking pots. During the summer of 1940, Irmgard Porcher and I were married in Munich in the Bonifazius Basilika, which is decorated with frescos of Heinrich von Hess (1798–1863). A great-grandson of the painter was married to the sister of my father, which made for a distant relationship. The biochemist Benno Hess is descended from the same forebear.

Kuhn's Institute was the home of chromatography and aluminum oxide. I soon became involved with this technique, being interested in its application in aqueous solutions. In Munich, Georg Maria Schwab had discovered that aluminum oxide, a commercially available adsorbent, could be made an anion exchanger by treatment with a mineral acid, preferably hydrochloric acid. We found the 'acid column' well-suited for the separation of organic anions - pantothenate from tunafish liver extracts in my case. With the aid of this technique, we soon obtained pure preparations from which, after hydrolysis with dilute sulfuric acid, β-alanine and D-α-γ-dihydroxy-β,β-dimethylbutyric acid (R-configuration) as lactone, could be isolated. These two substances were already known as building blocks of pantothenic acid. Thus it was shown that the B-vitamin from fish liver, which served as a growth promoter not only for certain yeast strains, but also for *Streptobacterium plantarum*, was identical with pantothenic acid.

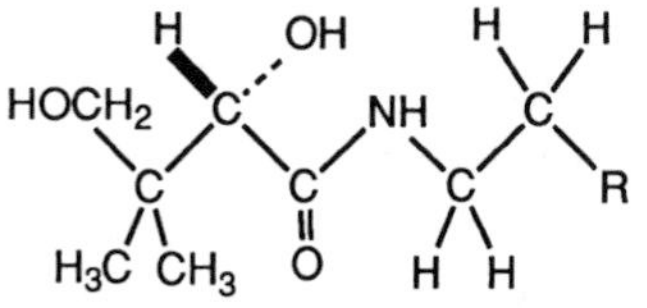

D-Pantothenic acid   R = CO_2H

D-Sulfopantothenic acid   R = SO_3H

Lactobac.bulgar.-factor   R = $CONHCH_2CH_2SH$

Fig. 2. Pantothenic acid, sulfopantothenic acid.

After 1935, when the discovery of the bacteriostatic effect of sulfonamides (F. Mietsch, G. Domagk, J. Trefouel) and its suppression by p-aminobenzoic acid (D. Woods, P. Fildes, 1940) was made, R. Kuhn was also interested in the competitive inhibition of bacterial growth by analogues of active agents. Consequently, we synthesized the two 'sulfopantothenic acids' from taurine, instead of β-alanine, and from the S-, as well as R-dihydroxycomponents. Indeed, the S-compound, corresponding to the natural pantothenic acid, totally inhibited the multiplication of lactic acid bacteria when present in a 1000–fold excess over pantothenic acid. The antipode was ineffective at this concentration. Depending on concentration, the inhibitory action could be abolished by pantothenic acid [8]. A review of my pantothenic acid related work up to 1974 can be found in ref. [9]. Later on, we also scaled the heights of coenzyme A which contains pantothenic acid as a building block (see p. 85).

The study of bactericidal compounds led Kuhn also in other directions, one of which had a favorable result for the Institute. With halogenated salicylic acids as a 'Leitmotif', Leonhard Birkofer synthesized 5,5'-dibromosalicil. This compound was found to be highly effective in an *in vitro* inhibition test for streptococci and staphylococci. In the jargon of the Institute it was called 'superpenicillin'. The responsible American authorities had heard about this, and therefore, might have not ordered a closing of the Institute after the occupation of Heidelberg in 1945. Unfortunately, the substance turned out to be ineffective in animal tests, since it was firmly bound by blood protein and, therefore, prevented from action on the pathogenic germs.

Kögl's tumor theory

Towards the end of the thirties, a theory concerning the nature of cancer caused quite a sensation: Fritz Kögl and Hanni

Erxleben from Utrecht had reported that they had found considerable quantities of D-amino acids, especially D-glutamic acid in the hydrolysates of malignant tumors. As had to be expected, in the protein hydrolysates of normal tissues only small amounts of D-amino acids occurred, racemized under the hydrolysis conditions. The regulating enzymes, specific for the L-configuration, would lose their effectiveness on the D-building blocks, thus explaining the uninhibited cell growth. Of course, there was no lack of experiments by other laboratories to confirm or contradict the findings of Kögl. The specific separation of acid amino acids with the 'acid' Al_2O_3 column (p 36) strongly suggested itself, to clarify the situation. We were able to adsorb the total amount of glutamic acid from protein hydrolysates on the acid Al_2O_3 and elute most of it in pure form with barium hydroxide. As determined by specific rotation, the glutamic acid from a Brown-Pearce tumor of the rabbit contained only 2% of the D-form. This amount was found in the hydrolysate of normal proteins also, due to the chemical racemization of the L-form.

American investigators also could not confirm the findings from Utrecht. Rittenberg and Foster had used the so-called isotope dilution method. But they isolated the glutamic acid by using the old precipitation procedure. Therefore, Kögl could claim deviating conditions for the lack of D-glutamic acid. Using glutamic acid which contained the isotope ^{15}N, we now combined chromatographic purification with the isotope dilution analysis. We had obtained the very precious ^{15}N-enriched ammonium salt, needed for the synthesis, from Professor Klaus Clusius in Munich, a pioneer of isotope separation. The analysis of the ^{15}N-isotope content of the glutamic acid, isolated from the tumors, was carried out by Wolfgang Paul in Göttingen, on a mass spectrometer which he had constructed himself. (Paul won the Nobel prize for physics in 1989. He died in 1993 in Bonn).

The results of our efforts with D-glutamic acid was a confirmation of the previous negative results [10]. Whether positive or negative, checking the results of other authors is always less than satisfying. Nevertheless, I considered the technically instructive effort justified, even though the problem appeared to be an artificial one.

The war only slightly touched Heidelberg and our family, into which daughter Sibylle was born in 1943. The town did not suffer any heavy bombing attacks. Rumors claimed that the Americans had chosen it for their future headquarters, which, indeed became true. Still, we had to suffer air-raid alarms day and night. The frequency increased steadily, and so we had to spend many hours in the air-raid shelter. During the day, these interruptions took place in the deep basement of the Institute. Frequently, we whiled away the time with chess games. My partners were Richard Kuhn or Heinz Maier-Leib-nitz who worked in the Physics Department of Professor W. Bothe. The former was an excellent chess player who only rarely gave me the pleasure of a remis.

Not many cities were spared by the bombs. The town of Pforzheim, the home of my parents-in-law and also relatives of my father's, was largely destroyed by a devastating attack at the end of February 1945, shortly before the end of the war. The parents of my wife were able to escape from the flames, but their house as well as those of other relatives who had moved out before, were completely demolished. In Munich, too, our former home together with the complex of the Chemical Institute fell victim to the bombs. Before this, my parents had retired to their house in Starnberg, taking along their library and most valuable furniture.

The end of the war in Heidelberg was impressive. On Good Friday 1945, a sunny, early spring day, the American troops moved into an almost completely undamaged Heidelberg without any resistance. Only the bridges over the Neckar had been senselessly blown up by the German army. The period after the war began with a feeling of liberation. Since the Institute was initially occupied by the Medical Branch of the U.S. Army, we used the free time to procure necessities for survival. We soon got into the possession of an old, three-wheeled, pick-up truck. Through connections of long standing with the Badische Anilin und Soda Fabrik (BASF) in Ludwigshafen, we also obtained combustible solvents which the indestructible motor swallowed, stuttering and groaning. With Günter Quadbeck, I went on many an adventurous foraging expedition. On one very hot day in July 1948, we had dropped in at the BASF plant office of Dr. Heinrich Hopff, a faithful benefactor. Suddenly, it turned dark, and an irresistable air pressure pushed the door to the hallway as well as its frame into the office and hurled the windows outside, as if moved by a ghostly hand. A few hundred meters away in the plant area, a great amount of dimethyl ether (which boils at −25°C under normal pressure) had escaped from the valve of a pressure tank car. The ether-air mixture had ignited, causing a tremendous, burning pressure wave to which more than a hundred people fell victim.

Research and teaching

Still during the war, in 1942, I obtained the Dr. habil. degree from the University of Heidelberg, which meant that I was appointed a docent (equivalent to an assistant professor) with the privilege to present lectures at the University. Then, in 1947, I was made an offer by the University of Mainz, which had been re-established by the French. There, with great enthusiasm, the buildings of a large army barracks were converted into a university including all disciplines. Fritz Strassmann was the head of the inorganic department as well as of the whole Chemical Institute. Werner Kern, a student of Hermann Staudinger, who had done important studies at Hoechst on the redox polymerization of unsaturated compounds, had already been appointed as one of the two professors of organic chemistry. I soon accepted the offer, since Kuhn gave me the assurance that I could keep my laboratory in the Heidelberg

Institute until satisfactory accommodations were ready in Mainz. This occurred after 2–3 years. During this time I gave courses in Mainz. W. Kern and I remodeled the artillery garages into laboratories and offices. In the Heidelberg Institute, after it had been vacated by the U.S. Army and later on in Mainz, I resumed my scientific research with good and eager young people who had returned from the war or captivity.

Reagents and equipment were at the lowest level, and so were the demands which one made. Physics had not yet invaded the chemical laboratory. Vacuum distillation apparatus, polarimeters, and microscopes were considered expensive instruments; spectroscopy in the visible light was reserved for specialists; UV and infrared spectroscopy were to be found only at a very few places. The automatic analysis of amino acid mixtures, described by Stein, Moore and Hirs not long before, was considered an unaffordable masterpiece. I remember a conversation with Adolf Butenandt in which we considered providing a few locations in the Federal Republic with such instruments where, on demand, analyses could be carried out for the whole country. In retrospect, one sees not only an economic miracle, from 1950 onwards, but also a laboratory instrumentation miracle, which soon made the traditional laboratory rooms so overcrowded that hallways had to be used for locating laboratory equipment.

Analytical studies

Paper chromatography had been invented in England. Reverting to the use of simple filter paper, Gordon, Martin and Synge had made a virtue out of necessity by creating an ingeniously simple, super-productive system to separate amino acids and other classes of substances. In my laboratory in Heidelberg, Edgar Fischer developed filter paper electrophoresis, applying it to amino acids first. This idea had already been pursued in other laboratories, but with relatively unwieldy apparatus.

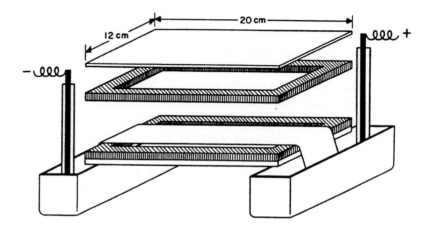

Fig. 3. Simple apparatus for electrophoresis on filter paper (ref. 7).

Our simple arrangement consisted of a wet chamber (Fig. 3), and two Petri dishes, filled with buffer, into which carbon electrodes were dipped [11]. At 110 V, depending on the pH, the acidic and basic amino acids were separated on the buffer saturated filter paper. Soon thereafter, the easy-to-handle system was used for protein mixtures, also by other investigators. The separation of protein mixtures by free electrophoresis by different migration rates in an electric field, in a U-tube, had been perfected earlier by Arne Tiselius [12] in Uppsala. In order to keep the moving boundaries undisturbed, the device had to be properly tuned.

Paper electrophoresis of blood plasma proteins was introduced into clinical laboratories as well. It was developed further to high voltage electrophoresis under refrigeration, and adapted for preparative use, preferentially with Gerhard Pfleiderer to other carriers such as starch, acetyl cellulose [13] and polyacrylamide, or agarose gel for analytical use on small glass plates. Today, molecular biology laboratories could not do without it for the separation of nucleotides, or of proteins, denatured by sodium lauryl sulfate.

Under Richard Kuhn, Pfleiderer had worked on a doctoral thesis on the use of tetrazolium salts for the reductive formazan staining of tissues. Therefore, he was biochemically interested, and I had planned to have him head a common biochemical laboratory.

Before the First World War, biochemistry, as the chemistry of the processes of life, was the domain of the medical faculties at German universities. Under the name of physiological chemistry some fundamental research was carried out; one only has to think of names such as Franz Hofmeister (proteins), Albrecht Kossel (nucleic acids), Franz Knoop (fatty acid metabolism), and Gustav Emden (glycolysis). Later on, the emphasis shifted more and more to the research institutes of the Kaiser Wilhelm Society (Carl Neuberg, Otto Meyerhof, Otto Warburg). In the natural science faculties of most universities, biochemical research played no role, even though almost one hundred years ago, a chemist, Eduard Buchner, laid the cornerstone of enzymology with cell-free fermentation. Exceptions were the chemical institutes at the Universities in Berlin (Emil Fischer, enzymes cleaving polypeptides and polysaccharides), and in Munich (Richard Willstätter, studies on purification of enzymes), and, Heinrich Wieland. Only at the end of the forties in Munich, the first chair of biochemistry, in the framework of chemistry instruction, was given to Feodor Lynen, who later on moved to the Max Planck Institute for Biochemistry in Martinsried.

In my future place of activity, I planned to anchor biochemistry in the discipline of chemistry under Gerhard Pfleiderer. This coincided with his wish to train further in a recognized enzyme laboratory after his doctorate. For this, he chose Theodor Bücher, the young disciple of Otto Warburg, who headed the clinical chemistry laboratory in the Hamburg-Eppendorf hospital. At the end of the war, Bücher had salvaged a few of the remains of the Institute of Warburg, which had been relocated from Berlin to Liebenberg in Brandenburg Marchess. Fleeing from the Russians, he had brought the remains to the

West. Bücher had learned the art of crystallization of enzymes of the carbohydrate metabolism of yeast and muscle tissues in Warburg's laboratory in Berlin-Dahlem. There, he had isolated and crystallized a 'phosphate transferring fermentation enzyme' of yeast (phosphoglycerokinase) [14].

Lactic acid dehydrogenases

Shortly after my move from Mainz to Frankfurt, where I was offered a chair at the Department of Chemistry in 1951, Pfleiderer joined me with many insights and the ability to crystallize lactic acid dehydrogenase (LDH) from rabbit muscle, and we soon took up studies with this enzyme. Early on, we observed that LDH, isolated from beef heart muscle, exhibited apparently different enzyme activities, depending on the origin of the dihydronicotinamide-dinucleotide (NADH, then called DPNH) which was used as a reductant of pyruvate to lactate. Crystalline LDH showed optimal conversion rates with NADH, prepared enzymatically. NADH preparations, made by reduction of NAD with dithionite, gave conversion rates only one half as fast.

We suspected sulfite ion, SO_3^-, to be the inhibitor, and so it was. Normally, the amount of the addition product of sulfite on NAD, already observed by O. Meyerhof, is relatively small, and can hardly be seen spectroscopically at 320 nm. We ob-

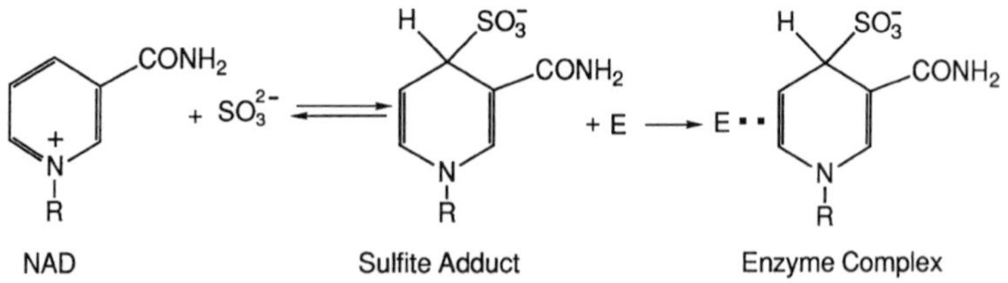

NAD   Sulfite Adduct   Enzyme Complex

Fig. 4. Addition of sulfite to the pyridinium part of NAD and binding of the adduct to lactate dehydrogenase (E).

served that in the presence of the enzyme, its concentration increased very strongly. The adduct which is similar to NADH, is firmly bound to the enzyme as a pseudo-substrate, as a matter of fact, so firmly that it protects the enzyme from denaturation by urea, proteolysis or heat [15].
This could be used to advantage in the purification of the enzyme from organ extracts, since companion proteins which do not bind NAD-sulfite or NADH, can be denatured and precipitated by heating.

Multiple forms

In 1956, using the electrophoretic separation method on paper or starch, we started comparative investigations of lactic acid dehydrogenases of different organs from various animal species. To our surprise, in our first experiments with rat heart

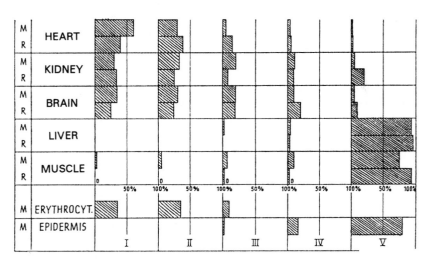

Fig. 5. *Paper electrophoretic pattern of amounts (in percent of total activity) of the LDH isozymes in organs of man (M) and rat (R). Visualized by spraying with pyruvate and NADH: the fluorescence in UV of it disappears in the presence of enzyme activity.*

LDH, we clearly observed 5 different bands, moving to the anode. These turned out to be lactic acid dehydrogenases, since at each band after spraying the paper with pyruvate and NADH, NADH-fluorescence disappeared (NAD^+ formation). The enzyme bands (I and II) which had moved the farthest towards the anode, were the most intensive, the three following ones were much weaker.

Then we read that J.B. Neilands, using moving boundary electrophoresis of beef-heart LDH, had crystallized beside the main boundary, a second 'impurity' for which he demonstrated LDH activity. In 1956, using starch block electrophoresis, Vesell and Bearn separated human serum into its protein components and found LDH-activity in 3 different zones [16]. At the same time, G. Pfleiderer, D. Jeckel and I had started to compare the LDH of rabbit or rat liver with that of skeletal muscle by electrophoresis. On cellulose acetate strips, 5 bands were visible. In the liver preparation, band V, which had moved the least, contained the most enzyme activity. LDH from brain, kidney and erythrocytes were more similar to the heart muscle type with the main activity in the fastest moving band I [17].

Pfleiderer and Wachsmuth found that the greatest LDH-activity of the human embryo resides in bands II, III and IV. In the course of development, this changes to the typical electrophoretic organ patterns of adults. Normally, blood serum contains only a tiny amount of LDH-activity in bands I, II and III. However, after damage to the heart muscle or liver, the corresponding pattern of LDH isozymes can be clearly observed in serum. Today, electrophoresis can look back to forty years as a clinical-diagnostic routine.

At about the same time, N.O. Kaplan and co-workers were able to insert 3-acetyl pyridine, instead of nicotinamide, into NAD [see ref. 17a]. With this analogue, they examined the dehydrogenation of lactate, derived from a variety of enzyme sources. They found that LDH from beef heart muscle reacted much more slowly than LDH from rabbit skeletal muscle.

They initiated a comprehensive series of comparative studies of different LDHs from numerous organisms. In 1959, C.L. Markert and F. Møller proposed the name 'isozyme' for the differing molecular forms in which proteins of the same enzymatic activity could exist [18]. In this publication they cited the ten enzymes, known at the time, which could be cleaved into components. They added isocitric acid dehydrogenase as a further enzyme. In our later studies of pantothenic acid synthetase from *E. coli*, we also found multiple forms [19].

Zymograms were the simplest way to discover isozymes, the number of which has increased tremendously in the meantime. To obtain larger amounts of lactic acid dehydrogenases, we soon also used chromatography. Towards the end of the 1950s, the ion exchange reagent, DEAE-Sephadex, had been developed together with Jerker Porath by the Swedish company Pharmacia. This reagent was ideally suited for the separation of isozyme mixtures [20].

The continuous differences between the five isoenzymes, e.g. the different affinities to the substrates lactate or pyruvate became clear when, in 1961, Appella and Markert proved that LDH consists of four subunits of which 2 types exist. These, in pure form, are present mostly as isozyme I in heart muscle (H), and as isozyme V in liver and skeletal muscle (M) [21]. These forms can associate to 5 types of the tetrameric enzyme: H_4, H_3M, H_2M_2, HM_3 and M_4.

The time in Mainz

Mainz, an old town of Roman origin, had a university which was endowed by Diether von Isenburg in 1477. It was closed in 1798 after the French Revolution, except for the Medical Faculty which continued to exist until 1877. On the initiative of the French occupation army, especially General Raymond Schmittlein, the University was reopened in May 1947. In the beginning, the limited facilities did not permit a well-directed research effort. Therefore, during the first years, I spent only part of my time in Mainz, giving lectures and procuring equipment, chemicals, and books. Research and training of students took place at the Institute in Heidelberg.

In the course of time, the bond to Mainz grew stronger. Collaborators, eager to learn, could begin their work there. At the end of 1950 my family, which had increased by the birth of our son Heinrich in 1947, transferred to Mainz into a house, pleasantly situated 'Am Rosengarten'.

Almost 80% of the inner city of Mainz was destroyed during the war, yet it demonstrated a new will to live, especially after the currency reform of 1948. The population overcame the deep depression with refreshing, inborn courage. Early in 1948, or, perhaps, already a year earlier, the famous 'mardi gras' parade took place. Wine, to create the necessary mood, was available in this center of the German wine business, although it was difficult to come by. If one succeeded in smuggling the popular beverage across the zonal border (Rhine), one could make favorable barter deals to obtain laboratory equipment, gasoline, food or the desired cigarettes. In a modest way, the author of these lines was successful in obtaining this latter 'currency' from American GIs in the castle of Heidelberg by exchanging his own water color sketches.

Biochemical research after the war

Through the emigration of Jewish and many liberal-thinking scientists, such as H.A. Krebs, O. Meyerhof, F. Lipmann, K. Bloch, R. Schönheimer, F.F. Nord, F.L. Breusch, and many more, biochemical research in Germany, obviously, lost almost all that had been built up in the ten years before the Hitler regime. After the war, in a relatively short time, in view of the advantageous position of American and later on also British sciences, an atmosphere of interest and enthusiasm developed among the young chemists and physicians who had survived the war. It is no exaggeration to state that Feodor Lynen played a central role in this development. He provided impetus with results in his investigations of fatty acids, sterols, and terpenes. Numerous, successful students of his spread his work style to other locations.

Other biochemists who also contributed to reconstruction were Carl Martius [2], Gerhard Schramm, Hans Friedrich Freska, and Paul Ohlmeyer. Theodor Bücher and Kurt Wallenfels established contact with a biochemical-pharmaceutical company which started the commercial production of enzymes. Fritz Turba from Prague, a student of Ernst Waldschmidt-Leitz showed up in Mainz, where he could resume his elegant studies of the analysis of proteins, peptides and amino acids. In Munich, Helmut Holzer, the first student of F. Lynen, pursued the phosphate metabolism in yeast, and Otto Wieland the pathology of phalloidin poisoning. Otto Westphal had established a little research group on bacterial carbohydrates in Säckingen/Rhine.

On my instigation, these and other young scientists met a few years after the war in a small circle for presentations and discussions, partaking of bread and wine at suitable places such as Bingen and Deidesheim. As a sponsor, Dr. Ernst Boehringer has to be mentioned with gratitude. He had done his doctoral thesis under Richard Willstäter in Munich, and, after Willstätter's resignation, had taken his doctoral exam under Heinrich Wieland, Willstätter's successor. He had now been the captain of his father's chemical-pharmaceutical firm in Ingelheim/Rhine. The symposia (Binger Tage) were enriched by the participation of foreign guests. Lester Krampitz, on sabbatical leave, came from the laboratory of Lynen, and once we were honored by a visit of Fred Sanger from Cambridge.

For organizing biological chemistry in West Germany after the war, Ernst Auhagen deserves great credit. In 1987 he summarized '40 Years of History of the Society for Biological Chemistry' in the 368th volume of the venerable *Hoppe-Seyler's Zeitschrift*.

Peptides

At the Chemical Institute of the University of Mainz which
slowly gained in scope, we turned our attention to the chemistry of peptides which was as good as forgotten in Germany at
that time. As is well known, the synthesis of peptides was
started in Germany at the beginning of the century by Theodor Curtius and, especially, Emil Fischer. After the First
World War, Emil Abderhalben in Halle, Fritz Wessely in Vienna and, especially, Max Bergmann in Dresden were successfully engaged in this field. But after the emigration of
Bergmann during the Hitler regime, it was totally neglected.
After the Second World War, the observations increased that
biological effects could be ascribed to several natural products,
composed of a few amino acids. The investigations of glutathione by F.G. Hopkins [22] had given a widely noticed signal.
For the organic chemist, the time seemed ripe to pursue peptide synthesis.

With Richard Sehring, I built on an old observation of Th.
Curtius (1881) that by reacting benzoyl chloride with silver
glycinate, a series of homologous benzoyl glycine peptides are
formed in addition to hippuric acid. We suspected the intermediate formation of mixed anhydrides of the peptide carboxyl
group and benzoic acid as the peptide-forming species and prepared such compounds from N-benzyloxycarbonyl amino acids
and benzoyl or acetyl chloride in the presence of tertiary bases.
Without isolation, they could be coupled to amino acids or peptides (mixed anhydride method) (Fig. 6) [23]. As often happens
in science, similar ideas were also developed in other laboratories (Vaughan in USA, Boissonnas in Switzerland). Anyhow,
we now had a voice in the matter of reawakening peptide
chemistry.

The identification of 'activated acetic acid' as S-acetyl coenzyme A by F. Lynen in 1950 not only was a milestone of biochemical metabolic research, but also in many ways stimulated preparative organic chemistry. For the peptide chemist

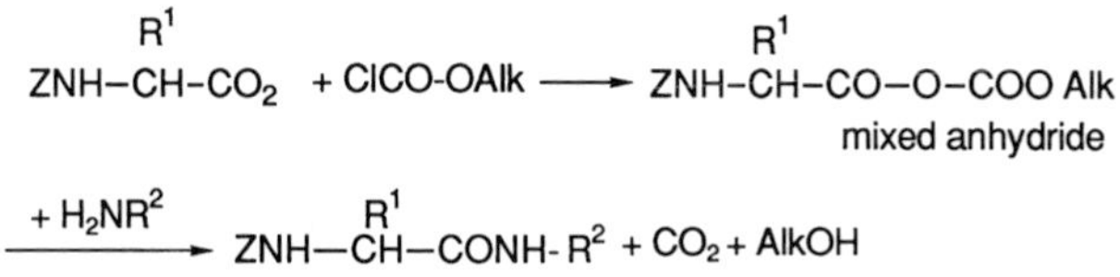

$$\text{ZNH-CH-CO}_2 \quad + \text{ClCO-OAlk} \longrightarrow \text{ZNH-CH-CO-O-COO Alk}$$

$$\overset{R^1}{} \qquad\qquad\qquad\qquad \overset{R^1}{}$$

mixed anhydride

$$\xrightarrow{+ \text{H}_2\text{NR}^2} \text{ZNH-CH-CONH-R}^2 + \text{CO}_2 + \text{AlkOH}$$

$$\overset{R^1}{}$$

Fig. 6. Formation of a mixed anhydride from a benzyloxycarbonyl amino acid and alkylchlorocarbonate and its coupling with an amine forming an amide (peptide) bond.

the question arose immediately, whether amino acid radicals bound to thiol groups would be suitable for the formation of peptides. Therefore, in 1951, I started, mainly with Ekkehart Bokelmann [24], a series of experiments with S-aminoacylthiols, trying thiophenol first. Using the mixed anhydride method, just described, N-Cbo-glycine could be coupled with thiophenol yielding the thioester (Fig. 7). The same was also possible with Cbo-di- and tripeptides. The aminoacyl compounds, thus activated, easily reacted with amino compounds to amides (peptides) [25].

Somewhat closer to nature, these reactions could be extended to cysteamine and to glutathione which at the time held center stage again. We succeeded in obtaining S-acyl derivatives of glutathione from thiophenyl compounds by S-S transfer. At the time, the mechanism of protein biosynthesis was still unknown. Therefore, besides phosphoric acid anhydrides and esters, amino acids bound at the cystein-sulfur could be considered for the formation of 'activated' amino acids. We wrote at the time: 'The question arises whether sulfur-bound aminoacyl radicals could act as energy-rich aminoacids in the biosynthesis of peptides...'.

Almost 20 years later, this question was answered positively by the elucidation of the biosynthesis of gramicidin S by *B. brevis* in the laboratories of Laland in Oslo, Kurahashi in Osaka, and, definitively, by Fritz Lipmann's group in New York [26]. It was shown that non-ribosomal peptide synthesis uses amino acids which are bound to the thiol groups of the

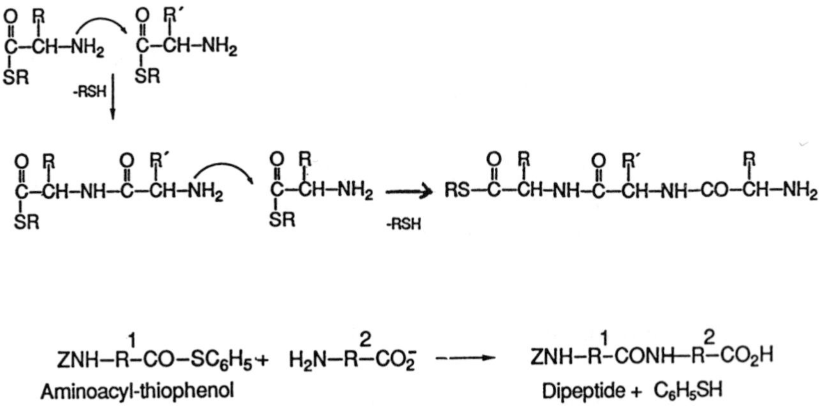

$$
\underset{\overset{|}{SR}}{\overset{\overset{\displaystyle O\ \ R}{\|\ \ |}}{C}}-CH-NH_2 \overset{\displaystyle O\ \ R'}{\underset{\overset{|}{SR}}{\overset{\|\ \ |}{C}}}-CH-NH_2 \quad _{-RSH}\Big\downarrow
$$

$$
\underset{\overset{|}{SR}}{\overset{\overset{\displaystyle O\ \ R}{\|\ \ |}}{C}}-CH-NH-\overset{\overset{\displaystyle O\ \ R'}{\|\ \ |}}{C}-CH-NH_2 \quad \overset{\overset{\displaystyle O\ \ R}{\|\ \ |}}{\underset{\overset{|}{SR}}{C}}-CH-NH_2 \quad _{-RSH}\longrightarrow \quad RS-\overset{\overset{\displaystyle O\ \ R}{\|\ \ |}}{C}-CH-NH-\overset{\overset{\displaystyle O\ \ R'}{\|\ \ |}}{C}-CH-NH-CO-\overset{\overset{\displaystyle R}{|}}{C}H-NH_2
$$

$$
ZNH-\overset{1}{R}-CO-SC_6H_5 + \ H_2N-\overset{2}{R}-CO_2^- \ \dashrightarrow \ ZNH-\overset{1}{R}-CONH-\overset{2}{R}-CO_2H
$$

Aminoacyl-thiophenol   Dipeptide + C_6H_5SH

Fig. 7. Aminoacyl thiols for peptide synthesis.

cysteine side chains of the synthesizing enzymes. (See my article on the role of sulfur in biomimetic peptide synthesis [27] and similar suggestions for prebiotic syntheses by C. de Duve [28] in the same F. Lipmann-Memorial Volume.)

The rapid development of peptide chemistry may be illustrated by briefly mentioning the European Peptide Symposia. By invitation of Joseph Rudinger, chemists from Poland (Emil Taschner), the German Democratic Republic (Manfred Rothe), the Soviet Union (Shemyakin), England (Geoffrey Young), Switzerland (Max Brenner) and the German Federal Republic (Stephan Goldschmidt, Erich Wünsch) met for the first time in 1958 in Prague to exchange ideas in the recently revived field of peptide chemistry. From then on, this arrangement was repeated annually, and later on every two years, in various European countries. As with all scientific congresses, over the years this developed into a large meeting. But it kept its attraction for us, since it led us to almost all European countries with their peculiarities and attractions. To talk about this in detail would fill a whole volume: I only want to mention the third symposium which took place in Basel in 1960. Klaus Biemann, coming from Boston, reported on the application of the then still very young mass spectroscopy of amino acids and peptides. Shemyakin and his wife also participated. On the wine-filled return from a trip to the Tessin on a 'train speciale' of the Swiss Federal Railway, Mrs. Shemyakin contributed to the merriment with lively Russian dances. To the 5th symposium in Oxford in 1962, Shemyakin sent the young Ovchinnikov, who just had concluded a sabbatical with V. Prelog at the ETH in Zürich.

As a ghost-participant, the 'old chemist from Westside' attended the symposia. I introduced him:

> There was an old chemist in Westside
> Who never obtained a pure peptide
> But nevertheless / At every congress
> He said next slide, please next slide, please next slide.

He showed up again in other limericks at several of the following meetings, e.g.

> Our peptide chemist is clever
> He does not prepare them as ever
> In bacteria he grew / A chain sixty-two
> But nobody knows what for – never.

Glutathione

Biochemists and organic chemists alike have been captivated by this wondrous peptide and thiol which is distributed universally and widely in aerobic nature. When we entered the chemistry of the S-acylthiols, it had been known for a long time (H.D. Dakin, as well as C. Neuberg, 1913) that glutathione is a coenzyme of 'glyoxalase' which converts methylglyoxal into D-lactic acid. In 1951, E. Racker found that the enzyme reaction is completed in two steps: glyoxalase I converts methylglyoxal and glutathione to S-lactoylglutathione, which is hydrolyzed to glutathione and D-lactic acid by glyoxalase II. C. Neuberg ascribed to methylglyoxal an important role in the metabolism of glucose, but later findings did not

support this idea. According to present-day knowledge, glyoxalase serves as a detoxification system for the very reactive
2-oxoaldehydes. To glutathione with its SH-group is given the
role of a protective, general scavenger, e.g. of formaldehyde,
oxygen radicals, heavy metal ions, and many more.

Forty years ago, this peptide could be believed capable of
other yet undiscovered abilities, among them a participation
in the biological peptide and protein synthesis (see p. 51).
Therefore, we synthesized for the first time several S-aminoacyl and other S-acyl derivatives of glutathione. Soon, I received an invitation, the first one from overseas. It came from
Heinrich Waelsch to attend a symposium on glutathione –
mainly organized by him – in the fall of 1953 in Ridgefield,
Connecticut.

First journey to the USA

In November, my wife and I boarded the 'Mauretania' of the
Cunard line, a medium large ocean liner, in Le Havre. She was
only sparsely booked and set her course for Southampton
where she took on several Irish passengers who intended to
visit relatives in the USA. The small number of passengers
and the unaccustomed life on board in old-fashioned elegance
made the 6-day crossing into an impressive voyage, marred
only by 1–2 uncomfortable days of 'rough seas'. But then the
slow appearance of the skyline of Manhattan and the majestic
docking at the pier made an unforgettable impression. In
those days, the Empire State Building dominated; less attractive skyscrapers, especially the huge blocks of the World Trade
Center had not arisen yet. We stayed with a female friend of
the family in then still idyllic Staten Island, and we again and
again enjoyed the 5-cent crossing on the Staten Island ferry, a
cheap and most interesting 'sea voyage', passing the impressive Statue of Liberty.

Heinrich Waelsch, who researched the glutamic acid metabolism of the brain, showed me his laboratory and office in a building at Columbia University Medical School. For the first time, I saw the cramped, even primitive conditions under which American scientists had to work, in comparison to our mostly older, yet more spacious laboratories. I had the same impression when I visited R.B. Woodward at Harvard. He worked under somewhat more comfortable conditions than H. Waelsch or H. Clarke, but his laboratory, except in instrumentation, did not measure up to the possibilities of our still new Kaiser-Wilhelm Institute in Heidelberg. On this first stay, we also visited friend Bernhard Witkop and family. He had just established himself at the National Institutes of Health in Bethesda. Among other marvels, he showed us the construction site of Building 10, which was just being put up. Twenty years later, during the time of my Fogarty Scholarship, I would become very well acquainted with it.

For the real purpose of my visit to the USA, I took the train from Grand Central Station in New York to New Haven where one of the participants of the symposium met me with a car. We soon reached a charming, small town, Ridgefield, where I was housed in an attractive, wooden building in colonial-English style. The symposium took place in the same building. By present-day standards, the number of participants was very small, indeed; about 50 competent researchers had assembled to discuss on one compound for 2 days [29].

Among others I reported on our synthesis of S-lactoyl-glutathione which proved to be identical with that formed by glyoxylase. We also elucidated the mechanism of the intra-molecular Cannizzaro reaction of α-ketoaldehydes to α-hydroxycarboxylic acids, which was shown to be a hydride migration of the aldehyde hydrogen atom. This led us to an attractive novel synthesis of α-amino acids: when the spontaneous internal redox reaction of the α-ketoaldehyde at the thiol was allowed to proceed in the presence of ammonium ions, the corresponding α-aminoacyl thiol, and thus the α-amino acid, was formed

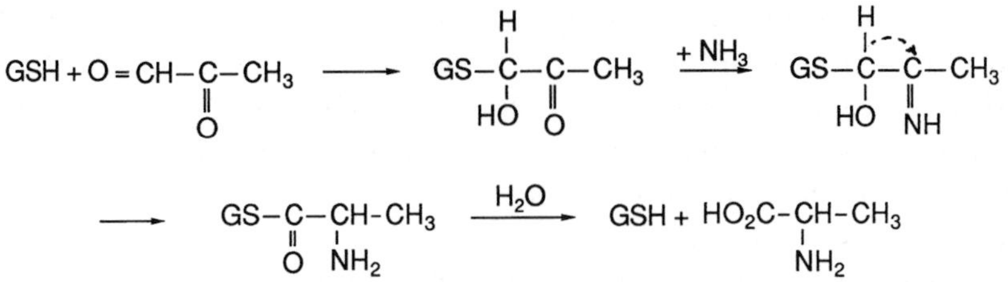

$$GSH + O = CH-C-CH_3 \longrightarrow GS-\overset{|}{\underset{HO}{C}}-\overset{||}{\underset{O}{C}}-CH_3 \xrightarrow{+NH_3} GS-\overset{|}{\underset{HO}{C}}-\overset{|}{\underset{NH}{C}}-CH_3$$

$$\longrightarrow GS-\overset{||}{\underset{O}{C}}-\overset{|}{\underset{NH_2}{CH}}-CH_3 \xrightarrow{H_2O} GSH + HO_2C-\overset{|}{\underset{NH_2}{CH}}-CH_3$$

Fig. 8. α-Amino acids from 2-oxo aldehydes and ammonia via S-aminoacyl compounds.

in good yield, with ketamine most likely an intermediate [30] (Fig. 8)

Among the speakers of the symposium were many scientists who later became famous, such as E. Racker [31], E.R. Stadtman, M. Calvin, H. Waelsch, K. Block, C.F. Anfinsen, W.W. Kielly, and D. Mazia. The atmosphere of the symposium was very congenial, and many of the friendships then formed have lasted to the time of the writing of this article. Unfortunately, Professor H. Waelsch died in 1965. We spent several enjoyable summer vacations in the Ticino with him, his wife Salome, and their children. To this day, we have maintained a cordial relationship with Salome who is a distinguished developmental physiologist and embryologist at Albert Einstein College (student of Spemann in Freiburg).

Johann Wolfgang Goethe University in Frankfurt a. M.

No sooner had we moved from Heidelberg to Mainz when I received an inquiry whether I would like to take over the vacant chair for organic chemistry at the University of Frankfurt. Although I knew the situation in Frankfurt, which was one of the cities most damaged by air raids, I did not ponder very long whether I should accept the offer. The chair was a prestigious one through Julius von Braun, Eugen Müller, and

Walter Borsche. Alexander Todd had done his doctoral thesis there, Louis Fieser was 1925/26 assistant of J. von Braun, and B. Helferich and K. Ziegler had worked there as associate professors. The medical-biological field had been greatly influenced by the accomplishments of the great Paul Ehrlich (1854–1915).

I accepted the offer and was asked to start giving lectures already during the summer semester 1951. Almost 1/3 of the Chemical Institute in the Robert Mayer Street, built in 1913, had been destroyed by aerial bombardment. Under the guidance of the undefatigable young assistant, Walter Ried, the students had repaired the worst damage with their own hands. Thus, laboratory practice could be started in the makeshift laboratories, and lectures could be given in the large, elegant auditorium which had just been rebuilt. Imperfect building conditions which were continuously improved, but never reached the ideal state, remained with me throughout my stay in Frankfurt. In spite of this, I felt comfortable there, and I became very attached to the old building and its inhabitants.

The University of Frankfurt was founded as late as 1914 as an endowment of the City under its able Lord Mayor Franz Adickes. It evolved from the Academy for Social and Economic Sciences by amalgamation with the buildings of the 'Physikalische Verein' and the 'Senckenbergischen Naturforschenden Gesellschaft'. What was, and is self-understood in the USA, namely, that citizens and private enterprises support the universities generously, is still the exception in Germany. But in Frankfurt, scientific interest, and from it the University, developed with the generous help of entrepreneur families such as Metzler, Grunelius, Passavant, Mumm, Rothschild, Lucius, and Meister (founder of the Farbwerke Hoechst). The University was supported by the City of Frankfurt until after the Second World War when its finances were incorporated into the budget of the State of Hessen.

Together with my Organic Chemistry colleagues, Walter

Ried and Manfred Wilck, I installed myself in the old, poorly
reconstructed building as well as possible. On the top floor,
under the roof, additional laboratories were established, especially for the biochemical department into which Gerhard
Pfleiderer was to move. To my great satisfaction, the interest
in biochemistry also spread to the neighboring Physicochem-
ical Institute of Hermann Hartmann, a gifted theoretical
chemist of comprehensive, humanistic general education.
(Originally, the journal, *Theoretica Chimica Acta*, founded by
him, took publications in German, English, French and
Latin.) In the same Institute, Joachim Stauff, Professor of biophysical and colloid chemistry, was active. In his laboratory,
Rainer Jaenicke developed a physical interest in the molecules of lactic acid dehydrogenases, which were investigated
in our Institute.

In Frankfurt, my exacting work load began with giving the
introductory lecture of organic chemistry. Each day, for one
hour, I had to present organic chemistry, accompanied by experiments, to an audience of all kinds of students, chemists,
biologists, future teachers and medical doctors. Starting with
elementary carbon, the course covered hydrocarbons, alcohols,
aniline, dyes, and many more ending with proteins. During
the first years, the rush of students was so heavy that, in spite
of the 400 seats in the auditorium, each lecture had to be given
twice. Of course, particularly popular demonstrations were
the distillation of brandy from cheap French 'grand filou', with
sampling of the product, or the fermentation of sugar solutions, or the coupling of an azo dye from a diazonium salt and
a corresponding phenol, or the production of nitroglycerol
which exploded with a loud bang when dripped on a hot plate.

Scientific research with local students started gradually.
Some students still worked in Mainz; others came with me to
Frankfurt. I worked on fields as different as chromatography,
electrophoresis, isozymes, peptide syntheses, model reactions
of oxidative phosphorylation, and natural substances like curare and poisonous Amanita mushrooms.

Half a century of mushroom research

The study of the deadly mushroom poison, amanitin, from the poisonous Amanita mushroom had its antecedent. In the first decade of this century, W.W. Ford in Baltimore, USA, worked on the enrichment of the poison from a white mushroom, probably *Amanita virosa*. However, he did not advance very far. At the beginning of the thirties, H.A. Raab in Munich, inspired by Heinrich Wieland, took up the theme. He turned to the domestic mushroom, *Amanita phalloides*. In his first publication, he reported a modest enrichment of a poisonous principle, and, importantly, recounted the diverse pre-history of amanitin research. In Munich in 1937, Feodor Lynen and, my cousin, Ulrich Wieland, succeeded in crystallizing the first homogenous substance, phalloidin [32] which, when injected, quickly caused the death of an experimental animal (white mouse). In 1940, R. Hallermayer isolated the slower acting 'amanitin' [33] in crystalline form. Amanitin and not phalloidin is the poison deadly after ingestion. In 1941, Bernhard Witkop, also a collaborator of Heinrich Wieland, was able to isolate from the hydrochloric acid hydrolysate of phalloidin *allo*-hydroxyproline (4-*trans*-hydroxy-L-proline), a hitherto unknown component, and 'Oxytryptophan' [β-(2-oxindolyl)-alanine].

At the beginning of molecular genetics oxytryptophan had become of great interest through investigations by A. Butenandt and A. Kühn in Tübingen. It had been found that the eye pigments (ommatins) of flies are formed from kynurenine, a physiological oxidation product of tryptophan. Oxytryptophan had been postulated as an intermediate in the oxidative pathway of tryptophan to kynurenine. Now it was available as a hydrolysis product of phalloidin. A sample from the Munich laboratory gave a positive effect in the eye test of mutant insects, but later on the authors found that the substance was not a normal product of tryptophan metabolism, but that kynurenine, arising from it by autoxidation, gave the positive ommatin effect.

Soon after the war we were able to resume the exploration of the components of *A. phalloides*, which had been started in Munich. First we tried to collect the necessary fresh mushrooms ourselves; later on, collaborators from the pharmaceutical-chemical firm, mentioned earlier, organized collection points in several forest rangers' houses in a wide area. There, mushroom collectors could hand in their unusual harvest against good pay. Additional mushroom hunts of our own with numerous students of the Frankfurt Institute, with a success party in a forest inn afterwards, gave welcome relief from the work in the laboratory. There was no lack of other diversions, such as annual boat rides on the Rhine, doctorate parties, etc.

Equipped with new analytical and preparative methods (paper electrophoresis, chromatography on cellulose), we could continue the Amanita research in Frankfurt. We soon recognized that the Munich 'Amanitin' consisted of two components which we called α-amanitin and β-amanitin, and that *Amanita phalloides* contains other 'amatoxins' (γ-amanitin, ε-amanitin, amanullin), all of which we isolated. Here, Angeliki Buku should be meritoriously recognized. My collaborator and colleague of many years, Heinz Faulstich, should also be singled out among the many other helpers whose important contributions are enumerated in a monograph [34].

Clearly, two families of toxic peptides could be distinguished: the amatoxins and the phallotoxins. Common to both of them is their constitution as cyclic peptides, the ring of which is crosslinked by an indol-sulfur bridge. Both also contain hydroxylated amino acids. However, their toxic action on the molecular level is fundamentally different. In the white poisonous mushroom, *Amanita virosa*, Faulstich found and characterized the virotoxins, another class related to the phallotoxins [35]. And finally, among other cyclic peptides, we discovered antamanide in *A. phalloides*, a peptide with antitoxic properties.

Amatoxins

Amatoxins are the components of the green Amanita mushroom, called the 'death cap', responsible for the still frequent deaths which occur after a meal of mushrooms. Descriptions of mushroom poisoning go as far back as antiquity. For instance, the Roman Emperor, Claudius, is said to have died from being poisoned by a meal of mushrooms which his wife and niece, Agrippina, had served him. She was afraid that he might withdraw the succession of his throne from her son, Nero, in favor of his son, Britannicus. Since death from poisoning sets in only after several days (up to eight days) the impossibility to prove its presence has made it into a sinister and horrible instrument both in legend and *de facto*.

Apart from the green 'death cap', the amatoxins are also found in the white *Amanita virosa*, which occurs more frequently in North America, where it is called 'destroying angel'. Considerable quantities of toxins were also discovered in *Galerina* and ('Morgan's') *Lepiota* species.

Even before the investigations of the amatoxins led to structural formulas, I developed a color test with E. Fischer and H. Dilger: when the dried mushroom juice is spotted with concentrated hydrochloric acid on newspaper, due to the presence of lignin, a clearly visible blue color appears. Today, the identification of amatoxins in blood or urine of a patient is possible with other very sensitive tests, for example a radioimmunoassay, devised by H. Faulstich, or by chromatographic methods.

As is known from fatal mushroom poisoning of humans, amatoxins develop their toxic effects slowly. Death may occur as late as 8 days after ingestion. Eight to ten hours after the meal, at the earliest, there is a severe intestinal reaction, followed by pseudo-recovery. On the third day, necrosis of the liver sets in that may lead to death within several days. An important measure of therapy, among other approaches, is the administration of silymarin, the active principle of the thistle *Silybum marianum*, which inhibits entrance of the toxin into the liver cells. If the therapy is not started before the third day, the prognosis is unfavorable. Other therapy measures have been proposed repeatedly over the last decades. A survey, still valid to-day, was presented by competent researchers at a symposium in Heidelberg in 1978 [36].

Unsuccessful experiments were undertaken to treat amanitin poisoning with an immune serum, specific for ama-toxins, similar to the treatment of snake bites. On the con-trary, amanitin became twice as toxic when bound to its im-munoglobulin. It became 50 times more toxic when complexed with a Fab-fragment of a monoclonal antibody specific for amatoxins [37]. The experimental animals died of kidney fail-ure while their livers were undamaged.

For the elucidation of the mechanism of action of amatoxins, Wieland and Dose [38] in 1954 applied paper electrophoresis to serum proteins of poisoned rats and mice. In the grass frog, we found that in a few days after injection of α-amanitin, the serum proteins, mainly albumin, were drastically reduced. In those days, the biogenetic connection between nucleic acids and proteins was not yet known, and the finding of an inhibi-tion of protein synthesis could not be further explored. Only ten years later, Luigi Fiume in Bologna and his co-workers began successful investigations on the pathology of amanitin poisoning. The first change in the cell nucleus of liver and kidney of mice, observed with the electron microscope already after 30 minutes, is a fragmentation of the nucleoli (Fiume, Laschi [39]), coinciding with an *inhibition* of RNA-synthesis. This was the fundamental observation which was then further elaborated and refined at several places.

The further story has been described in a monograph [34] and references cited therein. It led to the realization that in the cell nucleus of all eukaryotes the DNA-dependent RNA-polymerases II (or B) is inhibited by 80%, already in 10^{-8} M concentration, by α-amanitin and almost all other amatoxins. This enzyme is responsible for the synthesis of messenger-RNAs. Eukaryotic RNA-polymerase I (formation of ribosomal

RNAs) is practically resistant, as is the RNA-polymerase of prokaryotes (bacteria etc.). This enzyme is also not inhibited in mitochondria, which is one of many clues that these organelles are originally of bacterial nature. RNA polymerase III (responsible for transfer-RNAs) is inhibited by amanitin concentrations 1000-fold those effective at polymerase II.

The strong inhibition of m-RNA synthesis explains why amatoxins induce cessation of protein synthesis and thus necrosis of liver and kidney, and why the deadly effect does not occur at once, but only after a fairly long delay. Twenty years later, we closely co-operated on this subject with Pierre Chambon in Strasbourg, after the elucidation of the chemical structure of the amatoxins had advanced enough to be able to incorporate tritium into the molecule, and thus make possible quantitative studies. Today, α-amanitin is commercially available as a specific bio-reagent, to detect by its inhibitory action all events in which transcription of DNA into m-RNA participates. In contrast to direct inhibitors of protein synthesis, such as puromycin and cycloheximide, which inhibit the elongation of the polypeptide chain, α-amanitin specifically inhibits the formation of RNAs, the initiators of protein synthesis, by preventing the translocation of the RNA polymerase on the DNA strand to the site of the next phosphodiester formation (A.C. Vaisius [40]).

With tritium-labeled amanitin, one can explore to which bio-molecules – beside RNA-polymerase II – α-amanitin is bound. Amatoxin, containing fluorescein, has also been used in cell research [41], and amatoxin, bound to a solid phase, has served to separate RNA-polymerase II specifically [42]. The possibilities for application in molecular biology are by no means exhausted.

Toxic effects of the phallotoxins

In 1937, phalloidin was obtained from A. phalloides as the first homogenous and crystalline compound [32]. It belongs to the

fast-acting poisons of the mushroom. A dose of 150 μg kills a
white mouse of 20 g in a few hours when given intraperitone-
ally (not orally).

After a few not very informative experiments with mice by
M. Vogt in W. Straub's laboratory in Munich, the elucidation
of the mode of action was pursued again by my younger
brother, Otto Wieland at the end of the forties. The difficult
experiments first demonstrated no drastic deviation of the
various parameters from the normal. But then it was recognized that, under the influence of phalloidin, the livers of mice
and rats swelled to almost twice their size, and became dark
red and brittle. The increase in volume was caused by the infiltration of blood into the liver. The increase of blood inside the
liver led to a decrease in the periphery, and thus to a 'bleeding
to death into the liver'.

In the following years, physiologists and pharmacologists
learned to keep heart and intestine, as well as liver, of experimental animals functional outside the body for several hours.
The search for the cause of phalloidin poisoning continued in
perfused liver preparations. In 1960, Otto Wieland and F.
Matschinsky observed an instantaneous stop of the flow of bile
and the formation of vacuoles which enlarged in the course of
the experiment. In experimental animals the vacuoles also
contained solid blood components, such as erythrocytes. This
indicated endocytosis through invagination of the plasma
membrane.

In 1965, the pathologist Liugi Fiume of the University of
Bologna, to whom we owe the first observations of the toxic
effects of the amatoxins (p. 61), observed that newborn mice
and rats are resistant against a five-fold lethal dose of phalloidin in the first two weeks.

This phenomenon of the resistance of young rodents remained of interest, and was pursued further in Otto Wieland's
and, later, in our laboratory by my son, Eberhard and A. Walli.
Using radioactively labeled phalloidin, they found that the livers of 12-day-old rats incorporated the toxin considerably

more slowly than adult animals. Therefore, the excess was eliminated in the urine. They also observed that the young liver remained almost unchanged in structure when it had far surpassed the concentration of toxin, critical for that of grown animals. To the present day, a full explanation for this resistance cannot be given.

After the development of the perfused liver preparation, the next technical advance was the division of the organ into its functional hepatocytes with the aid of collagenase by the Americans, Berry and Friend, in 1969. Frimmer and co-workers were able to observe a direct action of phalloidin in such cells; in micromolar concentration it caused, already after ten minutes, soap bubble-like protrusions of the plasma membrane [43] (Fig. 9). The protrusions did not revert, nor did they influence the vitality of the cells, as was proven by exclusion of

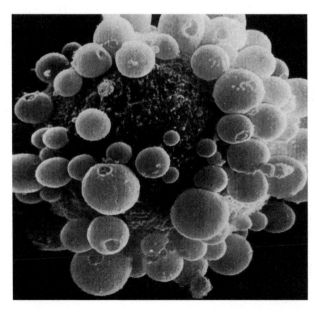

Fig. 9. Deformation of a hepatocyte as a consequence of phalloidin in vitro.

trypan blue. Thus, the pathological action of phalloidin could be explained as de-stabilizing the cytoplasmic membrane, which in the *in situ* organ permitted entry of the blood by endocytosis, and, in the isolated cell, led to the formation of blebs, caused by the pressure inside the cell. We found that death in mice by phalloidin, and blebs formation on hepatocytes could be prevented by antamanide.

Antamanide

When working up large volumes of methanol extracts of the green poisonous mushroom, minor components were found besides the toxins. In one of the fractions J. de Vries, detected an antitoxic component which later was crystallized from a greater amount of mushrooms [44]. We gave it the name antamanide ('anti-amanita peptide'). When 0.5 mg/kg of antamanide was injected into a white mouse, it prevented certain death from an injection of 5 mg/kg of phalloidin, given a few minutes later. Antamanide was shown to be a cyclic decapeptide, consisting of only 4 sorts of amino acids: cyclo-(ValPro-ProAlaPhePheProProPhePhe) (see Fig. 10).

The elucidation of the structure of antamanide took place in the years 1967–1968, the last years I spent at the University of Frankfurt.

As early as in 1966, Richard Kuhn had asked me whether I would accept an appointment as his successor at the Max-Planck-Institute in Heidelberg (as the former Kaiser-Wilhelm-Institute was now called). Kuhn was going to retire at the age of 68 in 1968. It did not take long reflection on my part to express acceptance, especially when taking into account the student unrest, just started, which forebode a heavy and irreparable impairment of university life.

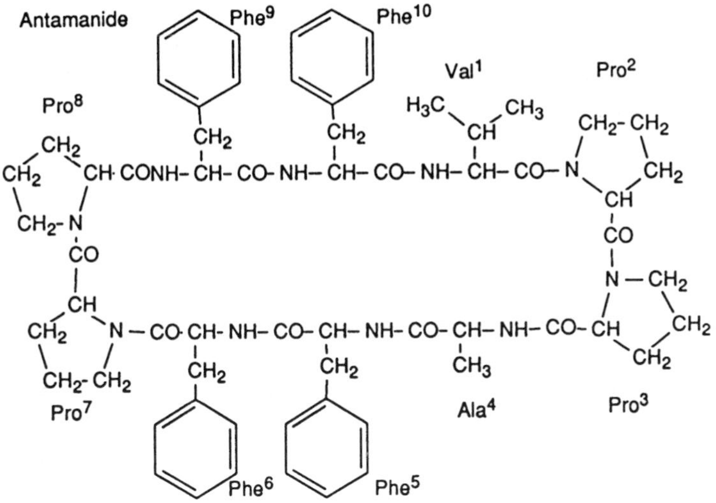

Antamanide

Phe^9 Phe^{10}

Pro^8 Val^1 Pro^2

Pro^7 Phe^6 Phe^5 Ala^4 Pro^3

Fig. 10. Antamanide.

An import system in liver cells

The synthesis of antamanide presented no difficulties to us as peptide chemists. It was followed by the synthesis of a few dozen analogous cyclopeptides in which building blocks were exchanged or substituted by others. Thus, the influence of structure on the antitoxic (or 'cytoprotective') activity could be elucidated. We found a variant in which phenylalanine no. 6 was replaced with ^{3}H-containing lysine and in which the ε-aminogroup of the lysine contained a photolabile anchor to be effective. When rat hepatocytes were illuminated in the presence of this radioactive antamanide, followed by electrophoresis of the proteins in the presence of sodium lauryl sulfate, the antamanide was shown to be fixed to two membrane proteins.

The same experiment was performed with a photolabile, radioactive phallotoxin with the result that the same polypeptides appeared as carriers of the toxin which had strong affinity to antamanide, and also for (radioactively labeled) bile

salts [45]. Thus a molecular proof was given of the cytoprotective property of antamanide as a competitor, and of the well-
substantiated postulation of Max Frimmer's group of the existence of a multivalent transport system in liver.

Amatoxins are transported into the liver cell by the same
pathway, but much slower. *In vivo*, their incorporation cannot
be prevented by antamanide which even inhibits their excretion by the bile. Here, many questions on bile formation and
secretion are still open, which, perhaps, can be brought closer
to an answer with the aid of these rare mushroom components.
After this intermezzo about antamanide which will be continued later, I want to return to phalloidin, amatoxins and the
Institute in Frankfurt.

Structures of the phytotoxins

The elucidation of the structure of the phytotoxins occurred
during my years in Frankfurt, after I had obtained first insights in Mainz. Both toxin families contain sulfur. Neuberger
[46], Cornforth, and Dalgliesh in London suspected that the
sulfur bond of *phalloidin* was located at the 2-position of the
indole ring of tryptophan, since in the ultraviolet spectrum of
phalloidin the absorption maximum of the indole ring was
shifted to higher wavelengths than in tryptophan. We were
able to confirm this, since after heating phalloidin with Raney
nickel (substitution of the sulfur atom by hydrogen) a normal
tryptophan spectrum was obtained. From β-indolylacetic acid
and methylsulfenylchloride, we synthesized 2-methylthioin-
dolyl-3-acetic acid as a model, the UV-spectrum of which was
identical with that of phalloidin. The desulfurized phalloidin,
dethiophalloidin, still was a cyclic peptide; therefore, phalloidin itself had to have a bicyclic structure. A peptide bond of
dethiophalloidin could be opened hydrolytically. The linear
peptide, thus obtained, was sequenced by W. Schön, using the
Edman degradation method. All details, as well as the detec-

tion and structure elucidation of additional phallotoxins, phalloin, phallacidin etc., are reported in the monograph [34],
mentioned above.
The structural formula of phalloidin is shown in Fig. 11a. The
space formula (11b) was already proposed in 1973 by D.J.

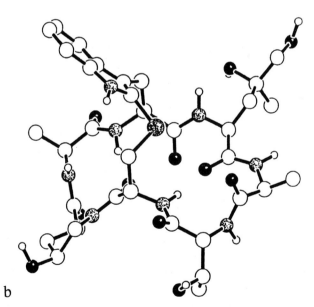

a   Phalloidin (H instead of OH: phalloin)

b

Fig. 11. Structural and space formula of phalloidin.

Patel, A.E. Tonelli and our laboratory, based on NMR data and minimal-energy calculations [47]. In the last two years, it was confirmed and further refined by H. Kessler with 2-dimensional NMR techniques and NOE effects.

The elucidation of the structure of α-amanitin presented greater difficultires. As with phalloidin, sulfur could be removed from the molecule, using Raney-Nickel. Thereafter, the UV spectrum showed the presence of a hydroxyl-substituted tryptophan. The position of the OH-group at C-6 could be deduced by UV-spectroscopy since the absorption curves of the isomeric hydroxyindoles are distinctly different.

The indole derivatives, hydroxylated in different positions, could also be differentiated by their color reactions. Treating a sample on filter paper with cinnamaldehyde and then hydrogen chloride vapor, was one of the most sensitive ones. 6-Hydroxyindole, and therefore α-amanitin, gave an intense blue-red color, while 5-hydroxyindole, and therefore bufotenin from the nontoxic, yellow *Amanita citrina*, gave a less intense brown-red color. Thus using chromatography it was easy to differentiate the two species.

In the first formula for α-amanitin in the mid-60s, U. Gebert and I postulated a thioether structure, as in phalloidin. Spectroscopic disagreements then induced H. Faulstich to correct the bridging to that of a S-sulfoxide. (Fig. 12)

W. Boehringer proved that α-amanitin contained the side chain of asparagine, $-CH_2-CO-NH_2$, and α-amanitin that of the corresponding, free aspartic acid, $-CH_2-CO_2H$. He succeeded in chemically converting β-amanitin into α-amanitin. In 1977, Kostansek et al., in the laboratory of W.N. Lipscomb, carried out the first crystal structure analysis on β-amanitin [48].

Antamanide formed a complex with sodium ions which showed up in the mass spectrum as ion m/e 1147 + 23 (= Na). In those days, compounds which complexed alkali metal ions were of the greatest biochemical interest, since in 1964 C. Moore and B. Pressman had discovered that the antibiotic valinomycin facilitated the uptake of potassium ions into mito-

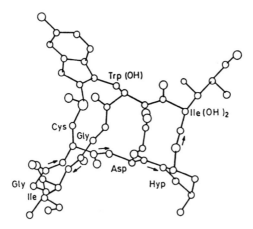

Fig. 12. Structural and space formula of α- and β-amanitin.

chondria. In the following years, many other ionophores were found which mediate the transport of alkali metal ions through natural and artificial lipid membranes [49]. The im-

portant Institute of the Academy of Sciences of the USSR in
Moskow was among the laboratories active in this field. There,
M.M. Shemyakin with many co-workers investigated the
structures and ion-binding capacities of valinomycin and
other cyclic depsipeptides.

An impending visit to this and other institutes in the USSR
prompted me to get in touch with Shemyakin, Yuri A. Ovchinnikov and V.T. Ivanov. Through further meetings in subsequent years, these contacts intensified. The Russians measured the Na-complex constant of our peptide by infrared, ORD,
and conductivity techniques, while we (with W. Burgermeister
and others) used measurement of potentials with ion-specific
glass electrodes, and vapor pressure determinations. The results were published in a modest joint publication [50]. The
structure of the crystalline Li^+-complex was solved later on by
Isabella L. Karle et al. [51] (Fig. 13).

The Russians found that antamanide was not a biologically
active ionophore. Its cytoprotective action is not directly connected with an ion-binding capacity.

Nobody could have foreseen the decline and end of the giant
Empire taking place only two decades later. So perhaps, a report on visits to the former USSR would be of interest, as it is
given in the Appendix on pages 90–96.

Oxidative phosphorylation

In the second half of the thirties, it was recognized, mainly in
the laboratories of O. Meyerhof (with K. Lohmann) and O.
Warburg, that adenosine triphosphate (ATP) is the general
energy carrier and transfer agent in the cell. In a memorable
paper of 1941, Fritz Lipmann explained the principle of 'group
potential' not only for phosphoryl, but also for other acylating
groups [52]. From measurements of ATP production, for example in mitochondria, as dependent on oxygen consumption, it
had become evident that ATP is formed in a more productive

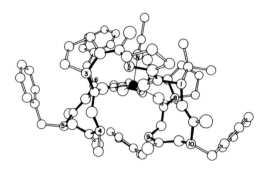

Fig. 13. Conformation of the antamanide-Li+-complex, crystallized from acetonitrile (from ref. 67).

pathway by oxidation of nutrients through the so-called respiratory chain. In this process the electrons are transferred from the substrate, to be oxidized to respiratory oxygen through many intermediate steps.

The unraveling of the main pathway of ATP formation constituted a great challenge for biochemistry. Modeling on the known mechanism of substrate phosphorylation, many biochemists considered analogous processes, by which oxidation of known or still unknown links in the respiratory chain might yield energy-rich phosphate bonds. As organic chemists, we also 'tuned into this channel'.

With Edmund Bäuerlein, who pursued this area by himself later on, we composed numerous *in vitro* systems, in which inorganic phosphate was converted to energy rich phosphate bonds by oxidation. With added ADP, these bonds reacted to form ATP in up to 25% yield. The simplest process proved to be oxidation of monophosphoric acid esters of hydroquinones which were activated as phosphorylating agents. More complicated was the system of thiols or disulfides, ADP and inorganic phosphate in which, on oxidation with hemin and oxygen, up to 17% of ATP was formed [53].

As with all oxygen-containing anions, in principle, the 'activation' of the phosphate ion consists of the elimination or displacement of electrons away from the central atom which becomes electrophilic (P^+ as the phosphorylating agent) and can react with a nucleophilic anion, such as ADP, to form ATP. Oxygen, as end oxidant, acts through a series of electron transferring intermediate compounds (pyridinium ions, flavines, quinones, Fe-porphyrin complexes) and reactive protein side chains (Fe- and Cu-sulfur clusters). All these were tried, theoretically and experimentally, in model attempts by us, and in other laboratories (for example V.M. Clarke in Cambridge and E. Lederer in Gif-sur-Yvette).

The function of a proton gradient as an energy source of a 'chemi-osmotic' ATP formation (in mitochondria), as shown by P. Mitchell [54], can be understood as an oxidative activation of phosphate. Protons, as well as other cations, are electrophilic particles which 'activate' by electron withdrawal, and therefore oxidize. (For example, protons oxidize metals to metal ions.) With the methods of the chemist, it is hardly possible to unravel and prove the molecular mechanisms which are involved in a cationically driven ATP-synthesis in the cell.

Conceptions exist, concerning how chemical energy is transformed into 'energy rich' bonds, or *vice versa*, and how the energy-rich bond in ATP is converted into mechanical work, such as muscle contraction. Generally, these theories rest on the assumption of conformational changes of the participating proteins. Can we expect that sometime in the future such, perhaps small, conformational changes can be 'seen' by modern, physical methods?

Enzymatic peptide synthesis

As is evident from the preceding narrative, we always were interested in the biological formation of the peptide bond. We were occupied with peptide synthesis via activated amino acids when H.G. Zachau, in Lipmann's laboratory, elucidated the nature of activated amino acids as being esters at the 3-

hydroxyl group of the terminal adenosine residue of transfer-RNA. Alkylesters had been known, since Th. Curtius in 1882, to form peptide bonds but so slowly that this type of reaction could be assumed to proceed in ribosomes only with the aid of a catalytic environment.

We could show *in vitro* that peptide synthesis with methylesters of amino acids is strongly accelerated by imidazole. In those days, histidine sidechains in the active center played a significant role in the explanation of enzymatic catalysis.

As early as 1958, Friedrich Jaenicke in my laboratory, synthesized the *primary* form of biologically activated amino acids, namely anhydrides with adenosinemonophosphoric acid. A review of the 'boom' period in synthetic peptide chemistry is given in comprehensive treatises, the last of which includes the period from 1956 to 1963 [55]. With the co-author of that review, Helmut Determann, I conducted several studies on enzymatic peptide bonding. In a series of subsequent studies, Determann thoroughly examined the problem.

The plastein reaction

At the beginning of this century, the Russian Savyalov coined the name 'plastein' for the insoluble precipitates which are formed by the action of proteinases on partially hydrolyzed fibrin and other 'peptones'. Informative investigations of this remarkable polypeptide synthesis were taken up by A.I. Virtanen and co-workers after 1948. They unequivocally elucidated the nature of this polypeptide mixture. To exactly define the conditions and sequence of the reaction, it seemed desirable to obtain defined peptides which, in the presence of proteolytic enzymes, could be polymerized to plastein-like macropeptides. We, therefore, began to isolate homogenous plastein active peptides from the very complex proteolysis products of a protein. For testing we used polymerization with pepsin at pH 4. The easily accessible Witte-peptone served as a starting material. In 1959, H. Determann and E. Albrecht began with 2.5 kg of Witte-peptone. The mixture of hundreds of components was subjected to precipitation with acetone, chromatography, thin layer electrophoresis, and further chromatography on ion exchange resins. Two peptides, consisting of 8 and 10 amino acids, were isolated which were highly active in the formation of plastein. The composition of these peptides was essentially determined by sequencing. Both carried lipophilic amino acids at the amino- and carboxyl-end.

Synthetic experiments soon brought forth the standard pentapeptide, tyrosyl-isoleucyl- (or leucyl-)glycyl-glutamyl-phenylalanine. Subsequently, while preserving the amino acids on both ends, the inner building blocks were varied. We found that the degree of polymerization was a function of the solubility of the product. Depending on the nature of the building blocks, it could already precipitate after doubling of the monomeric pentapeptide. However, when proline was substituted for glycine, homologous plasteins up to the pentamer (25 amino acids) were formed [56]. The tetrapeptide was found to have the minimum length for plastein active peptides to be polymerized.

The period of collaboration with H. Determann belongs to my particularly pleasant memories of the Frankfurt years. Later on in Heidelberg, I still tried to use this elegant method to connect, in a practical manner, peptide fragments which contained suitable terminal amino acids. In spite of partial successes, this effort was not continued, since classical peptide chemistry, including use of solid phase synthesis, had reached a level which could not be easily surpassed by the generally interesting biochemical method.

Again in Heidelberg

Richard Kuhn had definite plans for his activities before and after his retirement in 1969. After acceptance of the appointment as Director of the Department of Chemistry at the Max-Planck-Institute for Medical Research in Heidelberg, I was immediately to move into Kuhn's office in the northeast wing. He wanted to take over an office and a few laboratories in the northwest wing, to work on problems of his own choice, but also to collaborate with me.

During my frequent visits to the Institute in Heidelberg, already before receiving the offer, I had made close contact with Kuhn's last doctoral student, Hermann Kühn, who worked on crustacyanine, the greenish-grey protein pigment of crayfish and lobster shells. Unfortunately, this very promising biochemist, who was also an expert mountaineer (he took part in one of the Nanga Parbat expeditions), was killed by engulfment with a cornice in the Vallois Alps a few years ago.

The protein studies of Kühn, in Kuhn's laboratory, coincided with our interest in peptides. Kuhn had included this in his plans mentioned above, and he asked me to be of assistance to him and his student in the writing of his publication. This happened while Kuhn's state of health progressively deteriorated. From Frankfurt, I visited him several times in the Czerny Clinic in Heidelberg, to discuss the future of the Institute. To our great sorrow, already in the summer of 1967, Kuhn died after an operation of a carcinoma of the oesophagus. He was interred in a family grave on the mountain cemetery in Heidelberg. At the memorial service on November 15, 1967, in the aula of the New University of Heidelberg, Otto Westphal gave a moving review of Kuhn's character and life-work [57].

In May of 1968, I was appointed Scientific Member of the Max Planck Society, and Director of the Institute for Chemistry at the Max-Planck-Institute for Medical Research in Heidelberg, an institute which, then bearing the name 'Kaiser-Wilhelm-Institute', I had joined as an assistant in 1937.

Simultaneously with my taking charge, a thorough renovation of the laboratory facilities which dated back to 1930, the year of its opening was taking place. Now, greater demands were made on the purity of the laboratory air. The old hoods could not provide the amount of air circulation mandated now, and the pipes and conduits for gas, water and electricity had to be replaced. Also, much new laboratory equipment had to be purchased. I asked Dr. Christian Birr, a successful peptide chemist and one of the assistants who had come with me from Frankfurt to Heidelberg, to supervise these renovations. Without his talent for organization, we could not possibly have moved into the renovated wing during 1968.

Others who followed me to Heidelberg, were Dr. Angeliki Buku, the masterly mushroom poison investigator, Dr. Heinz Faulstich, the inspired co-thinker, so rich in ideas, and Dr. Edmund Bäuerlein who worked in a quite different area, model reactions of oxidative phosphorylation (p. 73). Also at the Institute in Heidelberg were several tenured co-workers from Kuhn's laboratory, e.g. Dr. Adeline Gaŭhe, Dr. Irmentraud Löw, Dr. Werner Jahn, with whom I had fruitful collaborations.

My family now included two sons. Eberhard was born in 1956 in Mainz. Towards the end of 1968, we moved from Mainz to Heidelberg into the house in Wilckens Street in which Richard Kuhn had lived with his large family. The house and two others were built in 1930 in red clinker brick style at the same time as the Institute and by the same architect, H. Freese. The house next to the Kuhn house initially was inhabited by Otto Meyerhof, the Director of the Department of Physiology, then by the physiologist Hermann H. Weber, and after his death by one of his successors, Wilhelm Hasselbach. In this sequence also, the exploration of the biochemical and physiological events of muscle action was continued, from the formation of lactic acid to the regulation by the Ca^{2+} level under participation of the sarcoplasmatic reticulum. The house, diagonally across, initially was inhabited by the Director of the Department of Physics. K.W. Hausser, and then, after his early death, by his successor, Walter Bothe (Nobel prize for physics in 1954), followed by Wolfgang Gentner. In

1975, we moved into our own house in wooded Schlierbach. Shortly thereafter, my wife, Dr. Irmgard Wieland, a trained physician, was able to pursue her profession again for ten years. She worked as a physician at the State Health Office in Heidelberg.

The Institute for Medical Research had quarters for small animals which could be used for pharmacological experiments. At the beginning of my time in Heidelberg, one could conduct toxicity experiments on mice and rats and also skillfully remove organs, such as liver, for transfusion experiments, without having to obtain written permission from the responsible authorities. This circumstance, as well as the not too petty rules for the handling of radioactive substances, had a very favorable effect on our investigation of the toxic and antitoxic properties of the mushroom substances. Many a project, which would have been much more difficult or impossible to work on in Frankfurt, was made possible in the well-equipped institute.

In Heidelberg we now turned more intensively to the search for molecular mechanisms underlying the effects of the peptides from A. phalloides. The structural characteristics, responsible for toxicity, were to be elucidated by modifications of the natural product and by synthesis of methodically modified amatoxin analogues. Also of interest was the binding site for the toxin at the complicated RNA-polymerase II, and the manner of its interference in the process of transcription. Since toxicity was a function of the inhibitory action on RNA polymerase II, isolated from cell nuclei, toxicity determinations on experimental animals were not required, and our studies were monitored with the in vitro test of inhibition of transcription.

During one of our stays in Bethesda, Maryland, we walked through the forests of West Virginia with Bernhard Witkop and found a number of the white 'destroying angels' which are very rare in Germany. The specimens were frozen immediately after harvest and sent to Heidelberg in insulated containers in dry ice. They were worked up by A. Buku and H.

Faulstich [58]. In chromatograms of the total extract, no α-, β-, or ε-amanitin were found. Instead, they discovered an equally toxic component which was identified as amaninamide. Amaninamide differs from α-amanitin only through the lack of the 6'OH-group at the indole ring.

Since the sulfoxide-oxygen plays no role in the toxic activity, synthetic experiments in the amatoxin series could be limited to the use of tryptophan which is more easily accessible than 6-hydroxytryptophan, and to the use of an oxygen-free thioether bridge. Here, Giancarlo Zanotti both in Heidelberg and in Rome did some excellent work.

The analysis of the components of the white Amanita from the USA yielded, besides phalloidin, a further group of fast-acting toxins which we called *virotoxins* [59]. The formula of the virotoxins is shown in Fig. 14. The difference is conspicuous since the virotoxins are not built bicyclically. Instead of the thioether bridge, the indole part contains a methylsulfonyl residue. The other side contains the sidechain of D-serine. A second hydroxylgroup was found at the proline ring. Hitherto, 2,3-*trans*-3,4-*trans*-3,4-dihydroxyproline had not been found in nature. Most likely, the additional hydroxyl groups in the virotoxins are responsible for their toxicologically equivalent activities with phallotoxins.

The mechanism of the action of phallotoxins (and virotoxins) so far had been understood as follows: the phytotoxins induced rapid formation of vacuoles in liver cells, caused by the endocytotical penetration of blood through a destabilized membrane. A further inspection of the rat liver, damaged by phalloidin was now started by Dr. B. Agostini with the electron microscope. In many slices, numerous filaments, lumped in bundles, were observed. *In vitro* these filaments could be produced by the action of phalloidin on the membrane fractions of unpoisoned rats [60]. These filaments, although not soluble in 10% KI solution, consisted of F-actin which was changed in its characteristics by the presence of phalloidin.

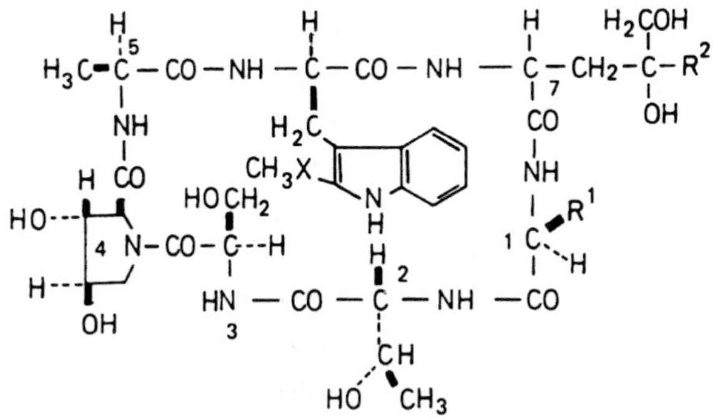

R^1: CH_3 or CH_2OH R^2: CH_3 or $CH(CH_3)_2$ X: SO or SO_2

Fig. 14. General formula of the virotoxins.

Phalloidin and actin

In the following years the action of the toxin on actin was studied extensively. We used an actin preparation from rabbit muscle. The 'genius loci' was favorable. In the neighboring Department of Physiology there was a long tradition in this field, since H.H. Weber, in 1954, had continued his investigations on muscle proteins there. His successor, W. Hasselbach, studied the regulation of muscle contraction. In his institute, the isolation of actin from rabbit muscle was a routine operation.

Actins form a large family of mostly similar proteins with a relative molecular mass (Mr) of about 43,000 Daltons. They occur in all eukaryotic cells as components of the cytoskeleton which pervades the interior of the cell as a mesh of several high molecular proteins. The reversible polymerization of G-actin to F-actin filaments, consisting of many hundreds of uniformly composed monomers, is a property of actins which determines their biological role. In the system of pure actin, with potassium, magnesium or calcium ions, and ATP, the equilibrium is almost completely on the side of F-actin. The concentration of unpolymerized actin then amounts to only 20 mg/l. Normally, below this 'critical' concentration no polymerization takes place, and F-actin completely dissociates into G-actin when a solution is diluted to such a low concentration. The kinetics of polymerization can be followed through viscosimetry or measurements of light dispersion.

We found that phalloidin induced the polymerization also in subcritical concentration. F-actin, produced in the presence of phalloidin, was stabilized. It did not revert to G-actin, even in several hundredfold dilution. We further found that phallotoxins stabilized F-actin filaments also against chaotropic agents, such as high concentrations of urea or potassium iodide. The stabilizing action on F-actin was also found against ultrasound, enzymatic proteolysis and heat. The dissociation constant K_D of the actin-phalloidin complex is rather small, about 10^{-8} M.

I was reminded of the heat stability of lactic acid dehydrogenases in the presence of NAD and sulfite (p. 44), or the stabilization of RNA-polymerase/DNA complexes by amatoxins (p. 63), examples which showed that the unfolding of protein structures is inhibited by specific, tightly bound molecules.

We studied the influence of changes in structure of the phallotoxin molecule on the affinity to F-actin on a large scale, after, in 1971, F. Fahrenholz synthetically had obtained norphalloin, a fully active compound, and after, in 1977, E. Munekata had obtained a natural phallotoxin, phalloin, by total synthesis [61]. All variations and syntheses cannot be enumerated here; they can be found in a comprehensive monograph [34].

I only want to mention the finding that the nature of the side chain no. 7 (Fig. 11) has little influence on the affinity of the molecule to F-actin. By chemical manipulation, reactive groups could be attached there, suitable for affinity labeling. Fluorescing molecules could also be attached which made phalloidin a sensitive reagent, with which one could visualize actin filaments in all kinds of cells. A timely review of these applications has been presented by H. Faulstich [62].

Kishino and Yanagida devised the following sophisticated way to measure the tensile strength of a single actin filament. Since, fluorescent phalloidin-like phalloidin prevented filaments completely from depolymerization, a filament stained with the fluorescent probe could be clamped between two moving glass microneedles and its stretching, until it tore, could be observed under a fluorescence microscope [63]. In our laboratory, Werner Jahn used the reagent to visualize the influence of phalloidin on actin in the liver cell. For this purpose, rat livers were perfused with a solution of phalloidin. After staining with fluorescent phalloidin, sections of such a liver exhibited the differences, shown in Fig. 15, as compared to normal liver [64].

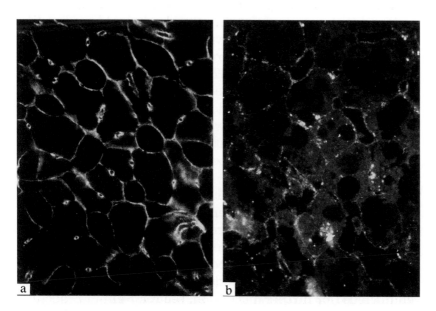

Fig. 15. F-actin filaments stained with fluorescent phallotoxin in (a) normal rat liver cell and (b) poisoned with phalloidin (from [64]).

Under the influence of phalloidin, the homogenous F-actin layer, beneath the lipid membrane, is torn, and the fragile membrane is exposed to the smallest changes in pressure. I postulate this to be the cause of the symptoms as indicated by all observations of liver poisoned by phalloidin. Microcystins, liver toxic peptides from blue-green algae (Cyanobacteria) also induce formation of protrusions on isolated hepatocytes, although morphologically different from actin blebs (Fig. 9). Microcystins do not react with actin in a stabilizing fashion; they reorganize the architecture of the cytoskeleton and, therefore, the pattern of the F-actin distribution [65], as a consequence of strong inhibition of protein phosphatases [66].

We used the technique of affinity labeling to elucidate the binding site of phalloidin to actin. Axel Deboben and Michael Nassal introduced tritium-containing anchor groups into phalloidin. These derivatives were incubated with F-actin, one of them under photolysis. Joe Vandekerckhove, in Ghent, determined the regions of the polypeptide chain where bonding had taken place, namely, at two different positions, at amino acids 115 and 355 which, therefore, had to be vicinal in the tertiary structure of F-actin [67].

Recently, the structure of the crystalline complex of G-actin with DNAse-I has been solved by X-ray analysis in the laboratory of Ken Holmes [68]. It was shown that the two regions to which phalloidin was fixed covalently, and which appeared to be far apart, in fact, in the tertiary structure were not very distant from each other.

This region is distant from the one where the myosin heads act at muscle contraction. Therefore, phalloidin has no influence on the contraction of glycerinized muscle fibres by ATP. Perhaps, I should also mention observations of a participation of actin, and hence the influence of phalloidin, in the process of transcription. There are considerable amounts of actin in the cell nucleus (for further discussion, see ref. 34).

Acylcoenzyme A without sulfur

Charles Jack Stewart, my first American co-worker (now professor at San Diego State University) came later on two sabbatical leaves to Heidelberg for a longer time. He had been familiar with coenzyme A from his earlier studies in Frankfurt, and once again chose it as the object of his studies. We were interested in the behavior of an *S*-acetyl-CoA in which the sulfur atom had been substituted by a methylene group. Such a compound could not act as an acylating agent, but, because of its analogous structure, might bind to acetyl-converting enzymes. For its preparation, one could not start with coenzyme A, but had to synthesize it 'from scratch' [69]. Acetono-CoA, as we called the compound, turned out to be a highly competitive inhibitor of the enzymatic synthesis of citric acid from *S*-acetyl-CoA and oxaloacetate as a pseudosubstrate. We did not investigate whether it reacts further with oxaloacetate to form a carba-analogue of citryl-CoA. Hermann Eggerer in Munich studied this system further. He alkylated the sulfur atom of CoA with the 3,4-dicarboxy-3-hydroxybutyl residue, corresponding to citric acid. 'Pseudocitryl-CoA', thus obtained, also proved to be a strong inhibitor of the enzymatic hydrolysis of citryl-CoA [70].

Acetono-CoA, as a pseudosubstrate, could be carboxylated by rat liver acetyl-coenzyme-A carboxylase which catalyzes the binding of carbonate to form malonyl-CoA. T. Shiba et al. proved that the carboxylation occurred at the methylgroup of the substrate, analogous to the natural substrate [71]. Further investigations with J. Knappe and H.P. Blaschkowski showed that acetono-CoA (and *S*-ethyl-CoA) also competitively inhibited the phosphotransacetylase from *Clostridium kluyveri* which catalysed the reversible formation of acetylphosphate from *S*-acetyl-CoA and phosphate. For allosteric activation (without co-valent participation), the enzyme pyruvate carboxylase needs *S*-acetyl-CoA. Here also, acetono-

CoA (and S-ethyl-CoA) proved to be an equivalent substitute [72].

The theme was continued by another guest scientist from America, Thomas Ciardelli. He mastered the difficult task of total synthesis of the carba-analogue of S-palmitoyl-CoA, heptadecane-2-onyl-dethio-CoA (here called HD-CoA). In the meantime, C.J. Stewart, again in Heidelberg, alkylated CoA at the sulfur atom to the heptadecyl compound [73]. Otto Wieland had demonstrated the inhibition of citrate synthesis by palmitoyl-CoA [74]. Now, both analogues exhibited a significant inhibition of the enzyme which transfers the palmitoyl residue from CoA to L-carnitine, thus providing a vehicle of transport of fatty acids through the mitochondrial membrane. Furthermore, HD-CoA was found to be a substitute for a natural substrate, palmitoyl-CoA, of the universal acyl-CoA dehydrogenase, a flavoprotein [75].

The methylene analogous acyl-CoA pseudosubstrates proved valuable in the laboratory of Janos Rétey who studied rearrangements dependent on coenzyme B_{12}. Certain mutases catalyze the exchange of a hydrogen atom for a group R at the neighboring carbon (Fig. 16). Methylmalonate-CoA-mutase rearranges the CoA-S-CO group and a hydrogen atom by a reversible reaction. Because the carba substrates cannot be hydrolyzed, the stereospecific cleavage and formation of the new bonds involved in the rearrangement could be observed for a longer period by NMR spectroscopy. Recently, J. Rétey wrote a review [76] about the mechanisms with radical participation, of this and other coenzyme A dependent reactions. With this review he clearly demonstrated again the fertility of the combination of problems of organic chemistry and biology. Very recently, HD-CoA performed another service to cell biology. According to James E. Rothman, the fusion of lipid double layer membranes which are necessary for the formation of vesicles for intracellular transport, requires several protein factors as well as ATP and

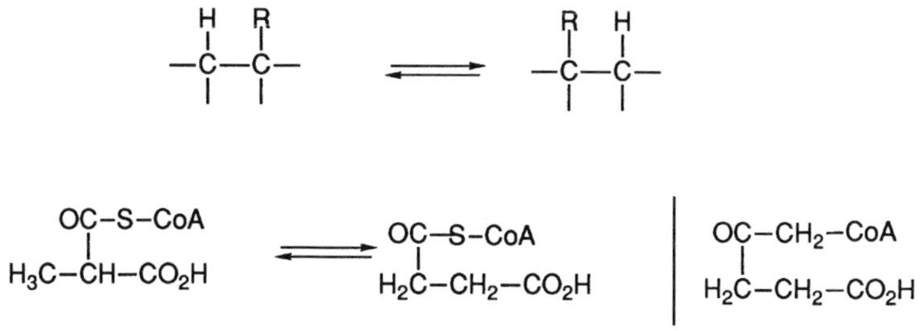

$$
\begin{array}{cc}
\text{H} & \text{R} \\
| & | \\
-\text{C}-\text{C}- \\
| & | \\
\end{array}
\quad \rightleftharpoons \quad
\begin{array}{cc}
\text{R} & \text{H} \\
| & | \\
-\text{C}-\text{C}- \\
| & | \\
\end{array}
$$

$$
\begin{array}{ccc}
\text{OC}-\text{S}-\text{CoA} & \text{OC}-\text{S}-\text{CoA} & \text{OC}-\text{CH}_2-\text{CoA} \\
| & | & | \\
\text{H}_3\text{C}-\text{CH}-\text{CO}_2\text{H} \rightleftharpoons & \text{H}_2\text{C}-\text{CH}_2-\text{CO}_2\text{H} & \text{H}_2\text{C}-\text{CH}_2-\text{CO}_2\text{H}
\end{array}
$$

Fig. 16. Mutase reaction. R can be CoA-S-CO or CoA-CH$_2$-CO as in B$_{12}$-catalyzed methylmalonate-CoA-mutase.

a long-chained acyl-CoA, such as palmitoyl-CoA. This was proven by the inhibition of the fusion process by HD-CoA [77].

Retrospection

Not long before I commenced my university studies, the first crystallization of an enzyme, urease, had been described by J. Sumner in 1926. This feat removed any doubt about the protein nature of enzymes. At that time, none of the 'additional nutritional factors', the vitamins, had been isolated in pure form or chemically characterized. The number of natural amino acids was not completely known, and since the era of Emil Fischer almost nothing new had been added to the field of peptides and their synthesis. In the subsequent 60 years, not only was there a revolution in the realm of physics, as well as technical development leading to air travel around the globe, but chemistry also made unimaginable progress, particularly in the physical methods of spectroscopy, X-ray structure analysis and application of computer techniques. In 1931, when I started my studies, I was a witness to the first success of X-ray analysis in the chemistry of natural products when the diffraction pattern of Rosenheim and King did not quite fit the structure of cholesterol suggested by Heinrich Wieland. Accordingly, the steroid structure was modified to its correct form (with Elisabeth Dane). Much later (1951) it was confirmed by R.B. Woodward in chemist's classical mode, by synthesis. He was awarded the Nobel Prize in 1965 'for his outstanding achievements in the art of organic synthesis'. He crowned his work with the difficult synthesis of vitamin B_{12} (with A. Eschenmoser) and died in 1979.

In the same year, Feodor Lynen, most tragically, had passed away. After an abdominal operation he succumbed to a postoperative complication in the summer of 1979.

During my career I was active as an academic teacher and researcher. At the beginning there was the classical effort to obtain all substances in crystalline form to guarantee purity. The delight of new, crystalline substances still prevails to this day: when the cytoprotective antamanide and its alkali metal complexes precipitated in beautiful crystals, when β-amanitin crystals permitted structure analysis by X-ray, and when, at the beginning of my biochemical interests, the isozymes of lactic acid dehydrogenase were obtained in crystalline form. My main pursuits were the development of methods of peptide synthesis which fortuitously supported me in my life-work of exploring the active principles of the white and green *Amanita* mushrooms. These interests have found fruition in two books: peptide chemistry in 'The World of Peptides', co-authored with Miklos Bodanszky [78], and mushroom research in the publication, named in ref. 33. The latter theme has been considered in a wider context in a brilliant essay dedicated to the author's 70th birthday by Bernhard Witkop [79]. Towards the end of my experimental work, in collaboration with biophysics, a new field arose: the exploration of molecular processes in the reproduction of viruses (nucleic acids) in prokaryotic and eukaryotic cells. This led to the current, routine manipulations of gene technology. With this, science entered the area of deliberate, aimed interference in life processes, formation of new forms and changes of organisms, theoretically also of human beings.

The fear of this and also of the creation of new, devastating viruses and microorganisms, unfounded as it might be, had political repercussions, differently articulated in various countries, depending on their national characteristics. To prevent the worst, namely total abstinence from undisputedly beneficial developments in biology and medicine, everything possible should be done to counter the regrettable lack of education in the natural sciences. In history, this once was called enlightenment.

Acknowledgements

I thank very much Dr. Klaus Florey, Princeton, for the translation and Mrs. Beate Isert-Hartmann for valuable support in preparing the manuscript.

Appendix

Foreign relations

Under the National Socialist regime and, naturally, during the war, almost all contacts with foreign countries ceased. Scientific relations were possible with Italy, however, for example with the, originally German, Stazione Zoologica in Naples where Kurt Wallenfels was permitted to elaborate on biological investigations of sea urchin pigments. Foreign journals reached our library through neutral countries or through France which was only partially occupied. They were eagerly perused, especially 'Nature' which was not entirely scientific, but also had political commentary. In the early summer of 1948, shortly after the war, three other West German professors and I were invited to take part in the First International Biochemical Congress in Cambridge, England. This was a great event for me. I made the acquaintance of colleagues with whom I had been familiar only through the literature. I also gained an impression of the dignified, but meager existence of academicians in post-war England.

Not much later, during the winter of 1952, Fritz Turba and I could participate in the first Ciba Symposium in London. There, we personally met well-known peptide and protein chemists who were leaders in the field, such as Chibnall, Porter, Martin, Synge, the Baileys, and Morgan from Great Britain, Cl. Fromgeaut (Chair), Roche, Desnuelle (worked with R. Kuhn before the War, see p. 32), and Acher from France, Fraenkel-Conrat and Craig from the USA, and Per Edman from Lund, Sweden. Dr. Wolstenholme, the Secretary of the Ciba Foundation, deserves great credit for his tireless efforts in support of the international medical and biological sciences.

International scientific mass tourism developed much later. Of course, I could not avoid being drawn into it on occasion. But I always preferred more intimate conferences, such as the European Peptide Symposia until recently. I also preferred contacts with colleagues on a personal basis.

Soviet Union

A very cordial professional relationship developed with the scientists of the Institute for Chemistry of Natural Products of the Soviet Academy of Sciences. The first visit to Moscow, in May 1969, was impressive in every respect. The airport, Sheremtjevo, had a suburban character, with very few take-offs and landings, and no crowds. We were received by Professor Mikhail M. Shemyakin and his assistant, Victor Tsetlin. As we experienced later on, arrival in the Soviet Union is connected with time-consuming passport, visa, and customs formalities. But Shemyakin only had to show his Academy badge, enclosed in blue leather, to wave us through. Already on our first stay in the USSR, we were surprised by the extent of the privileges of Academy members.

With our companions we stepped into a big, black limousine. The paved but rutted road to the city passed by isolated, wooden one story houses. Then we encountered a tank monument, marking the farthest advance of German troops in 1943. Soon thereafter appeared the first housing blocks which appeared shoddy according to our standards. After a drive of almost one hour, the city of Moscow with its beautiful large houses, palaces, churches, and theatres welcomed us. Here and there loomed the huge high-rise buildings of Stalinist architecture.

Already two years before this trip, my wife, by taking lessons with Russian emigrants and autodidactically, had learned enough of the difficult language to make herself quite well understood. The efforts of a foreigner to speak Russian were acknowledged with gratefulness and satisfaction. Our guide, Tsetlin, escorted us into the guest house of the Academy, a plain, new building with twelve floors. It was located at the beginning of the Leninskij Prospect. There was a lot of bureaucracy when we checked in. However, such inconveniences were more than compensated by the impressions already gained and those to be anticipated.

The Institute for the Chemistry of Natural Products, as it was still called, was located together with other institutes in an attractive, large building with six floors, probably dating from the end of the last century. It was situated in Uliza Vavilova, not far from the central building of the Academy. The Institute was crowded with personnel, furniture and equipment. The Director, Professor M.M. Shemyakin, resided in an attractive, large room, tastefully furnished. He was a friendly, well-educated man who exhibited great learnedness and open-mindedness in our conversation which was carried on in English-German, and also in Russian by my wife. His co-workers whom we met at the time and later on, also made a congenial impression. Yuri A. Ovchinnikov, my direct correspondent, happened to be in the USA. Therefore, the above mentioned V. Tsetlin and V.K. Antonov, the enzyme chemist, showed us through the spacious, but obviously too crowded Institute. The old steam equipment and unwieldy instruments caught the eye more than the new, interesting instrumentation, some of which still stood around in a court yard, packed in wooden boxes which had arrived from abroad, particularly from Japan.

Later visits arranged jointly by the German Science Foundation (DFG) and the Soviet Academy were focused mainly on studies of ion transporting antibiotics, such as valinomycin and other depsipeptides. The common interest in cyclic, biologically active peptides was the main reason for the visit. Of course, I also had to give lectures about our research projects.

On one of these occasions I also visited the distinguished, elderly V.A. Engelhardt, well-known through his muscle protein research at the Institute for Molecular Biology of the Academy. Unfortunately, Oparin, in the Bach Institute for Biochemistry, was out of town. Both institutes were solid edifices from the last century. I also paid a visit to A.N. Nesmejanov, the famous representative of elemental organic chemistry, in a 'new' building which, after only a few years, showed considerable wear. Wherever I went, the laboratory members were well-prepared for my visit with formulas, illustrations, and texts which quickly gave me an impression of the achievements of the laboratories. Finally, I briefly visited the Institute for Chemistry in the Lomonosov State University. The University was very impressive, if only because of its sheer size. In spite of the large proportions, it had a pleasing style, and was dramatically situated on a high bluff above the Moskva river.

On the 8th of May, the evening before the holiday commemorating 'Victory over Hitler Germany', Shemyakin escorted us to the train for Leningrad St. Petersburg, the Red Arrow. The elegant town surpassed our expectations. Beside science, there was the Hermitage, the Peter and Paul fortress, the city with the Neva and its many canals, the Summer Palace of

Peter I, and the Kirov Ballet. An excursion to the Finnish border brought us to the house of the famous painter, Repin.

The official purpose of my visit to Leningrad was a meeting with the creative and eminent patriarch of biochemistry and biophysics, S.E. Bresler. He was housed in a hall with a five meter high ceiling in an old warehouse at the harbor. In a corner behind a wall stood an ultracentrifuge, probably one of the first built by The Svedberg. The beam of the registering light had been conducted over several mirrors from the instrument to the very distant writing and measuring table of the experimenter. In the first years after its invention, the ultracentrifuge was a dangerous instrument. On occasion, the centrifugal force could hurl a part of the rotor with tremendous speed into the laboratory, like a bullet. With this instrument, Bresler tried to analyze in a D_2O-H_2O density gradient the mixture of polylysines which had been grown in ribosomes on a polyadenylic acid template.

Next I visited the peptide laboratory of Genadi Vlasov. His studies of nitrophenylsulfenyl derivatives were of interest to me, and I had intended to invite him to spend a sabbatical leave in Heidelberg at his convenience. However, this plan could be realized only much later when he came to my laboratory for several months in 1979. After the disintegration of the Soviet Union, travelling became easier. So he could spend the last quarter of 1993 in Germany, giving lectures and discussing joint interests in various institutes and laboratories.

The first visit to the USSR also gave us the opportunity of a trip to Taschkent. A four-hour flight transported us from spring in the overcrowded domestic airport, Domodedovo, to summer in the 'City of Roses'. Scientifically, not much can be said about the visit to the Uzbekistan Academy, but the town itself was unforgettable. It had been strongly shaken by a heavy earthquake years earlier. The original, one-floor adobe houses had suffered less than the cheap new buildings. Later, many old houses and also parts of the bazaar, had to make room for the monotony of modern buildings. In the open inner courtyard of one of the attractive older houses, we were invited to an intimate, late evening party, almost a revel, with the President of the Uzbekistan Academy, Sadikov, and his colleagues. It acquainted us with the merry disposition of these southern people.

On a day's excursion, we visited Buchara. It could have been called an oriental fairy tale city, had not the introduction of communism, about fifty years before, extinguished many memories of the pleasant arabic times, to which grand, decayed mosques and medresses, schools for students of Islam, gave testimony. On our return, we spent several more days in Moscow. There I met Yu. A. Ovchinnikov whose acquaintance I had made seven years before at the 5th European Peptide Symposium in Oxford. We not only discussed our scientific problems, but also visited the small chateau Archangelskoje and the famous State Circus. We also saw an excellent ballet performance in the Bolshoi Theatre.

This was not our first meeting with Professor Shemyakin. Already in 1958, I encountered him in Prague where Joseph Rudinger had founded the European Peptide Symposium. On this first and the many following European Peptide Symposia I already reported on pp. 52 and 53.

Unfortunately, Shemyakin died in 1970 at the age of 62, shortly after our last meeting at a scientific congress in Riga. Ovchinnikov became his successor as director of the Institute. As already planned by his predecessor, under his guidance a very generous institute complex was built on the outskirts of the city, the Shemyakin Institute for Bio-organic Chemistry. The first phase was erected in about ten years and dedicated to its purpose in 1984. On this occasion, scientists from other countries, active in the fields of interest to the Institute, were asked to plant little birch trees on the grounds next to the front of the entrance building. At a reunion in fall of 1989, I found that my birch, like all others, had grown well, but not luxuriantly, on the poor soil. Only the birch planted by Linus Pauling clearly towered above the others.

To deepen the contact between the Soviet Union and the Federal Republic a meeting of 25 Soviet and 25 German peptide and protein chemists was arranged in 1976. The symposium was to be repeated every two years and to take place alternately in a republic of the Soviet Union and in our country. For Moscow, of course, Ovchinnikov was the organizer who invited the participants from the Federal Republic to the first meeting at the Tadzhikic Academy of Science in Dushanbe. For this and subsequent meetings in Machackala at the Caspian Sea, in Odessa, as well as for congresses in Moscow and Alma Ata, we always arrived at the Sheremetjevo airport and were greeted by a delegation from the Academy. (At the 1989 meeting which took place in Dilizan in the Armenian mountains, Yuri Ovchinnikov, unfortunately, no longer was among us. The gifted, lively, and radiant man had succumbed to a blood disease during spring of 1988 at the young age of 52. His legacy is the large, bio-organic Institute of the Academy with its impressive twelve wings. For his successor, V.T. Ivanov this is a difficult challenge. As a further misfortune, V. Bystrov, one of the colleagues supporting him, died in 1990.)

In preparation for the Olympic Games of 1980 in Moscow, the arrival hall of the airport was rebuilt by a German firm. Feeder roads also were newly laid out. Generally, the growth of the city is modeled on the West European example. On arrival, each participant received a considerable amount of fresh ruble bills, and furthermore we had free bed and board. In a counter move, once at the Eibsee Hotel, once in Tübingen, the German peptide chemists as generously as possible returned the Russian hospitality with funds donated mostly from the German Research Association.

At all meetings, which were very productive because of the limited participation, excursions were organized which acquainted us with the geography, history, and culture of the meeting locations and surrounding areas. Hundreds of impressions could be related, beautiful and unforgettable, but also at times strenuous, tiring and depressing.

Traveling all over the globe as a scientist, one always finds people with common interests. This also held true for our relations with the Far East and America.

Japan

Our visits to Japan were most eventful and impressive. An academic lineage connected us with this country. In Freiburg, as far back as 1923 to 1925, Japanese postdoctoral fellows worked in the laboratory of my father. Another dozen followed in Munich from 1926 to 1937. This connection continued into the next generation. In Frankfurt and later in Heidelberg, I also had several Japanese co-workers with whom a lasting, mutual friendship developed. It all started with the close relationship between Munio Kotake and Heinrich Wieland. After his stay in Freiburg, Kotake, in his own country, became a professor of considerable renown at a university in Osaka. Many of his brilliant students obtained important posts at universities and in industry. He demonstrated a cordial veneration for his master (sense) by repeated visits to Starnberg, the retirement home of my parents. After the death of H. Wieland, Kotake continued visits to my mother and to my father's grave as well as to the Lynen family. He also maintained cordial relations with A. Butenandt and his wife which, undoubtedly, rested on common scientific interest, the study of tryptophan metabolism. Included in this circle was Bernd Witkop, the discoverer of hydroxytryptophan as a hydrolysis product of phalloidin (p. 109, 142).

Almost all our trips to Japan took place in the company of Bernd Witkop, to our joy and benefit. He was and is a well-informed, enthusiastic expert in Japanese culture and has a considerable knowledge of the Japanese language. His explanation for this rare gift is that he had harbored many Japanese guests in his laboratory at NIH in Bethesda who had great difficulty learning English. He, therefore, made the decision to learn Japanese, especially the language one uses to communicate with persons of higher standing.

The mutual, cordial veneration of Munio Kotake was transferred from father Wieland to the younger generation. Feodor Lynen was the first who, after the catastrophe of the World War, was invited to Japan for lectures during the early sixties. I received an invitation to give a plenary lecture at the IUPAC Congress in Kyoto in 1964. The strange beauty of this foreign country, but also its contrasts, immediately and strongly fascinated us. During our later visits, we observed the inroads of the western style of living into the traditional cultural life.

In 1964, almost no superhighways existed. One had to drive on the left on narrow roads. In the cities, slalom runs had to be executed around the interfering wooden utility poles. One frequently noticed kimonos being worn in the streets. Kyoto, with its hundreds of temples and shrines, was so overwhelming that the senses were not adequate to take everything in the first time round. Munio Kotake accompanied us in a taxi to several impressive temple sites, to the imperial villa (shagakuin), silver pavilion (ginkaku), to the great Buddha (Daibu tsu) in Nara, and to the more precious, small, black, wooden Buddha at that very place. Strange Japanese food, including the dreaded but tasty Fugu fish, in typical restaurants and festive geisha-parties rounded off the thousands of impressions.

In retrospect, the scientific events in the great Kaikan, which were introduced by samisen and koto sounds, almost faded into the background, since after the congress we went on excursions under the guidance of my faithful co-worker, Hiroshi Morimoto, now professor emeritus. In 1962/63, he had spent a longish study sojourn in Frankfurt. At that time, he was employed in a large pharmaceutical-chemical family enterprise. He had taken a few days additional leave which was rather rare. With Morimoto, we experienced an entertaining cruise on the Inland Sea from Kobe to Beppu. We visited the island of Kyushu with its thousands of monkeys, spoiled by the public, the thousand hot springs, and Mount Aso with the smoking crater of the Nakadake. Via Kumamoto, with its mighty castle and pagoda mountain, we came to Hakata, where I had to give a lecture. There, I met the colleague, Tanezo Taguchi, and his pretty wife for the first time. We have had the best of contacts with them ever since.

We continued by railway to Hikari to a branch factory of Takeda, passed Tokuyama (Kintai bridge), and returned to the Sea. Thousands of azaleas blossomed in Hiroshima, a memorial to the gruesome beginning of a new era in which, hopefully, world wars will be impossible forever. The enchanted Miyashima with the Tsukushima shrine helped to dispel these sombre thoughts. In Osaka I again was invited to give a lecture. There also, I had a reunion with Munio Kotake and made the lasting acquaintance of Tetsuo Shiba. In the Peptide Institute of the Protein Research Foundation, I met the elderly, dignified Professor Shiro Akabori and his young assistant and later successor, Shumpei Sakakibara; and I had an opportunity for discussions about peptide chemistry which was in a very active state of development. On the way I should mention an excursion to Sendai and to Matsushima, the pine island, also a lecture at the University Institute of Tetsuo Nozoe, a first meeting with Nakanishi (who soon thereafter emigrated to Columbia University in New York) and the mushroom researcher, T. Takemoto. Finally, we arrived in Tokyo.

In this strange megapolis we would have been lost without our guide who knew his way around. There was a confusing heterogeneity in the many high-rise buildings, going up, next to old and small wooden hovels in tiny gardens. The brightly colored, illuminated advertisements in the formidable Ginza, and the throngs of people in all streets and all locations almost made us faint. In Tokyo, we encountered a familiar face, Toshio Hoshino, one of the co-workers of Heinrich Wieland in Munich. He had become scientific director of a large firm,

Toray, which, among many other products, also produced synthetic fibers. Hoshino invited us to visit Toray in Kamakura for the next day. We made an interesting visit to the workshop of the research and development department, designed by a student of Le Corbusier. This allowed us also to admire the most famous large Buddha in Kamakura, albeit in a crowd of about one thousand schoolboys and girls as well as many international tourists.

We finished this visit to the Far East with a trip around the world, traveling on to Hawaii, California, and across the USA to New York. We had further opportunities to visit Japan. Only a few dates shall be mentioned. One was an invitation to Bernd Witkop and myself to present plenary lectures to celebrate the 77th anniversary of the birth, the so-called Kiju, of Munio Kotake at the end of October 1970. (We call it the 76th birthday.) After our lectures (mushroom poisons; frog poisons, Witkop in Japanese!), first in the evening in Osaka, the next day in the luxury mountain hotel, Hari Han, great geisha parties were given. We all were guests of the important beverage company, Suntory. Its President Saji intoned 'Am Brunnen vor dem Tore'. The party had been organized by the amiable Dr. Shiro Seno, one of Witkop's students, who had a leading position in the firm. With him and his wife, we had quite a few memorable parties in later years. After Kotake's Kiju we participated at a natural product congress in Fukuoka (Kyushu). As an interlude we shared unforgettable impressions with H. Morimoto and his wife (Nikko, Hakone, and a beautiful view of Mount Fuji from the Otome pass).

Finally, I want to recount the last adventure we had in Japan. After Hiroshi Morimoto, I was privileged to have further Japanese co-workers. In Frankfurt it was Akio Ohsuka, one of the last students of Kotake; in Heidelberg Masaaki Ueki, a peptide chemist from the school of Mukaiyama and Eisuke Munekata. Munekata with his family had spent $3\frac{1}{2}$ years in Heidelberg, and had been very successful with synthetic studies in the field of phalloidin (p. 69) [53]. These gen-

tlemen, together with Morimoto, planned my Kiju (1989) in
Osaka, in connection with a celebration of the opening of a
research institute in Noda, on the new campus of the Science
University of Tokyo, where I also was to present a plenary
lecture. There, it was my pleasure to encounter again my colleague T. Mukaiyama. I stayed with my wife and daughter in
a guest house of the Science University in Tokyo-Shinjuku. We
went on excursions to the new research and university town of
Tsukuba where E. Munekata held a professorship. With him
we traveled to Hokkaido (Sapporo), and with Professor Yuichi
Kanaoka we went to the young volcano Showa Sinzan, to the
Ainus and the bears, and then on to Osaka.

There, after my lecture at the Peptide Institute, mentioned
above, the Kiju ceremony took place at the Hangkyu Hotel in
great style and splendor in the presence of the chairmen, H.
Morimoto and Tetsuo Shiba, together with A. Ohsuka. There
also was singing. I had to sing 'Heideröslein' (one verse) in
duet with Tetsuo Shiba. My listeners and I enjoyed slides with
which I could recall the old times of the Japanese in the Institute of Heinrich Wieland in Munich. Shiro Seno also helped to
organize this party. A few days later, M. Ueki once more celebrated my kiju in a smaller circle at 'my' university. I shall
forever be grateful to all the sponsors, friends and colleagues
who made these unforgettable events possible for us.

China

A stay of three weeks in China helped to expand my scientific,
social, and political experiences. In 1976, the Max Planck Society had started to participate in an exchange program of
German scientists with the Chinese Academy of Science. F.
Lynen and H.A. Staab belonged to the first small group which
encountered – shortly after the death of Mao Tse Tung – a
politically unsettled China, deeply disturbed by the 'Gang of
Four'. Our trip in October 1977, after the long 'Cultural Revo-

lution' and after the fall of this dictatorial, anarchistic group, experienced an euphoric country, breathing freely again. Hope and rationalism dawned in all areas, including science. Under the effective guidance of the intelligent, German speaking Yang Dawei, an official of the Academia Sinica, our small group (five members) visited chemical, biochemical, and medical institutes of the People's Republic in Peking, and in Shanghai.

In both cities I had long-standing, personal relationships which had been established before the Second World War. My co-student and co-author, Hsing Chi-yi (see p. 31) had returned in 1938 from Munich to China, and we had lost contact during the war and post-war years. Thirty years later I encountered him again, almost by accident; he now was a professor at Peking University, and had survived the difficult times of the 'Long March' of Mao and the Cultural Revolution. He received me at Peking University for a lecture. Unfortunately, I had to give it under adverse conditions. Since the lecture hall could not be darkened, the projector gave very weak light, and was not adapted to my slides, I had to present the lecture on a gray blackboard with chalk. The institutes on the campus presented a rather desolate picture; it was the phase of resumption of normal academic life, as we know it. Somewhat more successful was a visit to the Institute for Biophysics where the Peking Insulin Research Group had established the structure of the zinc complex of insulin by X-ray analysis, independent of the English (Dorothy Hodgkin). The Chinese were very proud of their insulin research, and rightly so. The spatial formula has even been depicted on a postage stamp.

The Insulin Research Group of the Shanghai Institute of Biochemistry consisted of almost 50 scientists. In 1965, their intensive work on the hormone culminated in the internationally acclaimed, total synthesis of beef insulin. The driving force was Professor Wang Yu who now directed a large Institute for Organic Chemistry with 1200 co-workers. They were, partially very successfully, engaged in six fields of different

orientations. Wang Yu still spoke very good German. In 1937/
38, in the laboratory of Heinrich Wieland in Munich, he had
completed his doctoral thesis under Professor Elisabeth Dane,
and then had worked for 8 months in the laboratory of Richard
Kuhn where I came to know him quite well. The extended
reunion was the first after the war. I met him again several
times later on, also in Munich, where, in 1989, he received the
50 year renewal of his doctoral diploma.

The China visit brought us not only the acquaintance of numerous colleagues, it also afforded us insights into the life in
the cities and under the cities (the extensive network of underground shelters of Peking). A scenic high point was a river
cruise on the Li Kiang through the enchanted mountains of
Kweilin.

America

The first trip to the USA in 1953, to attend the Glutathione
Symposium, was followed by many more. Therefore, we became very well acquainted with this country. Both our physician sons and their young families spent 2 years in the USA,
Heinrich, the older one, in Oklahoma City, and Eberhard in
La Jolla. Visits there, followed by sight-seeing tours by air-
plane and car, opened to us the spacious, interesting land-
scapes, many National Parks and many cities. Other occasions
were congresses, lectures and visits with colleagues and former students, some of whom were spending post-doctoral
years there, and some of whom had taken permanent employ-
ment.

Bernd Witkop played a pivotal role in our visits. He had
studied chemistry in the Institute of Heinrich Wieland in Munich and then had done research in the 'Private Laboratory'
(see p. 124). After the war he left Germany, and soon became
director of a chemical laboratory at the National Institute of
Health (NIH) in Bethesda, Maryland. It was, and still is, ad-

vantageous for German students to work for a time in an American laboratory, after having obtained their doctoral degree. Following this trend, about a dozen of my co-workers spent fairly long post-doctoral periods in American laboratories, five of them in the laboratory of B. Witkop. Three of those stayed in America; tragically, Erhard Gross, the co-inventor of the cyanogen bromide cleavage of peptides, suffered a fatal accident on a visit to his native land. He was one of the driving forces of the American Peptide Symposium, and together with Johannes Meienhofer, he deserved much credit for the publication of peptide research results.

Another, early connection was that with Charles Jack Stewart of San Diego. In the mid-fifties, he came to Frankfurt on a Fulbright Fellowship and started to work in the field of S-aminoacyl compounds. He synthesized S-alanyl-coenzyme A with which he converted glutamate to the dipeptide alanylglutamic acid. As discussed on p. 85, we followed a wrong lead concerning biological protein synthesis, even though, later on, the phosphopantothenyl part of coenzyme A was recognized as the acyl-transferring agent in nonmicrosomal peptide synthesis, for example in the synthesis of Gramicidin S by *Bac. brevis*.

The opportunity to become acquainted with colleagues in the USA and to explore the country, was made possible for us through the good offices of B. Witkop. On his initiative, I was awarded a Fogarty Scholarship in 1973. Being a university lecturer (Adjunct Professor at the University of Heidelberg), and being responsible for the supervision of several graduate students at the Max Planck Institute, it was not possible for me to be absent from Heidelberg for a full year. Therefore, I divided the scholarship into three-month periods. In 1974, I moved with my wife for the first time into Stonehouse, Building 16, on the campus of NIH in Bethesda. The duties of a Fogarty Scholar included introducing oneself with a lecture in the central auditorium in building 1, presenting seminars in various institutes, and, where possible, conducting experiments in laboratories close to one's own field.

In building 4, where my former student, Wolfgang Burgermeister, worked in Witkop's department, I was able to participate in experiments to crystallize cyclic peptides. There, we met Isabella and Jerome Karle with whom we have been good friends ever since. Isabella had previously elucidated the structure of the crystalline Li-complex of antamanide by single crystal X-ray analysis [51]. Antamanide is a natural substance which inhibits the uptake of bile acids and other substances, including phalloidin, into the liver (p. 68). Since then, antamanide has led to the synthesis of other cytoprotective compounds, and to very refined conformational studies by NMR-analysis in the laboratory of H. Kessler.

One of the especially pleasant privileges of life as a Fogarty Scholar is the possibility of taking trips from Bethesda to chosen locations of scientific interest. We took advantage of this within reason. For example, I renewed the old friendship with H. Fraenkel-Conrat in the Stanley Laboratory in Berkeley, where, after leaving the Frankfurt Institute, Peter Duesberg had made fundamental observations about the changes of DNA in chicken fibroblasts, infected with Rous sarcoma virus. In San Francisco, I met Cho Hao Li in his Hormone Institute. At the time, I was a co-editor of his *International Journal of Peptide and Protein Research*. I remember with pleasure excellent seafood lunches with him.

Among the many other impressions I had, I want to single out a visit to Hans Neurath and his wife in Seattle. In his new, well-equipped laboratory, I had many interesting discussions with him and Ed Fischer. The Neuraths acquainted us with the charming city, situated on the Pacific coast, with its architectonically pleasing skyscrapers in the center and its beautiful surroundings. The area south of Seattle is dominated by the mighty volcanic cone of Mount Rainier. We did not only admire it from a distance. Hans Neurath owns a primitive cabin near the National Park. From there we could drive by car to the Paradise Inn and take exhilarating mountain hikes amidst the alpine flora.

REFERENCES

1   H. Wieland and R. Sonderhoff, Liebigs Ann. Chem. 499 (1932) 213.
2   C. Martius, How I became a biochemist in: G. Semenza (Ed.) Evolving Life Sciences Vol. 1 – Of oxygen, fuels and living matter, Part 2, Wiley Chichester, 1982 p. 1.
3   H.A. Krebs, see The Original 'Citric Acid Cycle' in M. Florkin and E.H. Stotz (Eds) Comprehensive Biochem. Vol 31, Elsevier, Amsterdam, 1975, p. 255.
4   Th. Wieland and Chi-Yi Hsing, Liebigs Ann. Chem. 526 (1936) 188.
5   A detailed story of the re-discovery of chromatography has been given by E. Lederer in G. Semenza (Ed.) Comprehensive Biochem. Vol. 36, Elsevier, Amsterdam, 1986, p. 437.
6   P. Desnuelle, Survey of a French Biochemist's Life in G. Semenza (Ed.) Comprehensive Biochem. Vol. 35, Elsevier, Amsterdam, 1983, p. 283.
7   Chen-Lu Tsou, The Highest Grade of this Clarifying Activity has no Limit-Confucius, in G. Semenza and R. Jaenicke (Eds.) Comprehensive Biochem. Vol. 37, Elsevier, Amsterdam, 1990 p. 357.
8   R. Kuhn, Th. Wieland and E.F. Möller, Ber. dtsch. Chem. Ges. 74 (1941) 698.
9   Th. Wieland and E.F. Möller in Vol. III Fermente, Hormone, Vitamine (R. Ammon and W. Dirscherl, eds.) Georg Thieme, Stuttgart (1974) 698.
10 Th. Wieland and W. Paul, Ber. dtsch. Chem. Ges. 77 (1944) 35.
11 Th. Wieland and E. Fischer, Naturwissenschaften 35 (1948) 29.
12 K.O. Pedersen, The Svedberg and Arne Tiselius. The Early Development of Modern Protein Chemistry at Uppsala in G. Semenza (Ed.) Comprehensive Biochem. Vol. 35, Elsevier, Amsterdam 1983, p. 250.
13 Th. Wieland and G. Pfleiderer, Angew. Chem. 67 (1955) 257.
14 Th. Bücher, Biochim. Biophys. Acta 1 (1947) 292.
15 G. Pfleiderer, D. Jeckel and Th. Wieland, Biochem. Z. 329 (1957) 104.
16 F.S. Vesell and A.G. Bearn, Proc. Soc. Exp. Biol. Med. 94 (1957) 96.
17 Th. Wieland and G. Pfleiderer, Angew. Chem. 69 (1957) 199.
17a N.O. Kaplan, Experiences in Biochemistry, in G. Semenza G. (Ed.) Comprehensive Biochem. Vol. 36, Elsevier, Amsterdam, 1986, p. 255.
18 C. L. Markert and F. Møller, Proc. Natl. Acad. Sci. USA 45 (1959) 753.
19 Th. Wieland, W. Löwe and A. Kreiling, Biochem. Z. 339 (1963) 1.
20 E.D. Wachsmuth and G. Pfleiderer, Biochem. z. 336 (1963) 545.

21 E. Appella and C. Markert, Biochem. Biophys. Res. Commun. 6 (1961)
    171.
22 N. William Pirie, Sir Frederich Gowland Hopkins in G. Semenza (Ed.)
    Comprehensive Biochem. Vol. 35, Elsevier, Amsterdam, 1983, p. 103.
23 Th. Wieland and R. Sehring, Liebigs Ann. Chem. 569 (1950) 121; Th.
    Wieland and H. Bernhard, Liebigs Ann. Chem. 572 (1951) 190.
24 Th. Wieland, W. Schäfer and E. Bokelmann, Liebigs Ann. Chem. 573
    (1951) 99.
25 Th. Wieland and W. Schäfer, Angew. Chem. 63 (1951) 146; Liebigs Ann.
    Chem. 576 (1952) 101.
26 W. Gevers, H. Kleinkauf and F. Lipmann, Proc. Natl. Acad. Sci. USA 63
    (1969) 1334.
27 Th. Wieland, in: The Roots of Modern Biochemistry (H. Kleinkauf, H.
    von Döhren, L. Jaenicke eds.) de Gruyter, Berlin (1988) p. 213.
28 C. de Duve, ibid. p. 881.
29 Chemistry and Properties of Glutathione. Proc. of the Glutathione
    Symposium in Ridgefield, Conn., 1953, Acad. Press New York (1954).
30 Th. Wieland, G. Pfleiderer and J. Franz, Angew. Chem. 66 (1954) 297.
31 E. Racker and F.W. Racker, Resolution and reconstitution: a dual auto-
    biographical sketch, in: G. Semenza (Ed.) Evolving Life Sciences Vol. 1
    – Of oxygen, fuels and living matter, Part 1, Wiley Chichester (1981).
32 F. Lynen and U. Wieland, Liebigs Ann. Chem. 533 (1938) 93.
33 H. Wieland, R. Hallermayer, Liebigs Ann. Chem. 548 (1941) 1; Th. Wie-
    land and H. Faulstich, 50 Years of Amanitin, Experientia 47 (1991)
    1186.
34 Th. Wieland, Peptides of Poisonous Amanita Mushrooms, A. Rich ed.,
    Springer-Verlag New York, Berlin, Heidelberg, London, Paris, Tokyo
    (1986).
35 H. Faulstich, A. Buku, H. Bodenmüller and Th. Wieland, Biochemistry
    19 (1980) 334.
36 Amanita Toxins and Poisoning, Proc. Internat. Amanita Symposium
    Heidelberg (1978); (H. Faulstich, B. Kommerell, Th. Wieland eds.), G.
    Witzstrock Baden-Baden, Köln, New York (1980).
37 H. Faulstich, K. Kirchner and M. Derenzini, Toxicon 26 (1988) 491.
38 Th. Wieland and K. Dose, Biochem. Z. 325 (1954) 439.
39 L. Fiume and R. Laschi, Sperimentale 155 (1965) 288.
40 A.C. Vaisius and Th. Wieland, Biochemistry 21 (1982) 3097.
41 E. Wulf, F.A. Bautz, H. Faulstich and Th. Wieland Exptl. Cell Res. 130
    (1980) 475.
42 A.C. Vaisius and H. Faulstich, Biochem. Cell Biol. 64 (1986) 923.
43 M. Frimmer, R. Kroker and I. Postendorfer, Naunyn-Schmiedeberg's
    Arch. Pharmakol. 287 (1974) 395.
44 Th. Wieland et al., Angewandte Chem. Int. Edn. Engl. 7 (1968) 204.
45 Th. Wieland, M. Nassal, W. Kramer, G. Fricker, U. Bickel and G. Kurz,
    Proc. Natl. Acad. Sci. USA 81 (1984) 5232.

46   A. Neuberger, An Octogenarian Looks Back in G. Semenza and R. Jaen-
     icke (eds.) Comprehensive Biochem. Vol. 37 Elsevier, Amsterdam 1990,
     p. 21.
47   D.J. Patel, A.E. Tonelli, P. Pfaender, H. Faulstich and Th. Wieland, J.
     Mol. Biol. 79 (1973) 195.
48   E. C. Kostansek, W.W. Lipscomb, R.R. Yocum and W.E. Thiessen,
     J.Amer. Chem. Soc. 99 (1977) 1273.
49   see W. Burgermeister and R. Winkler Oswatitsch in Topics Current
     Chem. 69 (1977) 93.
50   Th. Wieland, W. Burgermeister, W. Otting, W. Möhle, M.M. She-
     myakin, Yu.A. Ovchinnikov, V.T. Ivanov and G.G. Malenkov, FEBS
     Lett. 9 (1970) 89.
51   I.L. Karle, J. Karle, Th. Wieland, W. Burgermeister and B. Witkop,
     Proc. Natl. Acad. Sci. USA 70 (1973) 1863.
52   F. Lipmann, Advan. Enzymol. 1 (1941) 99.
53   E. Bäuerlein, M. Klingenfuss and Th. Wieland, Eur. J. Biochem. 24
     (1971) 305. Former review: Th. Wieland and E. Bäuerlein, Naturwis-
     senschaften 54 (1965) 80.
54   P. Mitchell, Bioenergetic aspects of unity in biochemistry: evolution of
     the concept of ligand conduction in chemical, osmotic and chemiosmotic
     reaction mechanisms in: G. Semenza (Ed.) Evolving Life Sciences Vol-
     ume 1 – Of oxygen, fuels and living matter, Part 1, Wiley Chichester,
     1981 p. 1.
55   Th. Wieland and H. Determann, Angew. Chem. 75 (1963) 539.
56   Last publication in the plastein series: H. Determann and R. Köhler,
     Liebigs Ann. Chem. 690 (1965) 197.
57   O. Westphal, Sonderheft der Mitteilungen der Max-Planck-Ge-
     sellschaft, 1967.
58   A. Buku, Th. Wieland, H. Bodenmüller and H. Faulstich, Experientia
     36 (1980) 33.
59   H. Faulstich, A. Buku, H. Bodenmüller and Th. Wieland, Biochemistry
     19 (1989) 3334.
60   V.M. Govindan, H. Faulstich, Th. Wieland, B. Agostini and W. Hassel-
     bach, Naturwissenschaften 59 (1972) 521.
61   E. Munekata, H. Faulstich and Th. Wieland, Liebigs Ann. Chem. 1758
     (1977).
62   H. Faulstich, S. Zobeley, G. Rinnerthaler and J.V. Small, J. Muscle Res.
     Cell Motility 9 (1988) 370.
63   A. Kishino and T. Yanagida, Nature 334 (1988) 74.
64   W. Jahn, H. Faulstich, A. Deboben and Th. Wieland, Z. Naturforsch.
     35c (1980) 467.
65   J.E. Eriksson, G.I.L. Paatero, J.A.O. Meriluoto, G.A. Codd, G.E.N.
     Kass, P. Nicotera and S. Orrenius, Exp. Cell Res. 185 (1985) 2815.
66   R.E. Honkanen, J. Zwiller, R.E. Moore, S.L. Daily, B.S. Kastra, M. Du-
     kelow and A.L. Boynton, J.Biol. Chem. 265 (1990) 19401.

67 J. Vandekerckhove, A. Deboben, M. Nassal and Th. Wieland, EMBO
    Journ. 4 (1985) 2815.
68 W. Kabsch, H.G. Mannherz, D. Suck, E.F. Pai and K. Holmes, Nature
    347 (1990) 37.
69 Ch.J. Stewart and Th. Wieland, Liebigs Ann. Chem. 57 (1978) 57.
70 H. Eggerer, W. Giesemann and H. Aigner, Angew. Chem. 92 (1980) 133.
71 J. Nikawa, S. Numa, T. Shiba, Ch.J. Stewart and Th. Wieland, FEBS
    Lett. 91 (1978) 144.
72 H.P. Blaschkowski, J. Knappe and Th. Wieland, FEBS Lett. 98 (1979)
    81.
73 Th. L. Ciardelli, A. Seeliger, C.J. Stewart and Th. Wieland, Liebigs
    Ann. Chem. 828 (1981).
74 O. Wieland, L. Weiss and I. Eger-Neufeldt, Biochem. Z. 339 (1964) 501.
75 C. Thorpe, Th.L. Ciardelli, C.J. Stewart and Th. Wieland, Eur. J. Bio-
    chem. 118 (1981) 279.
76 J. Rétey, Bio Factors 1 (1988) 267.
77 N. Pfanner, L. Orci, B.S. Glick, M. Amherdt, S.R. Arden, Y. Malhotra
    and J.E. Rothman, Cell 59 (1989) 95.
78 Th. Wieland and M. Bodanszky, The World of Peptides. A brief History
    of Peptide Chemistry, Springer-Verlag, 1991.
79 B. Witkop, Naturwissenschaftl. Rundschau 36, (1983) 261–275.

E.C. Slater, R. Jaenicke and G. Semenza (Eds.)
Selected Topics in the History of Biochemistry: Personal Recollections, IV
(Comprehensive Biochemistry Vol. 38) © 1995 Elsevier Science B.V.

Chapter 3

Stepping Stones - Building Bridges

BERNHARD WITKOP

Laboratory of Cell Biology and Genetics, NIDDK, National Institutes of
Health, Building 8, Room 403, Bethesda, MD 20892 (U.S.A.)

Between two wars

H.G. Wells did not use his time machine to travel back and revisit adolescence. If I were to do this, I would find a thin youth with a full head of hair look pensively into a future packed with more historic events than in all preceding centuries. Should I take this diffident youngster along on the machine and return further back, we would both meet a child happy in the belief that all people and phenomena exist only in the relationship to this contented little egotist. How fortunate that the various persons we are within the coordination system of time and space, never meet in the flesh; this saves the questing septuagenarian considerable embarrassment.

I can easily trace back the awakening of full conscience to an age of 2 to 4 years. The difficulty is that at that time dreams were as powerful as reality and, especially in retrospect, the two become unresolvable in the same way as in the Chinese fable of the man who woke up, not knowing, was he still the butterfly, he dreamt of, now imagining to be a man?

The end of the war (1918) was registered by indistinct commotions viewed from a perambulator moving in the busy streets of a small provincial town. The choice of a suitable birthplace is one of the most important decisions in our pre-

In Japan, tobi-ishi (literally: flying stones) serve not merely to form a path or a bridge but to guide the viewer's senses. For a time they lead him quietly past gently changing perspectives, but suddenly they will thrust up before him an unexpected scenic arrangement of startling force [1].

existence. I never regretted my selection of Freiburg, 'the pearl of the Black Forest'. The beauty of this medieval city, its intact atmosphere, its sense for tradition and continuity, in defiance of post-war penury and chaos, didn't fail to form and inform my infancy and adolescence.

The epicenters of this town, the Gothic cathedral and the old University, acted as magnetic poles: the people, like iron filings, moved within this powerful field in a foreseeable and decorous pattern. Decorous and foreseeable as this peaceful scenery may have seemed at the time, the shadows that were falling could be seen and described only by the poet:

> Things fall apart, the center can not hold;
> Mere anarchy is loosed upon the world,
> the blood-dimmed tide is loosed and everywhere
> The ceremony of innocence is drowned;
> the best lack all conviction, while the worst
> Are full of passionate intensity.

W.B. YEATS, 1920

Here is expressed a premonition of the day when in November 1944 I stood before the ruins of my birthtown, one of the saddest moments. I had to learn that the old houses and roads are as fugitive as the passing years.

The house of my childhood was no exception. The active link, my father Philipp Witkop (1880–1942), had become a member of the tightly knit and highly respected academic community in 1910. His literary interests and academic friends filled the house with a scholarly atmosphere that to a child became noticeable through the ubiquitous bookshelves and through recurrent breakdowns in communication. My brother and I spoke the local alemannic dialect. My father, a Westphalian, and my mother Hedwig Margarethe Hirschhorn (1888–1971), a native of the City of Mannheim in Northern Baden, both spoke an accent-free cultivated high-German. Fortunately, when entering the Real-Gymnasium, French was taught and helped to overcome any impasse at the dinner table by becoming the emergency vehicle of communication.

My father was a Professor of modern German literature. His guest-book, a leatherbound folio, is one of the few lasting possessions that accompanied me to the New World: It is a repository of autographs, epigrams and aperçus of all the luminaries and literati that shaped the unique cultural life of the ephemeral Weimar Republic (1919–1933). I remember the initial visitors as a child with surprising clarity. Hermann Hesse (1877–1962) who stayed in our house and read the beautiful fairy tale 'Piktors Verwandlungen' (Pictors' Metamorphoses), or the urbane and charming Stefan Zweig (1881– 1942) who was so amused, because the 8-year-old boy kissed his hand, not even knowing that, at least as a phrase, this custom existed in the old Vienna. Many years later, as an emigrant in Brazil, this sensitive recorder and monitor of history's and man's most subtle motions and motives took his life. He saw his 'world of yesterday' being destroyed by this murderous war.

Another more frequent guest was Thomas Mann who enjoyed taking long walks in the foothills of the Black Forest around Freiburg. My father, five years younger, and Thomas Mann had been fellow students in Munich at the turn of the century. In 1945, after the war, through his influence, he helped me to come to America. The problem of psychological connectedness and continuity within a person's lifespan is a dilemma that must have faced most creative minds, especially of many famous writers. Time, of course, has many faces: we are here concerned more with psychological time, our consciousness of the flow of events from the past to the future, than with time concepts of modern physics. More than a century ago Gottfried Keller, the Swiss poet, anticipated the relativity of time in his poem 'Die Zeit geht nicht':

Time Does Not Move

Time does not move - it's standing still;
We pass through it in turn.

It is a busy crossroad's inn
In which we pilgrims learn.

A something, form- and colorless
Acquiring shape and gain
As merging in it, sinking then,
We soon dissolve again.

Of morning dew a drop's ablaze,
Hit by a ray of sun:
One day may have a pearl's bright light,
While hundred years have none.

Time is a parchment, smooth and white
On which we write and scrawl
A solemn script with our blood
Till stream and tide move all.

To thee, oh wondrous, beauteous world,
Oh fairness without end
I write my busy eulogy
On this same document.

My happy self, it came to bloom
within thy orbed round;
My thanks increase thy limpidness
And let thy light abound

(my own translation)

The names of many of my classmates in elementary school and at the Real-Gymnasium as well as those of my friends in the Boy Scouts were taken for granted until it gradually dawned on me that these were the scions of luminaries who contributed to the eminence of Freiburg among German Universities: Hans Jürgen Staudinger (1914–1990), a dear friend for life, known for his contributions to the role of ascorbic acid in metabolic oxidation. Wolfgang and Theodor Wieland were neighbors and playmates in Freiburg until their father, my later teacher, Heinrich Wieland, moved to Munich where he promptly received the Nobel Prize (1928).

Professors who did and did not profess

The discovery of the true identity of one's parents usually comes at a time when they have departed from this world and when we are confronted with the same problems of advancing years and realize that they only preceded us, as expressed in the Latin tombstone lettering '*Eram quod es, eris quod sum*'. I find it easier to evaluate the literary legacy of my father than to judge and describe his personality without self-consciousness or embarrassment. Too many moments in old age reach back through a man's unresolved adulthood to the dim beginnings of his awareness as a child. The distance between generations then was much greater than now. In the meantime we have become conscious of this phenomenon. It no longer, at least in my own experience, is a 'chasm of silence which neither affection nor goodwill could bridge'.

Likewise, in my recollections he blends more easily with the seemingly serene background preceding World War I than with the gathering storm before World War II, the dissolution of his life, his family and his death at 62. Father came from Westphalian peasant stock, but his father sold his farm and moved to Gelsenkirchen, a city which soon was to become Germany's industrial heartland. My father hated this polluted, soot-infested industrial world, and we children were told time and again how fortunate we were to grow up at the border of the Black Forest in a landscape of unrivaled beauty. In Heidelberg he became Docent for aesthetics in 1909, in Freiburg Professor for contemporary German literature in 1910. His constant search for harmony and peace was ill-served by the events of the time.

More than half a century later, after the 2nd world war was over, historians tried to come to grips with the collapse of the University of Freiburg and most of their Professors who failed to 'profess' when they faced the onslaught of the Nazi takeover in 1933. The University, as an independent bastion of intellectual freedom, was in ruins eleven years before the en-

tire city would become a pile of pitiful rubble. In 1945 the
French commander of Freiburg asked the question: 'How was
it possible that the University surrendered, tolerated the loss
of academic freedom, did not protest the atrocities of the Third
Reich and actually aided and abetted the rise of totalitarianism?' The downward spiral started with Martin Heidegger
who as 'Rector Magnificus' of the University delivered an inaugural address on May 27, 1933, which welcomed Hitler and his
grab of power with open arms. As a 16-year-old, I witnessed at
the dinner table my father's consternation in response to this
surrender, but that did not prevent him from trying to avoid
any impression on the outside that he opposed the regime. By
contrast, my teacher Heinrich Wieland, exactly at the same
time, expressed his courageous attitude as follows: 'As early as
1933 I decided to follow a strategy of resistance that could be
carried right to the end' [2]. This lonely defiance at the University of Munich existed at the University of Freiburg in the
person of the historian Gerhard Ritter (1888–1967) and his
'Freiburger Kreis'.

Coping with a vanishing past

My mother's family can be traced to the 16th century, 1531 to
be exact, with Löb Oppenheimer zum Hirsch oder Hirschhorn
who moved from Heidelberg to Frankfurt in 1531. In the early
part of the 18th century Gabriel Hirschhorn, my great-great-
grandfather   founded   the   Tobacco   Company   Gabriel
Hirschhorn & Sons in Mannheim. Of his 17 children several
went abroad to England, France and Italy; two sons were involved in the revolution of 1848 and, like so many liberals of
those days, fled to the United States. My great-grandfather
bequeathed his flourishing tobacco business to my grandfather who married Wilhelmine Enthoven-Herschel, this way
spanning the net of the family to Amsterdam. This was a wise
decision, because later, in 1923, when the peak of inflation

(one dollar equal to 4200 billion marks) and the low of the postwar period was reached, the wealthy maternal dutch relatives invited my underweight brother and myself to the fleshpots of Holland for 3 months. So as a boy of six I got acquainted with the grachten (canals) and patrician houses of Amsterdam and Baarn where an uncle lived on a large estate in a rambling mansion with a whirl of social engagements. Twenty years later most of them were exiled to Theresienstadt, and some to Auschwitz from where they did not return. My mother fled to Holland in November 1938 and was kept hidden by the Dutch underground all through the war. She returned to Heidelberg in 1952 and died there in 1971.

Like most 'liberated' German Jews the Hirschhorns of Mannheim did not emphasize their 'jewishness' but took pride in being pillars of the community, such as members of the City Council, aldermen, confidants or friends of the Grand-Duke of Baden. Thus, when my mother got married, a telegram arrived from the Grand-Duchess with special congratulations:

'ihre königliche hoheit grossherzogin louise sendet ihnen zu ihrem vermählungstag die besten glückwünsche und lässt bitten auch ihren eltern höchstihren glückwunsch zu übermitteln = baronin rotberg', karlsruhe, den 5.3.1912.

Of these flourishing times of 'positive cooperativity' there are now only the memories left and there are scholars of Jewish culture and history who fault the German Jews for their excessive attempts to get assimilated into the local culture. Hannah Arendt comes to the pessimistic conclusion that the hoped for German-Jewish symbiosis depended on fallacies, if not deceptions, and from the very beginning was doomed to failure. In a similar sense, philosophers such as Hermann Cohen, Franz Rosenzweig and Walter Benjamin dealt with the German idealism as the last phase of a train of thoughts that started with the Aufklärung (enlightenment) and led to the attempted assimilation. Both Cohen and Rosenzweig, as they grew old, appealed to their contemporaries for a return to

Jewish piety and ethics and to a new synthesis of Jewish tradition with German cultural values. In this way they hoped to
relieve what Kafka called the 'burden of Jewish existence' and
the 'horror of solitude and abandonment' in favor of a novel
'sense of belonging'. Kafka went farthest in feeling a stranger
in the German language which, like a thief, he felt he had
stolen. Unlike Martin Buber, he was not proud of his achievement to sublime the German language, adopted or stolen as it
may have seemed to him, to new heights of style and expression. For many Jews 'in exile', paradoxically, the German language became their 'Heimat' in an alien land.

The painful transition from the fragile Weimar
Republic to the tyranny of the Third Reich

The nine years at a German 'Real-Gymnasium', from 1926 to
1935, provided ample experience for remembrance and rein-
terpretation of a time that still casts it shadows more than half
a century later. Unlike the Gymnasium with heavy emphasis
on the classics, the Real-Gymnasium tried to provide a balance between languages, the humanities and science, i.e.
physics, mathematics, chemistry and geology, each taught by
teachers of unusual ability. Especially Latin served as a challenge to the receptive young minds with all the grammatical
punctilios and finesses, woof and warp of the fabric of the texts
of Tacitus and the lyrical idiom of Horace and Virgil. The demand in English was for accuracy and precision which I picked
up so well that after my arrival at Harvard in 1947, I was
charged to provide courses for those students who failed their
chemical tests because of poor grammar, style or articulation.
Shakespeare was a subject both for English as well as German. There was more exposure to the dramas and sonnets of
the great bard than any of my three children had in the public
schools of Montgomery County, Maryland.

Physics and mathematics were taught by Karl Person, a

politically outspoken teacher who in 1933 paid the penalty of compulsory transfer to Karlsruhe just at the time when he moved into a house of his own in Freiburg. This kind of chicanery became popular after 1933 to neutralize opponents of the regime if they were lucky to survive. Person became President of the Parliament of Baden after 1945.

Friedrich Breusch taught chemistry in the same rigidly logical fashion as mathematics. He spent one full term in developing, step by step, the formulae of sulfuric and nitric acids; yet he was not only an unusual teacher but also a man of courage and principles. When in 1933 the fascist salute became mandatory in schools at the beginning of a lesson, he would refer to it as 'Aufhebung der Rechte', a pun meaning elevation of your right arm but also abrogation of rights. At age 62 he prematurely resigned in protest against fascism. This uncompromising courageous act made a deep and lasting impression on many of his students.

During the last four years my teacher in history and literature was Ernst Bender, one of the most dedicated pedagogs I have ever known. The quality of his instruction was on the College or University level and made us fall in love with literature, while commonly poems and dramas suffer irretrievable damage when offered through school systems. During the difficult transition time to a vicious totalitarian state in 1933 he made us read Marx' 'Das Kapital', stating that a critical judgement of this classical treatise required a thorough study. This was an interesting about-face: when fascism started, he introduced us to the origins of socialism.

The end of 9 years of gymnasium in those days was the 'Maturum' or 'Abitur' based on one week of written and three days of oral examinations, an event usually preceded and followed by intensive nightmares. There were reasons for more bad dreams, this time of escapes over the mountains into Switzerland, when the political situation brought me and my family into increasing jeopardy.

Chemistry in Munich: Richard Willstätter and Heinrich Wieland

Fortunately, the choice of a vocation or profession posed no problem. At age 15 I knew that chemistry attracted me most. At age 17, in December 1934, I traveled from Freiburg to Munich to visit Richard Willstätter, who knew members of my mother's family and who, after his resignation in 1925, lived in seclusion as a widower in a cultivated home in Möhlstraße 29. He resigned as early as 1925. His reasons and the circumstances in a German University, long before Hitler took over, are described in Willstätter's autobiography [3]. Felix Haurowitz (1896–1987), at this time staying in Willstätter's private laboratory, happened to be a witness of the students' demonstration where a number of colleagues implored Willstätter to reconsider his abdication as a Professor and as Director of the Institute. In a letter from December 16, 1978 to me, Haurowitz describes the historic event: 'At this convention Willstätter thanked the students and made the declaration – in his measured tranquil way of speaking – that he had no intention of staying with a Faculty in which Professors make decisions on candidates for vacant chairs not on the basis of qualifications but race. He then slowly rose and – without looking to the right or left – slowly moved out of the lecture hall. As this happened, all students rose. Nobody said a word, while Willstätter slowly left. It was like a funeral....'

When I visited Willstätter, his housekeeper, Elise Schnaufer, opened the door to the shy young man who then waited for the prescribed 15 minutes in the monastic Italian pews of the antechamber, until the 'Herr Geheimrat' opened the connecting door to his inner sanctum and welcomed the self-conscious postulant with open arms. The tenor and substance of that conversation is as fresh today as it was then; so is the picture of the hermit and scholar with his sad and kind blue eyes and parliamentarian manners.

He encouraged me not to seek the solution of my problems in

immediate emigration but promised to recommend my acceptance at the University of Munich to his successor and friend Heinrich Wieland. And, so it happened. During the subsequent months, one day I received a hand-written letter from Heinrich Wieland in which he promised to use his influence to secure admission for me at the University. I felt both honored and somewhat embarrassed that two such famous chemists devoted their time and energy to help pave the way for a chemical career of an unknown young novice. Later, off and on I had the pleasure of returning to Willstätter's precarious island of tranquility, the tranquility of resignation, to be sure, in a sea of trouble.

In the fall of 1935 I started my first semester in Munich. My teachers were Karl von Frisch in Zoology, Walter Gerlach in Physics, Otto Hönigschmid in Inorganic Chemistry and H. Wieland in Chemistry. It was the custom that the Director of the Institute, H. Wieland, would give the big introductory lecture, both in inorganic as well as organic chemistry, embellished and reinforced by many impressive experiments and demonstrations. Later at Harvard I regretted that the students there were not introduced to chemistry in a similar manner.

On September 30, 1938, together with other students and large crowds, I watched the motorcade, carrying Daladier and Chamberlain, moving down Ludwigstraße. The world held its breath then, believing to be on the brink of war. Little did they know that the conference meant nothing: all decisions had already been made in advance. 'Munich' soon thereafter became a symbol of appeasement and betrayal. Most disadvantageous peace is certainly not better than most just war. To the man in the street and to myself the outcome at the time meant a brief opportunity to live on borrowed time, until one year later 'dogged war snarled in the gentle eyes of peace'.

The elevation and relief at the time easily carried over into the kind of feeling that Goethe expressed (most probably much later, in 1820) on the evening following the 'Kanonade von

Valmy' 1792: 'From here and from today rises a new epic of
world history, and you can say, you were there'. Another quotation fits the years to come: 'What kind of times are these
when a conversation about trees becomes a crime, because it
implies a guilty silence about so many felonies' (B. Brecht).

'Kristallnacht' or the 'Night of Broken Glass'

The pursuit of scientific studies was rudely interrupted on November 9, 1938, the morning after the dramatic breakdown of
civilization not only in Munich but in all cities and towns of
Nazi Germany. On this morning I bought flowers and went to
Möhlstraße 29, which I was relieved to find intact. Willstätter
already had a visitor. In the familiar anteroom, against my
will, I became a witness to a conversation between Willstätter
and his guest whose voice I recognized as that of Heinrich
Wieland: 'Nobody will dare to touch you!' I heard Wieland say.
Willstätter countered by stating that his staying in Munich no
longer depended on good faith or good will. Wieland thought
that the Nobel Prize and the highest distinction, the Order of
'Pour-Le-Mérite' would offer some protection. When Wieland
left, I found Willstätter very tired. I said goodbye to him and
shook his hand. It was to be the last time.

Wieland, as an employee of the State, showed great forti-
tude by this visit that could have had very unpleasant consequences, disciplinary action or even dismissal. The same holds
for another courageous act, namely the removal of von
Baeyer's and his own bust from the place of honor in the laboratory entrance, when Wieland was told that Richard
Willstätter's bust was missing. The 'Kristallnacht' as a prel-
ude to the Holocaust was not duly recognized. Arthur von
Weinberg (1871– 1943), philanthropist and pioneer of the German chemical industry, at that time advised Paul Ehrlich's
widow that distinguished elderly persons would be exempt
from persecution. Under strong pressure from her family,

Hedwig Ehrlich left for Switzerland and the United States in the nick of time, while Arthur von Weinberg perished in Theresienstadt.

A few days later my mother, who after separation from my father, lived in Seeshaupt on Lake Starnberg in a house which I shared with her, was served notice to leave within six hours. She secured the help of Dutch diplomatic friends who ferried her across the border near Venlo in a diplomatic car. Less than two years later Holland was occupied and ceased to be a sanctuary. The reunion with my mother was in 1953, 15 years later. During the war there was no civilian traffic to Holland, and afterwards, with the British occupation in Northern Germany, the border was sealed. An appeal on humanitarian grounds was rejected by the British authorities shortly before I sailed for the States in June 1947.

The home in Seeshaupt, within the short three years of my mother's residence, attracted old family friends for unions and reunions that mostly were irretrievable, a fact that was recorded more subconsciously in those days. The full extent of the coming catastrophe was a reluctant premonition only. Elsa Bernstein, who as author of Humperdinck's 'Königskinder' had the pen-name Ernst Rosmer, visited several times. I had known the kind blind lady since the days of my childhood. She had a great gift of fascinating us children with wonderful stories and intricate paper cuts and fold-outs. Her blindness had become an asset. At this time she was the center of cultural and literary life in Munich, later she was the center of intellectual activity and compassion in the concentration camp of Theresienstadt where through her courage and inspiration, she attracted and organized young and old people, especially writers. The blind old lady survived Theresienstadt, where her younger seeing sister Gabriele perished.

The 'House of Prominent People' in Theresienstadt represented the second contingent of the intellectual elite that had failed to join the first contingent, while it still was possible to leave the country. In the same house there also lived and survived Leo Baeck. What an assembly of notables in the best of all possible concentration camps, where the 'final solution' left room for a delicate chance of survival for about 10,000 out of 40,000 inmates. 'Let the remembrance of the holocaust continue to infect the health of barbarism with debilitating dreams'.

Dissertation on the toxin of the deathcap

It was in the spring of 1938 that H. Wieland accepted me in his private laboratory to begin my thesis on the toxic principles of the deathcap, *Amanita phalloides*, the most toxic mushroom in Europe. This was a special privilege. Wieland, then 61 years old, accepted only a limited number of doctoral candidates. To be admitted to his private laboratory was good fortune in several ways: it provided a haven for privacy away from all other laboratories, close contact with the Master, and free chemicals and glassware. Although Wieland was only five years younger than Willstätter, and both had studied, directly or indirectly, under Adolf von Baeyer, the difference between the two of them seemed to amount to a full generation. Though both came from Baden, only the latter had some local color in his speech and was given to dialectic expressions. Wieland, in his informal way, liked the art of understatement with tongue in cheek. His skill of crystallisation was practiced with glass rod and test tube. When we apprentices blushingly produced a purified fraction not yet crystalline, he would dryly comment: 'Ein schönes Öl!' (A beautiful oil). The atmosphere in Wieland's laboratory, the constant contact with one of the old masters of the art of organic chemistry, the influence of this powerful personality, who worked more than he theorized, his fairness, his courage, his uncompromising attitude toward the totalitarian state are at the root of the definition of a great academic teacher [2].

A man who 'professes' his faith in times of adversity and persecution is a 'professor' and eventually becomes a martyr.

Wieland's colleague, the philosopher and musicologist Kurt
Huber (1893–1943) chose this destiny by supporting the
'Weiße Rose', the group of idealistic students who were the
nucleus of a resistance group at the University and who all
died for their conviction. His last statement at his trial illustrates both his character and his time: 'I do consider the return
to simple moral principles, to a lawful state, to mutual trust as
the restitution of legality. There is a last limit to formal legality when it is perverted into lie and amorality: When it serves
only as pretext for cowardice that is afraid of protest against
manifest violations of law. A government which forbids free
expression and criticism and considers suggestions for im-
provements as preparation for high treason and reason for severe punishments, such a state violates an unwritten law that
has always been alive in good citizens and will always have to
stay alive' [4]. Wieland's student Hans Carl Leipelt was executed on January 29, 1945, because he sympathized with the
'Weiße Rose'. Another student, Valentin Freise was put in
prison. Wieland showed great courage in trying to defend and
exonerate his students. At that critical time I felt ashamed of
being spared: 'As long as you don't feel ashamed of being alive,
while others are being put to death, you will remain what you
are, an accomplice by omission' (A. Koestler). Most probably,
we were saved only by the increasing confusion of the Gestapo
machinery caused by the relentless bombardment of Munich
and the advances of the Allies.

Wieland's laboratory at that time was less a nest of resistance than an oasis of decency. In the privacy of his laboratory
he freely expressed his antagonism to the murderous regime
and uplifted the four occupants, among them the later parliamentarian and minister in the Foreign Office Hildegard
Brücher. Thanks to his helmsmanship, scientific research continued and succeeded in spite of all handicaps. My task was to
process many hundred kilograms of *Amanita phalloides* on a

half-technical scale at the facilities of the C.H. Boehringer
Company in Ingelheim/Rhein in order to solve the structure of
the novel cyclopeptides responsible for the extreme toxicity of
the fly amanita [5]. After the war, the saga of the amanitins
and phalloidins was passed to the next generation: Theodor
Wieland solved their structures in a time span of almost 30
years [6]. While the winds of war still howled fiercely, but
some of the house of science was still standing, a new amino
acid from the hydrolysis of phalloidin (originally believed by
Butenandt [7] to be a 'pro-kynurenine' in the chain of events
that converts tryptophan to ocular insect pigments) was isolated [5], and its one-step synthesis, as well as a direct conversion of tryptophan into α-hydroxytryptophan was achieved [8–
10].

The end of the War

The end of the war was an experience not to be forgotten. Increasing dangers had forced my wife to be, Marlene Prinz, and
myself to go into hiding on a solitary farm ('Einödhof') called
Oberhof, in Southern Bavaria near Penzberg. When in April
1945 the last contingents of the ordinary army retreated from
the onrushing American armor, they left a heavy howitzer
right next to the farm. Ten hours later, SS-troops came by,
located some ammunition and started firing the gun. Within
minutes the response from advance American tanks hit the
gun and knocked it out of operation. The shelling lasted all
night until dawn. The farm people, the Polish helper, the
Ukrainian maid and we trembled all through the late afternoon in the basement in the expectation that the next shell
would hit the barn, part of the house, ignite the hay and straw
and incinerate the entire farm. We had already gone through
the saturation air-raids of Munich, and were resigned to losing
our last few belongings salvaged into our hide-out. Fortunately, there was only one hit on a neighboring farmhouse,

causing no fire. The next day we were between two fronts. More than one farm was ablaze, because SS-troops had stayed there and fired at approaching tanks. During an intermission of the shelling in the evening of the last day of April 1945, there was a knock at the door. When we opened there were Russian inmates from the concentration camp of Dachau asking for food and shelter in halting German. The night was bitter cold and snow was falling. We asked them in and gave them potatoes and skim milk. They looked pitiful, shivering in their ragged striped convict garb and showing all signs of starvation and malnutrition. Soon more of the hapless victims came. There was no longer room in the kitchen, so we moved them to the warm and comfortable stables, where next to the cows and horses they were delighted to rest on freshly piled up straw and eat what they had not had for many months, potatoes and skim milk. Their daily ration at Dachau had been 100 g of substitute bread and one potato, mostly rotten. As the evening passed, more and more refugees came. Finally, when we had about 150 persons, the Russian 'Kapo' told the next group to move on to the neighboring farms. Most farmers, that is farm women, because the men had been conscripted, understood the situation and asked the poor people in. There were only a few farms which were seized by force and where cattle was slaughtered. I mostly saw moving scenes of wordless gratitude. Some had tears, because they had forgotten that there can be mercy and compassion in this world. When we distributed the last cigarettes, there was no end of 'Spassibo' and we had to ask them not to kiss our hands.

Around noon the next day, a cold May 1st with fresh snow on the green meadows, the first American reconnaissance vehicles appeared. About half a dozen soldiers, all tall and wiry, crowded around the farm wife who kept shrugging her shoulders. They wore no insignia, but I spotted the Captain immediately: he was in his early forties, all the others in their twenties. He was taken aback when I addressed him in rather British literary English and thanked him for coming and ending my period of hiding. He looked up and down for quite a while, then he relaxed into a winning smile and shook my hand, an unheard of gesture of understanding and friendship, because Eisenhower's strict order of the day was no fraternization whatsoever with the Germans. There was another bilingual order of Eisenhower that was soon posted and promulgated everywhere: All Nazi laws on race and discrimination were declared null and void immediately. That meant, that after five years of illegal and dangerous togetherness we were able to get married.

During the afternoon we had our first GI's for tea, again an exception to the strict rule: 'Be wise, don't fraternize!' We talked about 'Gone with the Wind' and Rhett Butler's words: 'This is not the first time that everything in this world is upside-down. And, each time all the people lose everything, and there is complete equality. Then the whole game starts afresh with the stakes being a sharp mind and a strong hand. But there are always a few diehards who make it and eventually reach the position they had before the world went haywire'.

Few days in my life were as meaningful and concentrated as the days following the end of a war that had brought ruin and death to more than 50 million people. What is so memorable about these days is the depth that human contacts and relations reached among those that had suffered from oppression and totalitarianism. The good forces rallied immediately, found and embraced each other and marveled at the miracle of survival. The need to talk, report, evaluate and prognosticate was immense. Not only our bodies, but our minds were starved and needed freedom of thought and expression more than bodily food. It was at this point that we became fully aware of the burden of 6 years of war, now that it no longer weighed on us.

On May 7 came the announcement of the official armistice. The word 'peace' was apparently too much to ask for. That 'peace in our time' would remain an aim too ideal to be realized, we did not know then, nor were we anxious to know. The following days the American combat units arrived in full force.

They faced tremendous logistic problems, because, in addition to an intact German army corps returning from Italy, they had to collect and feed thousands of displaced persons and homeless concentration camp inmates who had been stranded in long box car trains on the way from Dachau to the Alps. What we had seen at the farm were only the strong survivors. Entire boxcars filled with emaciated cadavers were still standing around on the railroad tracks. The civilian population was ordered to file by these boxcars with their grisly contents and take due notice. Although most knowledgeable people had no illusions about the concentration camps, reality surpassed their worst fears.

Our wedding day later in August was the fitting climax of this period of liberation. There still was no mail and no train service. Through the office of the Bavarian Ministry of Education I was in the fortunate position to have the use of a car, an unheard of privilege in those turbulent days. Wieland's laboratory, after complete destruction in 1944, was relocated in Weilheim. He lived in Starnberg. The University authorities were in Munich. I worked in Freising. Trains did not operate. So the automobile was a godsend. This made it possible to collect my teacher, Wieland, and his wife from Starnberg.

The wedding lasted three days like in Grimm's fairy tales. Wieland in a whimsical toast compared the young couple to Pamina and Tamino who had to defy the powers of darkness and a villain, whose name was not Monostatos, and had to anneal their attachment in the fires of war without the benefit of a magic flute – although protected in some miraculous way. We were floating on the wings of music, conviviality and Wieland's best Moselle for many happy moments. The participants have all departed, the memories linger. The most beautiful festivals are celebrated on the brink of the nevermore.

Harvard University, 1947–1950

In 1939, before I had my thesis completed, I received an invitation to join the Chemistry Department at Harvard University. I arrived there only eight years later, crossing the Atlantic on the liberty ship Ernie Pyle. The advent in 1947 of a family coming from Munich was an unusual event. At the first reception Bob Woodward introduced us to James Conant who raised his eyebrows and asked: 'Did you serve in the German Army?' My answer: 'No, I was not considered worthy of such a distinction!' pacified him and he started sharing his memories of Hans Fischer and Heinrich Wieland and the honorary degrees that were offered to both of them at the Harvard tercentennial in 1936 [2].

Our contacts at Harvard in this first year were highly diversified. With Paul and Lou Bartlett there was a respectful relation, that later on, while skiing in Stowe with the Roberts, the Sheehans, Gilbert Stork, and others became extremely relaxed and cordial. The climax of winter fun was at the Ranch Camp near Stowe, Vermont. There we used to rough it on hard bunk beds, with Franklin stoves in wood cutters' lodges. In the beautifying mirror of retrospection those convivial and conbibulous evenings with Jack Roberts, John Sheehan, Paul Bartlett, Gilbert Stork and many other now famous pillars of science have an irretrievable charm which has not faded over the years. Memorable gatherings took place on the four-holers, a typical New England institution, probably devised originally by puritans anxious to continue group reading of the Bible even while obeying the call of nature. I remember that on such a tripartite session Paul Bartlett contributed an excellent idea to a mechanistic problem which subsequently fell into place like a Chinese jig-saw puzzle in ivory.

At Harvard I had a reunion with my closest friend Hans Heymann (1916–1979) who spent 1935–1938 in Wieland's Institute and then, in the nick of time, left Munich to obtain his Ph.D. at Harvard with Louis F. Fieser. His father, Bernhard

Heymann, (1861–1933) synthesized *Bayer 205* (Suramin) the famous chemotherapeutic agent effective against the trypanosomes suggesting the name 'Germanin' so that the new curative of sleeping sickness in his words, 'may do honor to the German name'. This great scientist and patriot would have lost his position as Director at the IG Farben in Leverkusen when Hitler came to power, had he not died in time.

Through Siegfried J. Thannhauser (1885–1962), who in Freiburg and later at Tufts College Medical School attracted wide attention as a biochemically oriented clinician, we met Kurt Meyer, whose titration of the enol tautomer of acetoacetate with bromine (1911) I had practiced in Wieland's laboratory. Wieland had told him once: 'This idea you only got because you were not familiar with the literature, otherwise you would have known that acetoacetate does *not* react with bromine!' In Thannhauser's laboratory in Freiburg, Hans Adolf Krebs had started work on the biosynthesis of urea [11].

In this world of cordial contacts, relaxed erudition, intact buildings and windows, enough food for three meals a day, we walked around like somnambulists, afraid that all was a dream. The contrast with war-torn and impoverished Europe was too great. That was just one generation ago!

In the laboratory there was a hectic activity. My graduate students, mostly GI's returned from active duty, were all outstanding. The change of academic climate between Munich and Harvard deserves some comment. In the exact sciences, especially under personalities such as Wieland, even under the trying conditions of totalitarianism, the student in Munich was free from grades and obligatory classroom attendance and there was a surprising degree of self-government in Wieland's department despite his precarious standing with the party and its officials. There were still free ideas – within certain limitations – minimal administrative interference in the research area, a trend of asceticism and self-denial in the service of science in the face of tremendous adversities. As a counterweight to a totalitarian government, the hierarchical struc-

ture of intellectual authority, as practiced by Wieland in his
institute, was probably the only organizational form to be able
to survive all the physical and mental hardships of war and
totalitarianism.

At Harvard I felt somewhat sorry for the students, mostly
combat veterans, whom it was my obligation to vex with cumulative, interim and final examinations. This was also the
time when James Conant implemented his idea of 'general
education' in rather forceful terms. Much later, when my children went through the American system of education and then
spent their Junior or later year abroad, the opportunity for
direct comparison came again. The more regimented college
education, at least in the areas where I had occasion for comparison, *viz.* literature, languages, forestry, and law, in terms
of factual knowledge or in the ability to organize an independent research project, compared favorably with the German
system of student independence, that had even become more
liberalized and, unfortunately, politicized by the seventies,
when in fact a decentralized democracy of specialists, most of
them mediocre, had been 'decreed' by the 'reform' of the German Universities, while in good old USA elitism was still practiced unashamedly at leading academic institutions, especially at 'fair Harvard' that emerged from the turbulent sixties
with hardly a scratch on its elitist escutcheon. New York City
College, by contrast, did not fare so well. Often called the 'proletarian Harvard', it had trained and produced notable artists,
journalists, scholars and many Nobel Laureates. The policy of
'open admission' ended this illustrious era. Clearly, there is a
wide gap between the exact and the social sciences: the former
try to aspire to Humboldt's unity of 'Forschung und Lehre' to
maintain intellectual mobility and versatility, while the latter
suffer from a sense of direction, a lack of serious competition
and misunderstood substitution of equality for quality. In the
'Matrix of Specialization' John Higham presents an expert
analysis of the American and European system of education in
the course of history.

In my meetings with Werner Jaeger (1898–1961), who had emigrated from Germany in 1934, he charmingly explained to me, the novice and chemist, his ideas of *Paideia*, namely the Greek ideal of lifelong transformation of the human personality, involving both conscious learning as well as letting the act of living inform your inner self, translate facts into values, processes into purposes, hopes and plans into consummations and realizations: man himself is the work of art that paideia seeks to form. Much later I followed this trail in philosophy to the 'Center of Hellenistic Studies', opposite Dumbarton Oaks in Washington, D.C., where Jeno Platthy, the poet-librarian, showed me the core of the library, the books of Werner Jaeger, acquired by Harvard.

Through the good offices of Robert Burns Woodward and Paul D. Bartlett my second year at Harvard introduced me to teaching duties both in the Laboratory as well as in the lecture room where I had to represent Bartlett during his sabbatical in California. Mechanistic organic chemistry was a great challenge for me who had been raised on an electron-free diet. My lecture was supposed to start at 10 in the morning, but the preceding teacher, Percy Williams Bridgman (1882–1961), as a rule used more than ten minutes of my allotted time. Organic chemistry at that time, before the arrival of Frank Westheimer and Konrad Bloch, was structural and static and hardly strayed into biochemical areas. Bartlett's favorite Wagner-Meerwein rearrangements stimulated my own research during this time. The importance of migrations of substituents or side chains in metabolic events, such as the formation of homogentisic acid from tyrosine or the migration of aromatic hydrogen in the 'NIH-shift' came much later.

Colleagues and friends

Robert Burns Woodward (1917–1979) [12]
The closest and most electrifying contact was with Bob

Woodward, then an extremely boyish-looking lean young man, exactly as old as I was. When I first knocked at Woodward's office on the first floor of the Converse Memorial Laboratory, I saw a student coming up and asked him, whether Professor Woodward was in. The student to my surprise said 'Let us see!' and opened the door. When I looked puzzled, the student smiled and commented: 'Now he is in!' We laughed, shook hands and immediately had a conversation about alkaloids that lasted for many hours.

Edward Crane in Woodward's laboratory worked on vomipyrine, a degradation product of Wieland's favorite alkaloid, vomicine, related to strychnine, at that time the most complex of all alkaloids. Almost simultaneously with Sir Robert Robinson (1886–1975), Woodward had figured out the structures of vomicine, as well as of strychnine which he synthesized later. My research on yohimbine in Munich [13] and at Harvard was the bridge to Woodward's interest in sempervirine that I had brought along as a gift of Prof. Janot in Paris, who together with V. Prelog had proposed a structure that did not account for its mystifying color and strong fluorescence. When I mentioned to Woodward Robinson's anhydronium bases derived from harmane, he immediately grasped the significance and saw the connection. Thus, sempervirine became the first naturally occurring anhydronium alkaloid [14]. Woodward's photogrammatic visualization of chemical structures never ceased to impress me. In that respect, whether structural, crossword or other puzzles, to him these were all games which he enjoyed and mastered more than we mortals. He, in my opinion, was the most gifted example of *homo ludens* for whom the science of chemistry became the supreme game.

In the long nights of protracted nocturnal discussions young Woodward's histrionic talents and his flair for special effects showed when we approached the solution of a problem and when he was ready for the dénouement with the words 'Elementary, my dear Watson'. He was not only a connaisseur and fancier of Sir Arthur Conan Doyle but, deliberately or subconsciously, used epithets and entire passages in his speech, be it informal or in lectures, that could have been lifted right from Sherlock Holmes: 'This case is exceedingly remarkable' or 'not devoid of a certain interest' or Holmes' characterization in Conan Doyle's words: 'He was a man, however, who, when he had an unsolved problem upon his mind, would go for days and even for a week, without rest, turning it over, rearranging his facts, looking at it from every point of view until he had either fathomed it or convinced himself that his data were insufficient'. Holmes and Woodward became interchangeable, and Woodward enjoyed quotations such as: 'Circumstantial evidence is occasionally very convincing, as when you find a trout in the milk'. Sherlock Holmes appealed to him, as he did to Paul Ehrlich, probably because Holmes displays the features of the amateur scientist serving society, while impressively enacting the very epitome of aloofness. Sir John Cornforth uses another picture: 'Woodward worked on this hard little nucleus (the synthesis of cephalosporin C) like a Japanese netsuke carver' [12].

When on November 9, 1979, Woodward's friends and colleagues gathered to memorialize him in the graceful chapel on the Harvard Campus, Lord Todd found the proper words to describe Woodward's dilemma [15]: 'Men like Bob Woodward, with a single-minded devotion to their science and a burning desire to excel are hardly likely to be good family men, much though they need companionship..... I confess that it was his passion to excel in all he did that at times disturbed me greatly, especially during the last few years, when his loneliness became increasingly evident'.

What a difference to Heinrich Wieland, who balanced his life and tried to serve nature, while Woodward, a genius in his own right, led a lonely battle to dominate it.

Otto Krayer (1899–1982) [16]

Wieland's work on the alkaloids from calabash curare was of great interest to Otto Krayer, Professor of Pharmacology, somewhat of a stepchild at Harvard, and pioneer on the hypotensive alkaloids from veratrum. More important perhaps for the history of the time was Krayer's unparallelled audacity in 1933 when he wrote a letter to the all-powerful Prussian Minister of Education in which he explained why he felt unable to accept the chair of pharmacology at Düsseldorf from which the Jewish incumbent, Philipp Ellinger, had just been removed: '.....the primary reason for my reluctance is that I feel the exclusion of Jewish scientists to be an injustice, the necessity of which I cannot understand, since it has been justified by reasons that lie outside the domain of science. This feeling of injustice is an ethical phenomenon. It is innate to the structure of my personality, and not something imposed from the outside. Under these circumstances, assuming the position would make it difficult to take up my duties as teacher with joy and a sense of dedication, without which I cannot teach properly..... The work to which I have heretofore dedicated all my strength, means so much to me that I could not compromise it with the least bit of dishonesty. I therefore prefer to forego this appointment, rather than having to betray my convictions; or that by remaining silent I would encourage an opinion about me that does not correspond with the facts'.

Krayer was summarily dismissed and informed that all German Universities henceforth were 'Off-limits' for him. In 1937 he joined the Faculty of Harvard Medical School as Associate Professor of Pharmacology and two years later succeeded Reid Hunt (1870–1948) as Head of the Department. As a member of the committee to help German university scientists, organized by the Quakers, he visited Wieland in Starnberg in 1948 and supplied him with cigars and new hope.

Percy Lavon Julian (1899–1975) [17]
'Studies in the Indole Series. VI. On the Synthesis of Oxytryptophan and Further Studies of 3-Alkylation of Oxindoles' [18] came to my attention as a student in Munich in 1938. In the same year, Percy Julian, a promising 'Afro-American' chemist

who had received his Ph.D. in Vienna in 1930, published 'The
Complete Synthesis of Physostigmine (Eserine)' [19], thereby
correcting a mistake of Sir Robert Robinson. At the time this
caused a major upheaval and provided fame and opportunities. Further intersecting interests on yohimbine, ketoyoby-
rine and yobyrine led to a life-long friendship. Julian made
medical history when shortly after Hench and Kendall's discovery of cortisone in 1948 he made available Reichstein's
Substance S in a process competing with the Merck Laboratories where Sarett had synthesized cortisone. Julian's physostigmine synthesis is still used in preparing the unnatural
antipode as a protective agent against organophosphors, as
well as in the study of derivatives, made by Arnold Brossi, now
in clinical trial to relieve the symptoms of Alzheimer's disease.

Amino Acids - Then and Now

The hydrolysis of phalloidin, besides the putative new amino
acid α-hydroxytryptophan, yielded allo-hydroxy-L-proline
which 14 years later was also isolated from sandalwood [20].
Sandalwood, in a stimulating collaboration with Ramadasan
Kuttan, yielded two more constituents of considerable biochemical interest: one more *sym*-homospermidine [21], the
other γ-L-glutamyl-S-(*trans*-1-propenyl)-1-cysteine sulfoxide
[22], in which the stereo-configuration of the sulfoxide is the

$$
\underset{H_3C}{\overset{H}{\diagdown}} C = C \underset{H}{\overset{S-CH_2CH(CO_2H)NH_2}{\diagup}} \quad (O)
$$

opposite of the same peptide isolated from onions, *Allium cepa*,
which presumably has the S-configuration of the sulfoxide. At
the time this was the first occurrence of sulfoxide diastereoi-
somerism observed in nature. Theodor Wieland had to cope

with the configuration of sulfoxides in his exhaustive investigations of the toxins of *Amanita phalloides* [6].

Another bridge to India, apart from sandalwood, was formed in my long association with L.K. Ramachandran (1928–1990) who joined my laboratory in 1958 to work on the protein from tobacco mosaic virus and later shared my interest in gramicidin.

Trans-4-hydroxy-L-proline was identified by Emil Fischer in 1902 in a hydrolysate of gelatin. The parent substance, collagen, contains also trans-*3*-hydroxy-L-proline, accessible by hydroboration of 3,4-dehydro-L-proline [23], as well as by a one-step ring-synthetic approach [24]. 3,4-Dehydro-L-proline was prepared by the action of hog kidney amidase on 3,4-dehydro-D,L-proline amide. By this method the optically labile amide yielded more than the theoretically possible optically stable 3,4-dehydro-L-proline, because the remaining labile D-amide spontaneously racemized simultaneously with the enzymatic cleavage of the L-amide. Here we have the first example of an enzymatic resolution coupled with the asymmetric transformation of an optically labile amide substrate [25].

3,4-Dehydro-L-proline proved its value not only as an antimetabolite of L-proline, but also as a synthone for several new natural amino acids, such as 2,3-*cis*-3,4-*trans*-3,4-dihydroxy-L-proline from diatomaceous earth, as well as for 2,3-*trans*-3,4-*trans*-3,4-dihydroxy-L-proline from the hydrolysate of virotoxin of *Amanita virosa* [26]. *Cis*-3,4-methylene-L-proline, a new natural amino acid from horse chestnuts, became accessible by the addition of carbene to 3,4-dehydro-L-proline [27]:

This new rigid boat-conformation amino acid is a powerful competitor for proline in many systems, such as bacterial proline permease where it inhibits proline uptake to the extent of 85%. Twenty years later, this amino acid would be a likely candidate for agonism at the NMDA (N-methyl-D-aspartic acid) receptor glycine modulatory site where a rigid analog of glycine, such as 1-amino-1-carboxy-cyclopropane shows a five times higher binding constant. The trans-lactone of γ-hydroxy-L-ornithine was prepared by perbenzoylation of L-histidine and catalytic reduction of the L-αδ-dibenzamido-γ-keto-n-valeric acid [28]. The erythro-γ-hydroxy-L-ornithine, corresponding to 4-hydroxy-L-proline [28], recently was located in a highly active antibacterial metabolite of Streptomyces griseorubiginosus [29].

While 3,4-dehydro-L-proline so far has not been found in nature, the homologous 4,5-dehydro-L-pipecolic acid or baikiain, a constituent of hardwood, became the starting material for two new amino acids: on hydroboration it yielded 70% trans-5-hydroxy-L-pipecolic acid which occurs in dates, in addition to 30% trans-4-hydroxy-L-pipecolic acid which is found in Acacia leaves [30]
.   In addition to trans-4-hydroxy- and trans-3-hydroxy-L-proline collagen contains δ-hydroxy-L-lysine whose configuration was determined by the application of the rule of Claude

Hudson. In its simplest form Hudson's Rule says that γ- or δ-(sugar)lactones in which the carbon carrying the hydroxyl group is related to D-glyceraldehyde, are dextrorotatory or $[d]_{D(lactone)} - [d]_{D(acid)}$ gives a positive value. In this way, the asymmetric carbon carrying the secondary hydroxyl group in lysine was shown to have the D_G-configuration and the open amino acid was *erythro-γ-hydroxy-L-lysine* [31].

The introduction of the hydroxyl groups into proline and lysine, i.e. the transition from pro-collagen to collagen, is a post-translational process. The enzymatic hydroxylation was studied in chick embryos with *cis* and *trans*-4-^{3}H-prolines [32]: the former retained 94% tritium, the *trans*-compound showed a loss of 98% tritium. The enzymatic hydroxylation, therefore, must proceed by front-side displacement of hydrogen by hydroxyl with retention of configuration.

Photochlorination of L-lysine yielded γ-chloro-L-lysine which was converted, *via threo-γ-hydroxy-L-lysine* to *threo-γ-hydroxyl-L-arginine*, a natural constituent of Lathyrus seeds [33]:

$$
\begin{array}{c}
NH \\
\parallel \\
NH - C \\
| \qquad \diagdown NH_2 \\
CH_2 \\
| \\
CH_2 \\
| \\
HO - C - H \\
| \\
CH_2 \\
| \\
H - C - NH_2 \\
| \\
COOH
\end{array}
$$

Neither *threo-* nor *erythro-γ*-hydroxy-L-lysine could be detected in hydrolysates of collagen or gelatin or of mediterranean sponge [34].

Proline as an archetype in interconversions carried out with A.A. Patchett in 1957 [35] resurfaced again 35 years later in the structure of *Vasotec* which was developed at Merck under Patchett's leadership and is one of the most important hypotensive drugs of the century [36]:

'*Amino Acids – Then and Now*' became the title of the Lecture delivered in Osaka late in 1993 to honor the memory of Shiro Akabori (1900–1992), Japan's leading pioneer in the chemistry of amino acids and proteins. In 1959 I met Professor Akabori who was a guest in the laboratory of Jesse P. Greenstein of the National Cancer Institute, the same year in which Greenstein's three-volume classic on the *Chemistry of the Amino Acids* appeared. A memorial issue for Akabori, containing this Lecture, entitled 'From Chemistry to Biology' will be published later this year by Kyoritsu Shuppan, Tokyo.

The reevaluation of old amino acids in the light of modern neuropharmacology and -physiology has lent new glamor to amino acids as common as glycine or glutamic acid as messenger molecules and neurotransmitters. The crowded program of the 1993 International Congress on Amino Acids in Vienna and the – regrettable – foundation of a new journal, 'Amino Acids' attest to the timeliness and significance of this field.

Much 'Trypto-Fun' and the International Study Group for Tryptophan Research (ISTRY) [9]

As early as 1889, Frederick Gowland Hopkins (1861–1947) took an interest in the yellow pigment of butterflies, a topic that was taken up again by Heinrich Wieland in 1925 and solved in 1940. The Sixth Meeting of the International Study Group for Tryptophan Research (ISTRY) in Baltimore, 1989, presented an opportunity to honor Hopkins' discovery of tryptophan in 1901, a molecule that half a century later emerged again as a fascinating research project for chemists, physiologists and pharmacologists, especially after the discovery of serotonin and its importance as a neurotransmitter and psycho-active agent. With the help of Arvid Ek, my first doctoral student from Norway at Harvard, the precursor of serotonin, the new natural amino acid 5-hydroxytryptophan became available in 1953 [37] and, with the expertise of Sidney Udenfriend at NIH, proved to be a natural substrate for 'aromatic amino acid decarboxylase'. Another solubilized specific enzyme converts tryptophan to 5-hydroxytryptophan, an important metabolic process which would be desirable to follow by a tritiated substrate, such as $5\text{-}^3\text{H-L}$-tryptophan or $4\text{-}^3\text{H-L}$-tryptophan [38].

There was no reason to suspect at the time that this kind of hydroxylation would involve a shift of hydrogen, deuterium or tritium. The retention of H, ^{2}H or ^{3}H in positions 4 or 5 on hydroxylation leads to 4-^{3}H-L-tryptophan with identical retention, viz. 85% for ^{3}H. This surprising discovery was termed the 'NIH-shift' [39] and in the last quarter of a century has blossomed to a fertile field of importance to drug metabolism, labile metabolites and interaction of DNA with carcinogenic metabolites of benzpyrene, an example of increasing sophistication in 'molecular' biology. Organic chemists are relieved that the epithet 'molecular' was never needed to characterize their discipline.

Tryptophan research on a regular international basis was organized in 1971, mainly as a result of the discovery of 5-hydroxytryptophan and the wider consequences for neurochemistry, psychiatry, cardiovascular studies and, more recently, immunobiology and neuroimmunobiology.

Almost every atom in free or bound tryptophan is capable of reacting selectively under appropriate conditions, both *in vitro* as well as *in vivo* [9]. That the primary attack of oxygen or oxygenases on tryptophan should occur at the 3-position was postulated as early as 1952, but the proof had to wait for more defined methods [40,41]: the elusive ring-chain tautomeric γ-hydroperoxy-ψ-tryptophan easily formed a tricyclic -hydroperoxyde that on standing or warming rearranged to formylkynurenine, a welcome analog to the metabolic process:

Formylkynurenine

Albert Hofmann's 'Problem Child' [42]

The golden anniversary of the accidental discovery of the halucinogenic effect of lysergic acid diethylamide (LSD) renewed interest in the report that Dr. Hofmann sent to his research director, Professor Arthur Stoll:

'Last Friday, April 16, 1943, I was forced to interrupt my work in the laboratory and proceed home, being affected by a remarkable restlessness, combined with a slight dizziness. At home I sank into a not unpleasant intoxicated-like condition, characterized by an extremely stimulated imagination. In a dreamlike state, I perceived an uninterrupted stream of fantastic pictures with intense, kaleidoscopic play of colors. After some two hours this condition faded away'.

The great potency of LSD with an effective dose of 0.3–1 μg/kg body weight seems to preclude metabolic studies *in vivo*. The discovery of the microsomal drug-metabolizing enzymes by B.B. Brodie and Julius Axelrod [43] offered an opportunity to simulate the process *in vitro*: indeed LSD on incubation with the oxidase from liver microsomes was inactivated and the product was identified by its reactions and by synthesis with the aid of sulfur dichloride *via* the disulfide [44]:

The oxindole nature of the metabolite was proven in the same way as in the case of α-oxytryptophan, namely opening of the lactam ring by heating with alkali, diazotization of the liberated anilino group with nitrite and acidification and coupling of the diazonium salt with β-naphthol in alkaline solution to the bright red azo dye. Wieland's classical chemical knowledge was passed on to his pupils not only in this case, but also when Feodor Lynen impressed Otto Warburg for the first time by demonstrating the labile S-acetyl coenzyme A thioester linkage with alkaline nitroprusside to surprise the delighted viewer by the purple color typical of free sulfhydryl groups.

From these studies it became clear that, as an exception, oxindoles or α-oxytryptophan may occur on the catabolic route of indoles.

The 4-position of (5-hydroxy)indoles is not only involved in LSD but also in dehydrobufotenin which arises in the toad by an enzymatic process which so far has not yielded its secrets:

Photocyclization of N-chloroacetyl tryptophan and protein-bound tryptophan (providing an easy access into the difficult 4-position) has been studied by my disciple and colleague Osamu Yonemitsu [45–47]. The analogous photocyclization with N-chloroacetyl O-methyl tyrosine leads to one of the most radical rearrangements ever observed and became a bugaboo in many cumulative exams [48]:

The bicyclic rearrangement product became a touchstone for Jerome and Isabella Karle's 'symbolic addition procedure' X-ray analysis which was honored by the Nobel Prize in 1985 [48].

Probing sequences of the gramicidins, serum albumins and TMV virus

When it comes to the selective cleavage of tryptophyl peptide bonds, two gaps have to be bridged: the brilliant work of Saul Winstein explored and popularized so-called 'neighboring group effects' and the phenomenon of 'anchimeric assistance'. But that was 'l'art pour l'art' with no attempt to yield to the temptation of straying into biochemical underbrush. The next bridge was cast to the Weizmann Institute, where, through the good offices of Ephraim Katzir-Katchalski at the International Congress in Zürich in 1955 I was privileged to secure the collaboration of Avram Patchornik who lost little time in discovering the cleavage of tryptophyl peptide bonds by N-bromosuccinimide (NBS), truly an applied neighboring group effect [49,50]:

Glucagon, a hormonal peptide containing one tryptophan was the first example which served to demonstrate the usefulness of this cleavage. Tobacco mosaic virus (TMV) protein (secured from Wendell Stanley, a postdoc of H. Wieland in 1930) was investigated by L.K. Ramachandran who explored the cleavage of tryptophyl peptide bonds in this exciting substrate [51]. The linear gramicidins A,B,C had to be separated first into pure components [52] by Erhard Gross (1928–1981) before the cleavage of the four Try-peptide bonds by NBS showed the liberation of one amino acid only. The subsequent synthesis of gramidicin confirmed the sequence studies [53]. The selectivity of the cleavage was shown with human serum albumin in which one peptide bond out of 564 was severed and a single new amino-terminal peptide was liberated [51].

Origin and fate of norepinephrine in man and in insects

In my student days the textbook version of the biosynthesis of the important neurotransmitter norepinephrine looked deceptively simple and convincing. Of course, the starting point was phenylalanine. But it required the discovery of the hereditary genetic defect, phenylketonuria, to focus attention on the soluble enzyme system, phenylalanine hydroxylase, to look for a rapid assay of this enzyme by offering it p-^{3}H-phenylalanine, in the same way as $trans$-4-^{3}H-L-proline served as a substrate for the determination of proline hydroxylase. But there was a difference: tyrosine was formed from 4-^{3}H-L-phenylalanine but with hardly any loss of tritium rather with more than 95% retention in the resulting tyrosine [39]:

Here we have another example of the 'NIH-shift'. The next step to dopa (dihydroxyphenylalanine) is a normal hydroxylation without surprises.

Siro Senoh, my first guest from Japan, in 1957 took up the next step by trying to model the introduction of a hydroxyl group in the β-position of the side chain, normally achieved by the enzyme dopamine β-hydroxylase, by the addition of the elements of water *via* a quinone methide [54]:

This model reaction, done with an N-acylated dopamine *via* dopamine quinone, has no application to man but to insects in which all steps, at all seven stages of sclerotization, are under enzymatic control [55].

At the quinone level, 1,4-addition to 6-hydroxydopamine [54,56], after removal of the protective group, takes precedence over 1,6-addition at the quinone methide stage. Preliminary experiments in dogs showed this amine to have unusual properties, but it took Hans Thoenen to establish that injected 6-hydroxydopamine selectively destroys catecholamine-containing nerve terminals. Siro Senoh was privileged to join Julius Axelrod in making available m-O-methylnorepinephrine (normetanephrine) [57], the chief metabolite of norepinephrine, an area for which Axelrod received the Nobel Prize in 1970.

Labile metabolites

The conversion of tyrosine to homogentisic acid involves hydroxylation with concomitant migration of the side chain, the same phenomenon that was later found with benzene homologs on hydroxylation with microsomal enzymes. A model reaction showed the base-catalyzed rearrangement of 4-carbomethoxymethyl-4-acetoxy-2,5-cyclohexadiene-1-one to homo-

CH_2COO CH_2—$COOCH_3$ $\xrightarrow{NaOH}$ OH CH_2—$COOH$ ... O ... OH

gentisic acid, a reaction whose steric course and mechanism was investigated in detail 15 years later [58].

Surprisingly unstable 4-imidazolones were encountered as rapidly disappearing precursors of formiminoglycinate [59], -aspartate [60] and -glutamate both in bacterial as well as mammalian metabolism.

An earlier encounter with a labile metabolite in the bacterial degradation of xanthine by a cell-free extract from *Clostridium cylindrosporum* led to the predecessor of the final product, formiminoglycine, *viz.* 4(5)-imidazolone, an unusually labile intermediate metabolite:

4(5)-Imidazolone   Formiminoglycine

The equally labile 4(5H)-imidazolone-5-acetic acid was characterized as an intermediate in the bacterial degradation of imidazoleacetic to formiminoaspartic acid by the action of *Pseudomonas fluorescens* [60]. The half-life time of this labile intermediate at pH 8 was less than one hour. The homologous N-formiminoglutamate, resulting from the breakdown of histidine by first histidinase, then urocanase and imidazolone propionate amido hydrolase, was studied at the NIH in 1955 by Tabor and Mehler, and its mechanism of formation clarified by J. Rethey in 1960 [62].

Finally, in 1970 the first labile arenoxide [61,62] was isolated as a fugitive intermediate in the microsomal oxidation of naphthalene [63]:

Just at the right time Emmanuel Vogel in Cologne had synthesized this tricyclic naphthalene 1,2-oxide, but expressed great disbelief that such an unstable molecule could ever be found in metabolism. A position paper on the importance of arene oxides in metabolism, long-range toxicity of drugs and etiology of certain cancers became a 'citation classic' [64].

Cyanogen bromide: from protein sequencing to protein accounting

The saga of the active one-carbon unit and transmethylation involves 'active' methionine, *i.e.* S-adenosylmethionine, and has fascinated investigators, such as Giulio Cantoni, Vincent du Vigneaud, William Stein and Stanford Moore. They all observed, but did not utilize, the intramolecular displacement of the sulfur function for the cleavage of methionyl peptide bonds. The key was cyanogen bromide which in the hands of Erhard Gross was used to 'audit' the sequence of pancreatic ribonuclease in 1961 [65,66]. The accuracy of the cleavage permitted for the first time sequencing of proteins as large as immunoglobulin by Gerald Edelman. The underlying mechanism points again to the principle of the neighboring group effect. The group in a methionine peptide that is being removed is the sulfur function:

In order to facilitate the departure of sulfur, the cyanosulfonium intermediate is formed by interaction with the pseudohalogen, cyanogen bromide. The displacement of the volatile and stable methyl thiocyanate CH_3-S-CN under the concerted attack of the imidolate anion leads to the unstable imino-lactone which spontaneously hydrolyzes to homoserine(lactone) and a new amino-terminal residue.

In 1966 Rolf Axen investigated the peptides released from plasma [67,68]. His results laid the ground for the coupling of substances containing primary amino groups to cyanamide-Sephadex, an idea carried out in Jerker Porath's laboratory in Uppsala. Surprisingly, the yield of coupled amino acid exceeded 100%. Blind experiments using unsubstituted Sephadex showed that, indeed, Sephadex after cyanogen halide treatment immobilized proteins in excellent yields! Thus, CNBr coupling replaced the isothiocyanate procedure for the synthesis of immobilized antigens and antibodies to be used in radio-immunoassays. The bridge between Bethesda and Uppsala is a nice footnote to useful events and discoveries.

A final note: the CNBr cleavage has been a key step in liberating insulin peptides from a precursor or chimeric protein made by *E. coli* through genetic engineering. A number of pharmaceutical companies produce more than 100 kg of human insulin a year by this method.

Photoreductions of uridine and thymidine [69,70]

Selectivity as a guiding principle in research and reactions has been a key issue for success. The temptation was overwhelming to transfer some of the experience on proteins to the less diversified RNA or DNA biopolymers. Coming from the photochemically active laboratory of Hans Schmid in Zürich, Peter Cerutti, in 1965 explored the photoreduction of uridine as well as of dihydrouridine with sodium borohydride. The starting point of this investigation was the observation that uracil and cytosine differ significantly in their electronic absorption spectra and in the π-electron density or spin density distribution of their first excited state as computed by HMO or SCF methods. In the event, photoreduction of a mixture of the four RNA components showed only the loss of uridylic acid. A light-independent secondary hydrogenolysis of hydrouridylic acid led to N-(β-D-ribofuranosyl)-N-(γ-hydroxypropyl) urea. Specifically the reduction to dihydrouridine was measured by the orcinol assay for ribose which is positive for the reduced molecule because of the lability of the β-glycoside linkage, and by the Archibald assay of the β-ureidopropionic acid formed on acid hydrolysis of the glycosidic bond and opening of the dihydrouracil by alkali:

When the dark hydrogenolysis was carried out with sodium borotritide, the incorporation of radioactivity into tRNA could be used as a rapid assay for dihydrouridine units.

By contrast to uridine, dihydrouridine is hydrogenolyzed faster in the light-independent reductive ring opening than it is formed. Dihydrothymidine, which may be a product of UV-irradiated DNA, is only a transitory intermediate in the photoreduction of thymidine in the presence of sodium borohydride. All products of this photoreduction have been isolated and the stereochemistry of the new asymmetric center which is formed by the reduction of the 5,6-double bond have been determined [71].

Thymine forms photodimers both from the excited singlet and triplet stage. There is now good evidence for the identity of the major photoproduct derived from irradiated DNA as the cis-syn dimer which was found to undergo easy hydrogenolysis in aqueous solution in the cold by NaBH$_4$ [72]. Hydrogenolysis of irradiated DNA with sodium borotritide should mark and modify areas of photodimerization with tritium, a method that has been suggested [73].

Wieland's legacy: from toads to frogs to birds [74]

The Wieland Centennial Lecture (1977) presented an opportunity to compare bufotoxin from the common toad, *Bufo vulgaris*, with batrachotoxin from the rare Columbian poison dart frog, *Phyllobates aurotaenia*:

Bufotoxin   Batrachotoxin

Bufotoxin was formulated by H. Wieland with the suberylarginine substituent in position 14 instead of 3, and the acetoxy group in position 5 instead of 16. Batrachotoxin's structure was deduced from Roentgen-ray diffraction data by Isabella Karle utilizing the 'symbolic addition procedure' developed by her husband, Jerome Karle [75].

John Daly [76] succeeded in exploring many species of Phyllobates and Dendrobates frogs, studying their habitats and crossovers, isolating or characterizing close to 300 different amphibian alkaloids and establishing their structures either by X-ray analysis through Isabella Karle or by sophisticated mass spectroscopy and NMR spectroscopy.

Takashi Tokuyama and Thomas Spande, my erstwhile associates, carried much of this now independent program, of
which epibatinine [77], an analgesic more potent than mor-
phine, as well as the occurrence of homobatrachotoxin in the
Pitohui bird of New Guinea pose new puzzles for herpetolo-
gists as well as ornithologists [78].

In the X-ray structure of the unusual histrionicotoxins [79]

Isabella Karle pointed to dimensions of the nitrogen and oxygen functions reminiscent of those of the neurotransmitter
acetylcholine. And, indeed, Edson C. Albuquerque [80] who
explored much of the electrophysiology of the frog toxins, clarified the effects of the histrionicotoxins on the ion conductance
modulator part of the acetylcholine receptor.

Epilogue

My entry into the eighth decade in 1987 appropriately marked
the end of a thirty-year helmsmanship, longer than my
predecessors, William Mansfield Clark, Claude Silbert
Hudson or Lyndon Frederick Small had ever served. The transition to scholarly pursuits was a relief.

In 1980 the appointment to membership in the Paul Ehrlich

Foundation in Frankfurt meant not only the selection of candidates for the most prestigious Paul Ehrlich prize, but also contact with the grandchildren of Ehrlich, the Ehrlich Archive and the history of Ehrlich and his times.

Ehrlich's definitive biography so far has not been written [81] and his fascinating letters have not been published. The best letters are the ones he wrote to his closest friend, his cousin Carl Weigert, in which he candidly talks about the many prejudices and setbacks which he encounters and the serious shortcomings of friends and colleagues, for instance Emil von Behring whose Nobel Prize would have been unthinkable without Ehrlich's guidance.

Ehrlich's own Nobel Prize was delayed until 1908, because Svante Arrhenius erroneously applied the principles of the mass action law to the quantitation of the immune response. Ehrlich's receptor concept was the beginning of molecular immunology. He was the first to recognize that antibody molecules 'amboceptors' have two different molecular combining groups, one with an affinity for the appropriate immunodeterminant on the cell surface, and the other with an affinity for complement. His keen insight was far ahead of the technical capabilities of the early 1900s. The biochemical tools to prove his foresight right did not become available until much later. That the 'amboceptor' has found its way onto the present DM 200 banknote may be visible only to the spectator forearmed with this knowledge.

The Heisenberg uncertainty principle applies not only to atomic but probably also to sociological particles. In both areas there are complexities so subtle that mere description may destroy them. When we try to analyze our own personal experience, especially from a position 'astride two cultures', 'joining two worlds' and attempt to attach special meaning to past events, we notice superior forces having molded the course of our lives and realize that there have been special and structural constraints influencing our development. Our effect on society rarely depends on our personal qualities alone but on the social energies with which our fellow humans charge us. Our personal talents are only the stimulants for the social dynamics to find a motor in us.

The reluctance to face this dilemma is more European, because the American attitude tends to vacillate between the assertion and suppression of self: From their point of view there exists an *invidious dualism* that contracts into *individualism* with some difficulty. This question goes deeper and concerns the essence of personal identity as well as the physical and psychological continuity throughout a human lifespan. Infancy, adolescence, manhood and senescence, are they all identical creatures of separate identities connected only by overlapping memories? This question has been raised by philosophers, but much earlier and more poignantly by poets. The cradle rocks above an abyss, and common sense tells us that our existence is but a brief crack of light between two eternities of darkness. Although the two are identical twins, man, as a rule, views the prenatal abyss with more calm than the one he is heading for.

In the year of the centennial, Albert Einstein's thoughts on man's existence are as timely as when he wrote them (1936):

'Of what is significant in one's own existence one is hardly aware of and certainly should not bother the other fellow. What does a fish know about the water in which he swims all his life? The bitter and the sweet come from the outside. The hard from within from one's own efforts. For the most part I do the thing which my own nature drives me to do. It is embarrassing to earn so much respect and love for it. Arrows of hate have been shot at me too, but they never hit me, because somehow they belong to another world with which I have no connection whatsoever. Today I live in the solitude which is painful in youth, but delicious in the years of maturity'.

Personal memories should be for use, not for show. They should not represent the past to serve the interests of the present. In some way the desire to write autobiographical notes is to yield to the earthly temptation to indulge in the memories of youth in a futile attempt to escape the passage of time and the advent of death, in another way it serves as a needed catharsis – all in all just feeble footprints on the sands of time. The worlds of childhood, of yesteryear and of a 'sunny past' are retrievable only as memories and then create a nostalgia which Proust described so well:

'Remembrance of a certain form or appearance usually coincides with the regret of the loss of a special moment and, in that respect, unfortunately, the houses, lanes and streets of our childhood are fugitives like the years that make up our life'.

Every life becomes history and merges into the legacy of the past where, in the end, its human contents is lost as soon as the relation to the living world and the contemporary scene ceases to exist. Most of its past, science discards forever, although in its time it was innovative and good science and blazed new trails. Like the aesculapean snake, science keeps young by shedding its skin as it keeps growing. Only professional historians succumb to the temptation to drag along a large cargo of objects, themes and their interconnections, a load the younger generation is unable or unwilling to shoulder. If, in a small way, memories do not only inform but also incite, rather than diversify conversation, then an individual's life, personal and unapologetic, may not have been lived in vain. Only the stems are broken in the act of threshing, as we are thrashed by a cruel destiny, but the grains stay intact and spring from the ears and husks and may go either to the mill or the field for future harvest. Some of our thoughts, random or inherited, scientific and otherwise, may have stayed at our doorstep as orphans of the mind, but we should not be worried lest they may find foster parents in time.

Science and humanism are indivisible. Some of the thoughts and principles, born by their union, guide us as lights, less like a beacon to a safe port, but more like a lodestar as a constant companion on a never ending quest.

REFERENCES

1 K. Tange, Katsura, Stepping Stones Along the Shore, Approach to Shokintei, Yale Unviersity Press, New Haven, 1960.
2 B. Witkop, Medicinal Research Reviews 12 (1992) 195–274. German version: Liebig's Ann. Chem. (1992) I-XXXII.
3 R. Willstätter, Aus meinem Leben, Verlag Chemie (1949).
4 Clara Huber, Kurt Huber zum Gedächtnis, Verlag Josef Habbel, Regensburg, 1947, pp. 25–26.
5 H. Wieland and B. Witkop, Liebig's Ann. Chem. 543 (1940) 171–183.
6 Th. Wieland, in: Springer Series in Molecular Biology, Peptides of Poisonous Amanita Mushrooms, Springer Verlag, Berlin, 1986, pp. I-XIV and 1–256.
7 A. Butenandt, W. Weidel and E. Becker, Naturwissenschaften 28 (1940) 447.
8 B. Witkop, Liebig's Ann. Chem. 556 (1944) 103–114.
9 B. Witkop, Heterocycles 20 (1983) 2059–2075.
10 B. Witkop in H.G. Schlossberger, W. Kochen, B. Linzen and H. Steinhart (Eds.), Progress in Tryptophan and Serotonin Research, Walter de Gruyter & Co., Berlin-New York, 1984, pp. 21–23.
11 H.A. Krebs, Reminiscences and Reflections, Clarendon Press, Oxford, 1981.
12 A. Todd and J. Cornforth, R.B. Woodward, Biographical Memoirs of Fellows of the Royal Society 27 (1981) 628–695.
13 B. Witkop, Liebig's Ann. 554 (1943) 83–126.
14 R.B. Woodward and B. Witkop, Am. Chem. Soc. 71 (1949) 379; cf. B. Witkop, J. Am. Chem. Soc. 75 (1953) 3361–3370.
15 Remarks by Lord Todd, in Robert Burns Woodward, A Remembrance, Harvard University Memorial Church, Friday, November 9, 1979, Harvard University Press, 1979.
16 A. Goldstein and O. Krayer, Biogr. Memoirs of the Natl. Acad. of Sci. 57 (1987) 150–225.
17 B. Witkop and P.L. Julian, Biogr. Memoirs of the Natl. Acad. Sci. 52 (1980) 223–266.
18 P.L. Julian, J. Pikl and F.E. Wantz, J. Am. Chem. Soc. 57 (1935) 2026–2029.
19 P.L. Julian and J. Pikl, J. Am. Chem. Soc. 57 (1935) 755–757.
20 A.N. Radhakrishan and K.V. Giri, Biochem. J. 57 (1954) 58–63.
21 R. Kuttan, A.N. Radakrishnan, T. Spande and B. Witkop, Biochemistry 10 (1971) 361–365.
22 R. Kuttan, N.G. Nair, A.N. Radakrishnan, T.F. Spande and B. Witkop, Biochemistry 13 (1974) 4393–4400.
23 F. Irreverre, K. Morita, A.V. Robertson and B. Witkop, J. Am. Chem. Soc. 85 (1963) 2824–2831.
24 K. Morita, F. Irreverre, F. Sakiyama and B. Witkop, J. Am. Chem. Soc. 85 (1963) 2832.

25 A.V. Robertson and B. Witkop, J. Am. Chem. Soc. 82 (1960) 5008–5009;
    84 (1962) 1697–1701.
26 A. Buku, H. Faulstich, Th. Wieland, J. Dabrowski, Proc. Natl. Acad.
    Sci. 77 (1980) 2370–2373.
27 Y. Fujimoto, F. Irreverre, J.M. Karle, I. Karle and B. Witkop, J. Am.
    Chem. Soc. 93 (1971) 3471–3477.
28 B. Witkop and T. Beiler, J. Am. Chem. Soc. 78 (1956) 2882–2893.
29 U. Schmidt, R. Meyer, V. Leitenberger and A. Liebernecht, Angew.
    Chem., In. Ed. Engl. 28 (1989) 929–930.
30 Y. Fujita, F. Irreverre and B. Witkop, J. Am. Chem. Soc. 86 (1964)
    1844–1846.
31 B. Witkop, Experientia 12 (1956) 372–374.
32 Y. Fujita, A. Gottlieb, B. Peterkovsky, S. Udenfried and B. Witkop, J.
    Am. Chem. Soc. 86 (1964) 4709–4716.
33 Y. Fujita, J. Kollonitsch and B. Witkop, J. Am. Chem. Soc. 87 (1965)
    2030–2033.
34 N. Izumiya, Y. Fujita, F. Irreverre and B. Witkop, Biochemistry 4
    (1965) 2501–2507.
35 A.A. Patchett and B. Witkop, J. Am. Chem. Soc. 79 (1957) 185–192.
36 A.A. Patchett, J. Med. Chem. 36 (1993) 2051–2058.
37 A. Ek and B. Witkop, J. Am. Chem. Soc. 75 (1953) 500; 76 (1954) 5579–
    5588.
38 J. Renson, J. Daly, H. Weissbach, B. Witkop and S. Udenfried, Biochem.
    Biophys. Res. Comm. 5 (1966) 504–513.
39 G. Guroff, J.W. Daly, D.M. Jerina, J. Renson, B. Witkop and S. Uden-
    fried, Science 158 (1967) 1524–1530.
40 A. Ek, H. Kissman, J.B. Patrick and B. Witkop, Experientia 8 (1952)
    36–40.
41 M. Nakagawa, H. Watanabe, S. Kodato, H. Okajima, T. Hino, J.L. Flip-
    pen and B. Witkop, Proc. Acad. Sci. 74 (1977) 4730–4733.
42 A. Hofmann, LSD, My Problem Child, McGraw-Hill & Co., New York,
    St. Louis, San Francisco (1980).
43 J. Axelrod, Trends Pharmacol. Sci. 3 (1982) 383–386.
44 K. Freter, J. Axelrod and B. Witkop, J. Am. Chem. Soc. 79 (1957) 3191–
    3193.
45 O. Yonemitsu, P. Cerutti and B. Witkop, J. Am. Chem. Soc. 88 (1966)
    3941–3945.
46 O. Yonemitsu, Electron Transfer Chemistry, Yakugaku Zasshi, 102
    (1982) 716–734.
47 T. Iwakuma, H. Nakai, O. Yonemitsu and B. Witkop, J. Am. Chem. Soc.
    96 (1974) 2564–2569.
48 R.M. Moriarty and Ch.W. Jefford, in W.A. Benjamin (Ed.), Organic
    Chemistry, A Problems Approach, Menlo Park, 1975, p. 53 and 155;
    X-ray analysis, cf. J. Karle, I. Karle and I.A. Estlin, Acta Cryst. 23
    (1967) 494–498.

49 T. Spande, Y. Degani, A. Patchornik and B. Witkop, Adv. Prot. Chem. 24 (1970) 97–260.
50 Patchornik, W.B. Lawson and B. Witkop, J. Am. Chem. Soc. 80 (1958) 4747–4748.
51 L.K. Ramachandran and B. Witkop, J. Am. Chem. Soc. 81 (1959) 4028–4032.
52 E. Gross and B. Witkop, Biochemistry 4 (1965) 2495–2501.
53 R. Sarges and B. Witkop, J. Am. Chem. Soc. 87 (1965) 2020–2027.
54 S. Senoh and B. Witkop, J. Am. Chem. Soc. 81 (1959) 6222–6231.
55 M. Sugumaran in: K. Binington and A. Retnakaran (Eds.), The Physiology of Insect Epidermis, Inkata Press, Victoria, Australia, 1991, pp. 143–170.
56 G. Jonsson, T. Malmfors and Ch.Sachs (Eds.), 6-Hydroxydopamine as a Denervation Tool in Catecholamine Research, North Holland Publishing Co., Amsterdam, 1975.
57 S. Senoh, J. Daly, J. Axelrod and B. Witkop, J. Am. Chem. Soc. 81 (1959) 6240–6245.
58 S. Goodwin and B. Witkop, J. Am. Chem. Soc. 79 (1957) 179–185. R. Leinberger, W.E. Hull, H. Simon and J. Rétey, Eur. J. Biochem. 117 (1981) 311–318.
59 K. Freter, J.C. Rabinowitz and B. Witkop, Ann. 607 (1957) 174–187.
60 H. Kny and B. Witkop, J. Am. Chem. Soc. 81 (1959) 6245–6251.
61 B. Witkop, Intra-Sci. Chem. Rept. Vol. 8, 4 (1974) 75–87.
62 Y. Kanoaka, The NIH Shift, Kagaku (Chemistry) 12 (1976) 920–925. H. Tabor and H. Mehler, Meth. Enzymol. 2 (1955) 231–232. J. Klepp, A. Fallert-Müller, K. Grimm, W.E. Hull and J. Rétey, Eur. J. Biochem. 192 (1990) 669–676.
63 D.M. Jerina, J.W. Daly, B. Witkop, P. Zaltzman-Nirenberg and S. Udenfriend, Biochemistry 9 (1970) 147–155.
64 J.W. Daly, D.M. Jerina and B. Witkop, Experientia 28 (1972) 1129–1149.
65 E. Gross and B. Witkop, J. Am. Soc. 83 (1961) 1510–1511.
66 B. Witkop, Science 162 (1968) 318–326.
67 R. Axen, E. Gross, J.V. Pierce, M.W. Webster and B. Witkop, Biochem. Biophys. Res. Comm. 23 (1966) 92–95.
68 B. Witkop, in M. Kageyama, N. Nakamura, T. Shima and T. Uchida (Eds.) The Chemist's Magic Bullets: Selectivity as a Guiding Principle in Biomedical Research, Science and Scientists, Essay by Biochemists, Biologists and Chemists, Scientific Societies Press, Tokyo, 1981, pp. 351–360.
69 P. Cerutti, K. Ikeda and B. Witkop, J. Am. Chem. Soc. 87 (1965) 2505–2507.
70 B. Witkop, Photochemistry and Photobiology 7 (1968) 813–827.
71 Y. Kondo and B. Witkop, J. Am. Chem. Soc. 93 (1971) 764–770.
72 T. Kunieda and B. Witkop, J. Am. Chem. Soc. 93 (1971) 3493–3499.

73   T. Kunieda, L. Grossman and B. Witkop, Biochem. Biophys. Res.
     Comm. 33 (1968) 453–456.
74   B. Witkop, Angew. Chemie, Int. Ed. Engl. 16 (1977) 559–572.
75   T. Tokuyama, J.W. Daly, I. Karle, J. Karle and B. Witkop, J. Am. Chem.
     Soc. 90 (1968) 1917–1918.
76   J.W. Daly, Progress in the Chemistry of Organic Natural Products (W.
     Hertz, H. Grisebach and W.G. Kirby, eds.) 41 (1982) 205–340, Springer
     Verlag, Wien, New York.
77   T.F. Spande, H.M. Garaffo, M.W. Edwards, H.J.C. Yeh, L. Pannell and
     J.W. Daly, JACS 114 (1992) 3475–3478.
78   J.F. Dumbacher, N.M. Beehler, T.F. Spande, H.M. Garaffo and J.W.
     Daly, Science 258 (1992) 799–801.
79   J.W. Daly, I. Karle, C.W. Myers, T. Tokuyama, J.A. Waters and B.
     Witkop, Proc. Natl. Acad. Sci. 68 (1971) 1870–1875.
80   E.X. Albuquerque, K. Kuba, A.J. Lapa, J.W. Daly and B. Witkop, Proc.
     Natl.Acad. Sci. 70 (1973) 949–953.
81   B. Witkop, 'Paul Ehrlich's Leitgedanken und lebendiges Werk,'
     Naturwissenschaftliche Rundschau 34, 361–379 (1981); also in Sci-
     ence, Technology and Society in the Time of Alfred Nobel (C.G. Bern-
     hard, E. Crawford and P. Sorbom, eds.) Oxford, Pergamon, 1981, p. 146.

E.C. Slater, R. Jaenicke and G. Semenza (Eds.)
Selected Topics in the History of Biochemistry: Personal Recollections, IV
(Comprehensive Biochemistry Vol. 38) © 1995 Elsevier Science B.V.

Chapter 4

Recollections: Vacillation of a Classical Enzymologist

ERNST J.M. HELMREICH

Medical Clinic, Division of Clinical Biochemistry and Pathobiochemistry,
The University of Würzburg School of Medicine, Versbacher Str. 5, Würzburg
(Germany)

There is the familiar notion that an autobiography is an impudent lie. And indeed, sticking to the truth does not necessarily make a story more exciting. Therefore, a reader of my recollections will soon find out that I have led an uneventful life in tempestous times and that the little that I might have contributed to biochemistry I owe to the most part to my luck in finding and working with gifted and interesting people. Moreover, I shall only recall my own scientific endeavours. I do hope that my friends from within as well as those outside science will forgive me for not mentioning many happy occasions and meetings.

A sense of humour, an inborn optimism and a certain generosity on my part I make responsible for the many gratifying personal interactions that arose from joint scientific ventures. Perhaps, a protected youth is not the best preparation for a self-centered and ambitious career as a scientist. On the other hand, what I might have been lacking in ambition I have gained in friendship and in a quite personal satisfaction in doing science. Last but not least, no autobiography without an epigraph. I have plagiarised mine from Friedrich Nietzsche

who is always good for an opinion fitting to every occasion. In *Menschliches – Allzumenschliches, Bd. I, p. 526* he says that someone used to thinking in abstract terms easily forgets what he has witnessed but firmly keeps in mind the ideas formed by his life's experiences.

The early years in Munich

Biographies of scientists quite often begin with reminiscences of early youth and chemists recall that already as kids they had begun to experiment in the basement of their parents' homes with sometimes devastating effects. I cannot say that any of my youthful activities were of that sort or gave a premonition that I would devote my life to research in the natural sciences. I was the only child. My father was engaged in banking and business interests prevailed in my family, though I received a sound education in an old-fashioned secondary school (gymnasium) with remarkably little political indoctrination. Moreover, a distant uncle on my mother's side was a professor of Latin and Greek and took an interest in my education. Therefore, if there had not been the political circumstances in Germany in the thirties inevitably leading to war, I might have studied the humanities, perhaps art history. An important influence in my life was my maternal grandfather who designed and manufactured custom-made furniture. Although he had become a businessman, he was always happy to leave his office and tinker with me at a workbench. Thanks to him I learned to enjoy working with my hands and accomplishing something tangible. I believe this is apparent in my scientific work since quite often I was occupied more like a craftsman in developing a technique, a new experimental approach, and/or the tools to tackle a problem. The ways and means to reach a goal interested me more than the goal itself. Actually, after I had reached a goal, it lost its interest and I often turned to something quite different. Not the insight fi-

nally achieved gives satisfaction, rather it is the process leading to it, is a quotation attributed to Edgar A. Poe.

Soon after my 18th birthday, in 1940, I joined the army.
However, my service was of short duration and I never saw
military action. This was due to septicemia, which I developed
following the army's routine anti-typhoid vaccination and became violently sick, which was not too surprising considering
that effective antibiotics with the exception of sulfanilamides
were not yet available. I therefore spent most of 1941 in the
hospital where I was first thought to be a hopeless case. But
quite unexpectedly I recovered, was released from the army
and judged unfit for further military service, and in the winter
of 1941/42 began to study medicine at the University of Munich. Thus I survived the war unscathed, as an uninvolved,
quite naive bystander. Later, I was deeply ashamed for not
having been touched by and cared more about the horrors
around me. To this day I have not lost this feeling of guilt. At
least in this latter respect I may perhaps be different from
many of my contemporaries who were able to repress the past.
Alexander Mitscherlich has characterised my generation as a
generation 'unfähig zu trauern', incapable of sorrow and regret.

After completing medical school, I worked as a physician at
the University of Munich Medical Clinic where I was encouraged by a cardiologist (H. Nowy) and Professor Walter Seitz,
the head of the outpatient clinic, to try my hand at research on
the action of digitalis glucosides [1]. At that time, I fell in love
with a young lady who was studying chemistry and later medicine and we married in 1949. This may have been the turning
point in my decision to become a biochemist and I began to
look for a place to learn chemistry. This turned out to be quite
difficult because most laboratories were still in ruins and students like myself who had already completed their studies
were discouraged from further competing for a place to study
with those who had returned from the war and were eager to
begin their education.

The fact that the newly appointed professor of organic chemistry at the Technical University of Munich (at that time still called Technische Hochschule), Stefan Goldschmidt, accepted me as a chemistry trainee and student was perhaps the most fortunate event in my life. My chemistry training gave me the tools to do biochemical research. Stefan Goldschmidt was one of the few prominent Jewish emigrants who returned to postwar Germany. He was successor to the Nobel prizewinner Hans Fischer. Professor Goldschmidt took a personal interest in my chemical education. Thanks to him I even became a 'Privatdozent' for biochemistry at the University of Munich Medical School in 1953 for work I had done in an organic chemistry laboratory at the Technical University. My 'habilitation' dealt with the metabolism of fructose [2]. Work in carbohydrate metabolism made me acquainted with papers of Carl and Gerty Cori which made such an impression on me that without ever having met them I wanted to work in their laboratory at Washington University in St. Louis as a postdoc. At the same time I was studying, together with Helmut Holzer who was in Feodor Lynen's laboratory, the relationship of NAD^+-NAD + H^+ levels in the liver to fatty acid synthesis and ketone body-formation and the derangement of fatty acid metabolism in diabetes [3]. I became acquainted with 'Fitzy' Lynen who was in charge of the biochemistry laboratory in the famous chemistry department of Munich University. The chairman there at that time was still Heinrich Wieland who had won the Nobel prize in 1927. Lynen had just gained worldwide recognition for his discovery that the acetyl residue in acetyl-CoA is linked to the coenzyme as an energy-rich thioester (Nobel Prize 1964) and when he met the Cori's (Nobel prize, 1947), he recommended me to them and they accepted me as a postdoc. I obtained a National Academy of Sciences fellowship and in the autumn of 1954, I went with my wife and my 3-year-old daughter Irene to St. Louis where we had the good fortune of being able to stay with Victor Hamburger in his house in University City.

St. Louis, 1954–1968

I arrived in the Cori's lab after R. Levine and M.S. Goldstein
had published their evidence that insulin and muscular activity made muscle permeable to sugars [4a]. Carl Cori was not
yet quite ready to accept that. He wanted to find the sugar
which had disappeared from the blood in the muscle cells. But
in order to assess the intracellular sugar concentration, the
extracellularly distributed sugar concentration had to be
known. Therefore, my first assignment was to design a method
to assess the extracellular space in the whole animal. This was
represented by the distribution of D-raffinose after its injection into nephrectomized rats. It was now possible to quantitatively follow the distribution of five non-metabolized or poorly
metabolized aldo-pentoses and of galactose between plasma
and tissues after injection of insulin and stimulation of mus-
cles, and compare it with raffinose which was confined to the
extracellular space. For that purpose the carcass of a whole rat
had to be quantitatively extracted with water in order to determine if and if so how much, pentose had disappeared
through utilization. I went about this task by boiling the dead
rat, putting the soft boiled rat through a meat grinder and
extracting the ground meat exhaustively with hot water. I recovered nearly 100% of the injected pentoses which pleased
Professor Cori, but not my colleagues in the lab, who insisted
that I go about my smelly business at night after they had left.
The results confirmed Levine's and Goldstein's findings that
insulin and muscular activity indeed increased the penetration of pentoses and galactose into the intracellular compart-
ment of muscle [4]. Although my findings were no surprise,
they marked a change in the direction which research on insulin action took in the Cori lab. The idea that insulin promotes
hexokinase activity had given way to the view that the rates of
penetration and utilization of glucose are somehow adjusted to
each other and that besides permeability of the muscle membrane factors controlling enzyme activities may limit the rate

of glucose metabolism in muscle [5]. These experiments marked the end of my personal involvement in research on insulin action, which I did not regret. The role of the insulin receptor and the mechanism of insulin-dependent recruitment of a glucose carrier in muscle and adipose tissue were eventually elucidated in other laboratories more than 15 years later.

As my two postdoc years in the Cori laboratory came to an end, the Cori's asked me whether I would be interested in staying on, provided a suitable job could be found. They made me acquainted with Herman Eisen who had just arrived from New York to take over a position as head of the Division of Dermatology at Washington University Medical School. I liked his research program, and when he offered me a position as assistant professor of biochemistry in medicine, I accepted without hesitation. But in order to obtain the immigration papers, my family and I had to return to Munich in December 1956 where my daughter Ilka was born. The time I subsequently spent with Herman Eisen was not only personally enjoyable, it was also scientifically rewarding. Our joint paper together with Milton Kern on the secretion of antibody by isolated lymph node cells [6] was immediately acknowledged and its importance recognized. It showed that isolated lymph node cells from rabbits immunized with dinitrophenylated bovine γ-globulins were capable of synthesizing and secreting antibodies *in vitro*. From the kinetics of secretion it followed that lymphoid cells committed to antibody synthesis survive the process of secretion which is restricted to γ-globulins. Subsequently, David Kipnis and I became interested in the kinetics of aminoacid transport. I contributed a viable preparation of isolated primary lymph node cells from guinea pig and rabbit, Dave brought his expertise with α-aminoisobutyric acid transport to bear on the problem, and Carl Frieden advised us on the kinetic treatment of our data. The result of this cooperation was one of the early reports indicating that carrier-mediated transport processes in the cell membrane are functionally asymmetric on the inner and outer surface with respect to the kinetic properties of the carrier. Both the results obtained by us with mammalian cells [7], as well as the data published by Horecker, Thomas and Monod [8] on galactose transport in *E. coli*, indicated a transition of the carrier from a state of high affinity for the penetrant on the outer cell surface to one of low affinity on the inner cell surface. Together with the late Eric Reiss, David Kipnis and I then examined the kinetics of amino acid transport, the appearance of amino acids in the intracellular amino acid pool, and their eventual incorporation into protein in mammalian cells. The kinetic analysis indicated functional heterogeneity (compartmentation) of the intracellular amino acid pool with respect to the precursor role of amino acids for protein synthesis. Moreover, the similarity of these observations with those obtained with bacteria and yeast again suggested that the functional compartmentation of intracellular amino acids is a fundamental biological property of all living cells (cf. Kipnis et al. [9]).

Work went well in the laboratory and my wife and I had found friends who are still good friends today; in short, we and our children liked St. Louis. I had no doubt that cellular immunology and cell biology were fields of research which offered many exciting and interesting problems to be attacked. So I made up my mind to stay with Herman Eisen as he became chairman of the Department of Microbiology at Washington University Medical School, succeeding Arthur Kornberg in 1961. But I had underestimated what an authority Carl Cori was for me. I was impressed, if not to say fascinated, by Carl Cori whom I admired not only as a scientist but as a cultured and erudite man who spoke Italian and German fluently and who could recite with perfect recall classical German poems, from Friedrich Schiller and Johann Wolfgang von Goethe to Heinrich Heine and Rainer Maria Rilke. So I returned in 1961 to biochemistry as a tenured Associate Professor. In the same year, my family and I became American citizens. For the next years I worked closely with Carl Cori until he left for Harvard in 1966; however, the close cooperation with Carl Cori was not without its problems. Although I did carry out independent research of my own, I wished to run my own show and build up my own laboratory, and had a hard time freeing myself from Cori's intellectual magnetism. Together with Bill Danforth, we worked on the role of metabolic interconversion of glycogen phosphorylase by reversible phosphorylation in contracting frog muscle. For these experiments it was necessary to find an antifreeze which would allow measurements of the *in situ* ratio of the phosphorylated (phosphorylase *a*) and the non-phosphorylated (phosphorylase *b*) enzymes without changing the temperature-dependent interconversion ratio in deep-frozen frog muscle. Bill Danforth and I still recall gratefully that as we were pondering the problem of which type of antifreeze we could use that would not denature the enzymes but would allow introduction of inhibitors of the interconverting enzymes into the frozen muscle, my technician Roger Sherman proposed glycerol which was immediately adopted and shown to work [9b]. For other experiments on the regulation of glycolysis during muscular activity, which were carried out together with Simon Karpatkin, I converted a Dubnoff shaker to a serial incubator with a sample changer for experiments on isolated electrically-stimulated frog sartorius muscle. This allowed control of contraction and unimpeded efflux of lactate, the end-product of glycolysis in the contracting frog muscle under anaerobic conditions (cf. Karpatkin et al. [10]). Our results demonstrated a coordinated activation of the chain of glycolytic enzymes which we termed a 'functional operon'. A great advance in understanding the functional organization of glycogenolytic enzymes was brought about by the subsequent characterization of a glycogen organelle and the enzymes attached to it, as well as the discovery in Eddy Fischer's lab in Seattle of 'flash' activation of this enzyme-assembly by calcium which couples metabolism and contraction in skeletal muscle [11]. The close relationship of our work in St. Louis to that in the laboratories of Edmond H. Fischer and Edwin G. Krebs, whose research on reversible phosphorylation was honored in 1992 with the Nobel prize, resulted in personal contacts which became mutual friendships. However, the work which influenced me most, was my work with Carl Cori on the role of 5'-AMP as an activator of phosphorylase *b* (cf. Helmreich and Cori [12]). This work led to my deep involvement in another fundamental aspect of metabolic regulation besides reversible phosphorylation, namely allosteric regulation, a concept which was developed at the same time by Jacques Monod and colleagues [12a], as well as by Daniel Koshland and his colleagues [13]. Actually, our experiments were carried out at the same time that Jacques Monod and colleagues developed their model. They used phosphorylase as a specific example illustrating the importance of conformational changes of enzymes for their catalytic activity [14]. After Carl Cori left for Harvard, I continued to work on allosteric regulation of phosphorylase, first with Maria Michaelides, Boyd Metzger and Luis Glaser, and later with L.L. and J. Kastenschmidt (cf. Michaelides et al., [15]; Metzger et al., [16] and Kastenschmidt et al. [17, 18]. In the latter work with the Kastenschmidts we used equilibrium binding of substrates and the cofactor AMP, rather than activity measurements and kinetics, to follow the allosteric transitions of the phosphorylase protein in the course of activation by AMP or phosphorylation.

In the fifties and sixties, Washington University in St. Louis was an exciting place to work. Arthur Kornberg had isolated the enzymes which synthesize DNA. Rita Levi-Montalcini had discovered NGF, and Stanley Cohen EGF, in Viktor Hamburger's laboratory. These are only a few of the most notable events associated with Washington University. At this time and in this environment my own scientific concepts were formed and the goals defined that I set for my research. Carl Cori's belief that an enzymologist should eventually try to find out how the enzyme he isolated and characterized *in vitro* functions in the intact cell was an intellectual guideline that I have followed ever since.

My interest in the conformational transitions occurring in enzymes after their binding of allosteric effectors led me to spend a sabbatical in Manfred Eigen's laboratory in Göttingen in 1967. He had developed a method for dissecting reaction sequences into elementary steps by temporarily disturbing equilibria through temperature jump and other means. As I arrived in Göttingen to work with phosphorylase, Manfred Eigen and Kasper Kirschner had just finished a paper which elegantly established the value of their method for kinetic studies of allosteric transitions of an enzyme, in this case yeast D-glyceraldehyde-3-phosphate dehydrogenase [19]. The method was not as successful for phosphorylase, which eventually turned out to have several transitions which were impossible to separate in time, at least when using the indicator bromothymol blue, the binding of which to phosphorylase b was shown to be temperature-dependent by Agnès Ullmann, Roy Vagelos and Jacques Monod [20]. But the sabbatical in Göttingen was to have long-lasting consequences for me, as I have recalled recently (1994) [21]. During this extended period of time in Germany, my friends and colleagues tried to persuade me to return to Bavaria where I was born and raised and where my mother was still living. They reminded me how much I loved hiking and skiing in the Alps. In Würzburg, the chair of physiological chemistry was vacant. Carl Martius had been chairman after the war, but had left for the Eidgenössische Technische Hochschule in Zürich and his successor, Fritz Turba, a protein chemist, died soon after he came to Würzburg. Würzburg is an old university, first founded in 1402, but closed in 1413, because the rector was assassinated by his manservant and then permanently established in 1575. Würzburg University is a scientifically respectable address, having at one or the other time men like Rudolf Virchow, R.A. von Koelliker, Adolf Fick, Emil Fischer, Wilhelm Conrad Röntgen, Wilhelm Wien, Theodor Boveri and Eduard Buchner on the faculty. Karl Landsteiner studied in Würzburg as a student of Emil Fischer. Aside from the university, Würzburg, with its surrounding hills rising above the Main river valley, is an attractive area where excellent wine is grown. However it was difficult for me to make up my mind. I vacillated and changed my mind a few times. The day on which I finally committed myself to take the Würzburg position happened to coincide with a lecture given by Manfred Eigen in Munich, just after he was informed of having received the Nobel prize in October 1967. After the lecture we met with Manfred to celebrate, and in the good spirit we all were I wrote a picture postcard from a well-known bar in Munich to my wife in St. Louis informing her that I had finally decided to go to Würzburg. This was not well received by my family as I found out after my return to St. Louis. I guess I never could convince them that I had made my decision to go to Würzburg sober and responsibly. In any event, the decision to return to Germany was made and I started in Würzburg in the summer of 1968. Later I have summarized my experiences as a re-migrant under the heading of Mark Twain's 'European travelogue', 'An innocent abroad'. Now, 25 years later, I can say that I did not regret to have returned to the country where I was born because in a way which I did not foresee German science has been reborn. I do respect the younger generation that now occupies chairs and responsible positions in German science for their talent, their commitment and their achievements.

Würzburg, 1968–1991

By then I had established a close and friendly relationship with Anne and Carl Cori and occasionally I came to Boston to pay a visit to them. At one of these occasions just after I had returned to Germany, Carl asked me what I planned to do in Würzburg. I told him that I wanted to find out how pyridoxal phosphate functions in glycogen phosphorylase and that I also wanted to study the mechanism of adrenaline action. Carl was in favor of the second choice because he had always been interested in hormone actions. But he was not much in favor of the first choice. Carl Cori considered it a secondary problem, since the primary role of pyridoxal phosphate as an essential cofactor of all enzymes involved in amino acid metabolism was clearly established by now, thanks mainly to the work of Esmond E. Snell in the USA and Alexander Braunstein in Moscow and their respective schools. He was right, of course; as Esmond Snell liked to say, glycogen phosphorylase was at its best only a honorary member in the club of pyridoxal phosphate-dependent enzymes; also, there is more pyridoxal phosphate bound to glycogen phosphorylase in man and mammals than in all other pyridoxal phosphate-dependent enzymes. In 1957, shortly after pyridoxal phosphate was found to be a stoichiometric constituent of muscle glycogen phosphorylase in Carl Cori's laboratory in St. Louis (cf. Baranowski et al. [22], Edmond H. Fischer and Edwin G. Krebs in Seattle used $NaBH_4$ to reduce the cofactor-enzyme bond (a Schiff's base) to a secondary amine, and found to everyone's surprise that reduced phosphorylase with irreversibly attached cofactor retained activity (cf. Fischer et al., [23]). Therefore, this brilliant and simple experiment indicated at once that if pyridoxal phosphate actually should play a role in phosphorylase catalysis, it would have to do this in a manner different from that of all other known pyridoxal phosphate-dependent enzymes. Other pyridoxal phosphate-dependent enzymes are inactivated on reduction because the 4-formyl group of pyridoxal 5′-phosphate is directly involved in transaldamination reactions in the course of catalysis. I started to work on the role of pyridoxal phosphate in glycogen phosphorylases right after I came to Würzburg. It was a long haul which took us seven years after Knut Feldmann, a physicochemist working with me, found the first clue to the solution of that puzzle. I had recognized quite early the value of NMR-spectroscopy in enzymology, and utilizing ^{31}P-NMR spectroscopy with one of the first 180 MH wide-bore instruments, Knut Feldmann and Bill Hull from the Bruker company showed that in all active forms of glycogen phosphorylases the 5'-phosphate of the cofactor was dianionic or became dianionic in the course of conversion of inactive to active phosphorylase (cf. Feldmann and Hull [24]). This was a breakthrough which immediately suggested a plausible explanation for the essential role of pyridoxal 5'-phosphate in glycogen phosphorylases, namely that of a proton-donor-acceptor shuttle in general acid-base catalysis [25–27]. However, our proposal was questioned by Neal Madsen who, like myself, was once a postdoc in Carl Cori's laboratory [28, 29]. Neal offered an alternative interpretation based on his own studies of the binding of glucose 1,2-cyclic phosphate to phosphorylase also by ^{31}P-NMR spectroscopy. The cyclic phosphate was thought to be a transition state analog of glucose 1-phosphate. The data allowed two interpretations, either the phosphate of pyridoxal phosphate becomes re-protonated, as we had postulated, or the phosphate of pyridoxal phosphate is constrained and tightly coordinated. Neal Madsen and colleagues [30] preferred the latter possibility and decided that the cofactor phosphate in the active form of phosphorylase is a tightly coordinated dianion, and furthermore suggested that such a distorted phosphate is an electrophilic species which could effect a partial withdrawal of electrons from the substrate phosphate, thus labilizing the glycosidic bond. An electrophilic (Lewis acid) mechanism, such as that proposed by Neal Madsen and S.G. Withers [31] would not involve proton transfer to or from the coenzyme phosphate but rather it would imply that the phosphate stays dianionic throughout the catalytic cycle. Such a mechanism would basically be an operationally irreversible phosphoryl transfer mechanism with formation of a pseudo-pyrophosphate bond. To the contrary, the Würzburg group always considered phosphorylases as glucosyl – rather than phosphoryl – transferases and we were able to prove it by studying the reaction of phosphorylase with glycosylic substrates [32, 33]. I would have probably never thought of using glycosylic substrates if Helmut Klein who was an organic chemist before he joined my group had not suggested them. Formation of a transferable glycosyl residue from a glycosylic substrate involves a stereoselective protonation and leads to products retaining the activating proton [34]; therefore, these compounds are ideally suited to the study of a reaction mechanism involving substrate protonation. We were lucky that these non-glycosidic compounds were readily accepted by phosphorylase as substrates. I have recently reviewed these results ([34a], containing citations of the relevant original literature). Among the reactions of phosphorylase with glucals studied by us, the reaction with D-gluco-heptenitol [35]) proved to be especially important because it turned out to proceed only in one direction, e.g. glucosyl transfer to phosphate yielding heptulose 2-phosphate. The fact that heptulose 2-phosphate formed from heptenitol is a 'suicide' inhibitor which binds with high affinity (K_i = 2–14 μM) and in the absence of a primer made it a welcome tool for the X-ray crystallographers [35, 36]. Thanks to the formation of heptulose, 2-phosphate in the crystal, interaction of the phosphate of the cofactor with the substrate phosphate could be demonstrated, which was consistent with the proximity of the phosphates already made apparent by Neal Madsen's and Toshio Fukui's work. The X-ray crystallographic structural determination of the rabbit muscle phosphorylase b-heptulose-2-phosphate complex brought to light that there are no positive charges positioned in such a manner that they could constrain and distort the 5'-phosphate dianion of pyridoxal-5'-phosphate as one might have expected if an electrophilic mechanism, such as that postulated by Madsen and his colleagues, was operable. This evidence together with the earlier discovery of the stereospecific protonation of D-glucal [37], provided the first direct proof of a general (Brønsted) acid-type activation mechanism for the phosphorylase reaction.

In the forward direction, phosphorolysis of α-1,4-glycosidic bonds in oligo- or polysaccharides is started by protonation of the glycosidic oxygen by the substrate orthophosphate followed by stabilization of the incipient oxocarbonium ion and subsequent covalent binding to form α-glucose-1-phosphate.

In the reverse direction, protonation of the phosphate of glucose-1-phosphate destabilizes the glycosidic bond and promotes formation of a glucosyl oxocarbonium ion-phosphate anion pair. In the subsequent step the phosphate anion facilitates nucleophilic attack of a terminal glycosyl residue on the carbonium ion bringing about α-1,4-glycosidic bond formation and primer elongation. Both in the forward and in the reverse reactions, the phosphate of the cofactor pyridoxal-5′-phosphate acts as a general acid ($PL\text{-}POP_3H^-$ or $PL\text{-}OPO_3^{2-}$) as predicted by the NMR-experiments, and functions as a proton shuttle to protonate the substrate phosphate. To accomplish this, the phosphates of the cofactor and the substrate must approach each other within a hydrogen-bond distance (2.8–3.1 Å) corresponding to a distance of the phosphorous atoms of 4.7–5.3 Å. The phosphorous atoms in the phosphorylase-heptulose 2-phosphate complex are actually only 4.8 Å apart. In the case of the phosphorylase-glucose 1-phosphate complex, the oligosaccharide bound to the catalytic site is thought to help in turning the phosphate of glucose-1-phosphate towards the cofactor phosphate, as in the case of heptulose-2-phosphate, so that pyridoxal phosphate can function in catalysis [38]. Moreover, in the case of the natural substrates glucose 1-phosphate or P_i, it is the allosterically triggered movement of Arg- 569 [38], which brings about the formation of the transition state intermediate and which ensures that the phosphate is able to act as a good base to extract a hydrogen from the oligosaccharide, whereas when the reaction proceeds in the direction of glycogen degradation, the glycosidic bond is weakened by direct protonation from the phosphate. However, until now, no structural information has come to light with respect to the disposition of the saccharide bound to the catalytic site. The anticipated role of the oligo- or polysaccharide bound to the catalytic site therefore remains speculative. It is the most important problem with regard to the catalytic mechanism which remains to be solved.

Dieter Palm, who had worked with me on phosphorylase from the beginning, continues with his group in Würzburg to work on the phosphorylase mechanism using site-directed mutagenesis. They have identified Glu-637 (in conjunction with Lys-533) as the amino acid side chains most likely involved in the control of the protonation state of the phosphate of pyridoxal phosphate in *E. coli* maltodextrin phosphorylase [39, 40]. Now, they are trying to distinguish different modes of oligosaccharide binding in the direction of synthesis and of degradation [41].

If one accepts my reasoning, it becomes apparent why nature uses pyridoxal 5'-phosphate in phosphorylase catalysis because no amino acid side chain could accomplish an equally well-balanced proton transfer in a reaction strongly biased towards phosphorolysis and against hydrolysis. Moreover, pyridoxal 5'-phosphate has a different positional mobility than amino acid side chains constrained by protein structure. This together with the versatile role of orthophosphate are cogent reasons why in the case of glycogen phosphorylases protonatable groups of a vitamin are more advantageous than those of amino acid side chains to facilitate catalysis [42].

It gave me great satisfaction that the catalytic mechanism proposed by us for glycogen phosphorylase including the unique participation of a vitamin B_6 derivative, pyridoxal phosphate, was shown to be compatible with the X-ray structural evidence obtained in Louise Johnson's laboratory in Oxford [43–45]. But quite apart from the catalytic mechanism, the beautiful work of the X-ray crystallographers has unraveled many new and interesting aspects, especially on the control of activity by allosteric activation. It may be anticipated that in the not too distant future phosphorylase may become one of the best studied allosteric enzymes deserving a place alongside hemoglobin, which thanks to the work of Max Perutz is the allosterically-regulated protein 'par excellence' [46].

My interest in studying the role of pyridoxal phosphate in

glycogen phosphorylase catalysis identifies me as a classical
enzymologist rooted in an era of biochemistry in which the
clarification of the role of small molecular weight cofactors and
vitamins was hailed as a great achievement. It was therefore
predictable that I would again use the techniques of classical
enzymology to study β-adrenergic signal transduction. What
originally attracted me to that problem was not only the challenge of finding out how hormones act, it was also a challenge
trying to purify membrane-bound enzymes present in minute
amounts. I had the good fortune of finding a gifted and hard-
working postdoc who was up to the formidable task that
awaited us. It was due to Thomas Pfeuffer's and his wife's
skills and tenacity that we had a good start in this competitive
field in the early seventies. We were among the first to make
use of metabolically stable GTP analogs, such as GTPγS, introduced by Fritz Eckstein, and Gpp(NH)p and Gpp(CH$_2$)p
which were 10–40 times more potent than GTP as activators
of isoproterenol-stimulated adenylyl cyclase from pigeon
erythrocyte membranes. Since treatment of membrane preparations with these stable analogs protected the hormonally
activated proteins against adverse detergent effects, we were
able to isolate and enrich a protein fraction which contained
about 90% of the radioactive metabolically stable GTP-deriva-
tive (Gpp(NH)p) which originally bound to membranes. The
solubilized GTP-binding protein fraction could then be separated from adenylyl cyclase by chromatography on Sepharose,
resulting in about a 40- to 80-fold purification from membranes. Removal of the nucleotide binding protein from adenylyl cyclase was also achieved by affinity chromatography
with GTPγS coupled to Sepharose via a spacer and resulted in
a 75% loss of Gpp(NH)p activation. This was the first evidence
indicating that adenylyl cyclase and the protein which binds
GTP are separate moieties [47]. Thomas Pfeuffer subsequently identified the GTP-binding moiety as a protein with a
Mr of ~ 42,000 and showed that activation of adenylyl cyclase
by hormone (isoproterenol) or fluoride was transmitted by this

protein [48]. Moreover, Dan Cassel and Thomas Pfeuffer [49] were able to show that the same Mr 42 kDa protein was also labeled on treating pigeon erythrocyte membranes with cholera toxin and ^{32}P-NAD$^+$. Cassel and Selinger [50] had already shown that cholera toxin inhibits the catecholamine-stimulated GTPase activity in turkey erythrocyte membranes, making GTP in cholera toxin-treated membranes as effective an activator as nonhydrolyzable GTP-analogs.

In the following years, Thomas Pfeuffer and my group advanced in different directions: Thomas went on to purify adenylyl cyclase, and I tried with Wolfgang Burgermeister and later with Mirko Hekman to purify the turkey erythrocyte β-adrenoceptor. Whereas Thomas made another important contribution and successfully purified bovine brain and rabbit heart adenylyl cyclase to homogeneity for the first time, cleverly making use of a forskolin affinity column [50], I could not keep pace with Bob Lefkowitz and his group, although we eventually also managed to purify the β-adrenoceptor to homogeneity by using quite effective affinity gels chemically different from those used by the Lefkowitz laboratory. Moreover, we synthesized useful photoaffinity labels [51]. Since we had managed to purify all the components of the β-adrenoceptor/G-protein-dependent adenylyl cyclase system, Thomas Pfeuffer, Alex Levitzki and I agreed in 1984 at a meeting in St. Paul de Vence to join forces and to collaborate in an effort to start with the separate purified components and reconstitute the whole signal transmission chain in lipid vesicles. This was achieved at about the same time as Alfred Gilman and his laboratory in Dallas reported on their reconstitution experiments [53]. In our case, success was largely due to the skillful and hard work of two young ladies, Debby Feder, a graduate student from Alex's laboratory in Jerusalem, and Mie-Jae Im, a postdoc from Korea, [54, 55].

In the course of this work, Mie-Jae Im had observed that separate $\beta\gamma$-subunits from trimeric G-proteins alone could bind to the β-adrenoceptor [56]. Moreover, when $\beta\gamma$-subunits

purified from bovine brain were added to purified β-adrenocep-
tor in lipid vesicles in the presence of G_s also purified from
turkey erythrocytes, the $\beta\gamma$-subunits promoted the interaction
of the activated β-adrenoceptor with G_s and increased its
GTPase activity [57]. These observations became the starting
point of a new line of research which I have followed ever
since. More recently, we found that $\beta\gamma$-subunits promote even
the binding of seemingly unrelated α-subunits to the non-activated β_1-adrenoceptor and that basal GTPase activity is stimulated even further under these circumstances (Kurstjens et
al. [58]). We have interpreted that to mean that the basal
activity in the hormonally non-activated state is due to a pre-
coupled G-protein-receptor complex, suggesting that β-recep-
tors and perhaps other receptors of this kind may form long-
lived complexes with $\beta\gamma$-subunits.

I find the specificity of protein-protein interactions and the
conformational transitions in G-protein-effector signalling
systems and their functional consequences intriguing. Usually, a well-defined chemical change in a low molecular weight
intrinsic or extrinsic ligand of a receptor makes the receptor
capable of productive coupling to G-proteins. The prime example is the 11-*cis* → *all trans*-transition of the intrinsic ligand
retinal in rhodopsin on bleaching. My personal view, if not to
say my personal bias that protein-protein interactions in
membranes, like those in the soluble state, are the decisive
specificity-determining factors, has been greatly influenced by
my work on the control of function and mobility of proteins by
membrane lipids. Therefore, I believe that heterologous binding interactions of $\beta\gamma$-subunits of trimeric G-proteins with receptors and effectors are as specific as their binding to their
respective α-subunits, and moreover, that protein–protein interactions in the membrane are as important as fatty acid
acylation and isoprenylation reactions for anchoring or target-
ing proteins, such as $\beta\gamma$-subunits to membranes. In contrast to
interaction with membrane lipids, interaction between proteins can provide the specificity required to distinguish among

the many receptor- and effector-subtypes. Such a sorting and editing function of the $\beta\gamma$-subunits might be important for the multiplicity of coupling reactions of G-proteins with receptors and targets. What is now needed are quantitative data of the binding of different cloned and expressed α- and $\beta\gamma$-subunits to each other and to receptors [59,60,61].

The interactions of membrane proteins, such as receptors, often require that proteins move laterally in the plane of the plasma membrane. In the case of the G-protein-linked receptors, an early step in the activation cascade requires the heterologous interaction of the receptor with G-proteins and adenylate cyclase. Actually, Alex Levitzki [62] postulated that the rate-limiting step in these signal-transduction systems is collision coupling between the components in the membrane. Therefore, to understand the mechanisms of these processes, it is necessary to characterize the rates of and constraints on the motions of the interacting membrane proteins. The experimental access for solving this problem is difficult because the proteins involved in signal transmission, except for rhodopsin, do not have a natural intrinsic chromophore. Therefore, for generating a measurable local concentration of signal, as is required for studying gradients introduced by photobleaching and for measuring their dissipation due to rotational and lateral diffusion, one must find the best suited optical probe emitting a sufficiently strong signal in order to selectively and specifically couple it to membrane proteins of interest which lack usable intrinsic chromophores. We are still trying to find suitable fluorescent ligands. With one of the first fluorescent β-adrenoceptor ligands which Mirko Hekman had synthesized in my laboratory, I went to Elliot Elson's laboratory in the winter of 1980/1981. Elliot had just moved from Cornell to Washington University in St. Louis. Thanks to Elliot, I had an interesting and stimulating sabbatical. We were joined by Yoav Henis, a former postdoc of Elliot who is now a professor at Tel Aviv University. With his skills and his excellent and patient experimentation, we were able to show by means of

fluorescence recovery after photobleaching that the majority
of antagonist- bound β-adrenergic receptor is immobilized in
the membranes of cultured secondary liver cells over a temperature range of 4–37°C. In view of the lateral immobility of
the major portion of the β-receptors, several possibilities could
be considered. For example, the G-protein and the target enzyme, e.g. adenylyl cyclase, may be sequestered in the same
local regions on the cell surface as the β-receptors. In this case,
lateral diffusion of the receptor over micrometer distances as
measured by fluorescence recovery after photobleaching
would not be required, and lateral or rotational diffusion over
submicron distances may suffice for activation [63]. The work
started in St. Louis is far from completed. We are still trying to
find an answer to the questions raised above, concerning the
mobility and distribution of the G-proteins and the target enzymes in the membrane and the forces and structures which
cause immobilization of β-receptors in the resting, nonacti-
vated state. In this context, I would like to recall some early
work on the role that the structure of the membrane lipid bilayer plays in the regulation of adrenoceptor-linked signal
transmission [64]. These and other data finally led us to suggest that the drastic change of β-receptors around the phase
transition in 'fluid' phospholipid-enriched cells is due to structural changes in the membrane resulting from the introduction of lipids rather than fluidization. Nowadays, with lipid
domains being popular and taking into account that the pack-
ing geometry in mixed lipid-protein systems is governed by the
spatial requirements of the lipids, a mutual interdependency
between lipid and protein structural order seems plausible
[65].

Retirement and new challenges

Obviously, there is still a long way to go following the road
signs established by previous work. Since the fall of 1991, I

have retired as chairman of the Department of Physiological Chemistry at Würzburg University Medical School. Thanks to the generosity of my colleagues, I am able to continue my research now in the Division of Clinical Biochemistry and Pathobiochemistry of the Medical Clinic of Würzburg University. Ulrich Walter, the head of the division, and his American wife Suzanne Lohmann came to Würzburg about 10 years ago from Paul Greengard's laboratory at Yale University and worked for some years in the Department of Physiological Chemistry. I feel at home in the new environment. I like to be with young people involved in hands-on experiments directly at the workbench. In 1990, I spent one summer and fall at the Max Planck Institute of Biophysical Chemistry in Göttingen with Manfred Eigen and Thomas and Donna Arndt-Jovin searching for suitable fluorescent dyes for monitoring β-adrenergic receptors in intact cells [66]. This work is continued by Fritz Boege and others. And in 1993 and 1994 I spent several months in the USA as a Fogarty scholar at the NIH. As long as I still find my experiments exciting and as long as graduate students share that excitement with me, I feel entitled to continue running a lab. Of course, such a decision is more easily made by a person like me who never developed a hobby amusing enough to divert him and turn him away from science. But why should one want to quit now, in times which may someday be judged as the most exciting period in the history of the biological sciences? Because now, for the first time, a successful attack on the most challenging problems of cell biology and their reduction to the molecular level has become feasible, problems so complex that nobody would have dared to try to challenge them only 20 years ago.

REFERENCES

1   Helmreich, E. (1950). Der Einfluß von Strophanthin auf das Cytochro-
    moxydase-Cytochrom-C-System. Biochem. Ztschr. 321 144–151.
2   Helmreich, E., Goldschmidt, S., Lamprecht, W. and Ritzl, F. (1953). Der
    Einfluß von Kohlenhydraten, insbesondere von Fructose, auf den
    Umfang und den zeitlichen Ablauf der Bildung von aktivierter Es-
    sigsäure und Brenztraubensäure in der Rattenleber. II. Vergleichende
    Untersuchungen über die Dissimilation der Fructose und der Glucose.
    Hoppe Seylers Z. Physiol. Chem. 292 184–206.
3   Helmreich, E., Holzer, H., Lamprecht, W. and Goldschmidt, S. (1954).
    Bestimmung stationärer Zwischenstoffkonzentrationen. II. Die Ent-
    stehung der Ketokörper und ihre Beziehung zur Glykolyse. Hoppe-Sey-
    lers Z. Physiolog. Chem. 297 113–126.
4   Helmreich, E. and Cori, C.F. (1957). Studies of tissue permeability. II.
    The distribution of pentoses between plasma and muscle. J. Biol. Chem.
    224 663–680.
4a Levine R. and Goldstein, M.S. (1955). Recent Progess Hormone Res.,
    11, 343.
5   Kipnis, D.M., Helmreich, E. and Cori, C.F. (1958). Studies of tissue
    permeability. IV. The distribution of glucose between plasma and mus-
    cle. J. Biol. Chem. 234 165–170.
6   Helmreich, E., Kern, M. and Eisen, H.N. (1961). The secretion of anti-
    body by isolated lymph node cells. J. Biol. Chem. 236 464–473.
7   Helmreich, E. and Kipnis, D.M. (1962). Amino acid transport in lymph
    node cells. J. Biol. Chem. 237 2582–2589.
8   Horecker, B.L., Thomas, J. and Monod, J. (1960). Galactose transport in
    Escherichia coli. II. Characteristics of the exit process. J. Biol. Chem.
    235 1586–1590.
9   Kipnis, D.M., Reiss, E. and Helmreich, E. (1961). Functional heteroge-
    neity of the intracellular amino acid pool in mammalian cells. Biochim.
    Biophys. Acta 51 519–524.
9b Danforth, W.H., Helmreich, E. and Cori, C.F. (1962). The effect of con-
    traction and of epinephrine on the phosphorylase activity of frog Sarto-
    rius muscle. Proc. Natl. Acad. Sci. USA 48 1191–1199.
10 Karpatkin, S., Helmreich, E. and Cori, C.F. (1964). Regulation of glyco-
    lysis in muscle. II. Effect of stimulation and epinephrine in isolated frog
    Sartorius muscle. J. Biol. Chem. 239 3139–3145.
11 Fischer, E.H., Heilmeyer, L.M.G. Jr., Haschke, R.H. (1971). Phosphory-
    lase and the control of glycogen degradation. In Current Topics in Cel-
    lular Regulation, Vol. 4 (Horecker, B.L. and Stadtman, E.R., Eds.), pp.
    211–251. Academic Press, New York.
12 Helmreich, E. and Cori, C.F. (1964). The role of adenylic acid in the
    activation of phosphorylase. Proc. Natl. Acad. Sci. USA 51 131–138.

12a Monod, J., Changeux, J.P. and Jacob, F. (1963). Allosteric proteins and cellular control systems. J. Mol. Biol. 6 306–329.

13 Koshland, D.E. Jr., Némethy, G. and Filmer, D. (1966). Comparison of experimental binding data and theoretical models in proteins containing subunits. Biochemistry 5 365–385.

14 Monod, J., Wyman, J. and Changeux, J.P. (1965). On the nature of allosteric transitions: A plausible model. J. Mol. Biol. 12 88–118.

15 Michaelides, M.C., Sherman, R. and Helmreich, E. (1964). The interaction of muscle phosphorylase with soluble antibody fragments. J. Biol. Chem. 239 4171–4181.

16 Metzger, B., Helmreich, E. and Glaser, L. (1967). The mechanism of activation of skeletal muscle phosphorylase a by glycogen. Proc. Natl. Acad. Sci. USA 57 994–1001.

17 Kastenschmidt, L.L., Kastenschmidt, J. and Helmreich, E. (1968). Subunit interactions and their relationship to the allosteric properties of rabbit skeletal muscle phosphorylase b. Biochemistry 7 3590–3608.

18 Kastenschmidt, L.L., Kastenschmidt, J. and Helmreich, E. (1968). The effect of temperature on the allosteric transitions of rabbit skeletal muscle phosphorylase b. Biochemistry 7 4543–4556.

19 Kirschner, K., Eigen, M., Bittman, R. and Voigt, B. (1966). The binding of nicotinamide-adenine dinucleotide to yeast D-glyceraldehyde-3-phosphate dehydrogenase: temperature-jump relaxation studies on the mechanism of an allosteric enzyme. Proc. Natl. Acad. Sci. USA 56 1661–1667.

20 Ullmann, A., Vagelos, P.R. and Monod, J. (1964). The effect of 5'-adenylic acid upon the association between bromthymol blue and muscle phosphorylase b. Biochem. Biophys. Res. Commun. 17 86–92.

21 Helmreich, E.J.M (1994) Recollections. Excursions in Biophysics by a classical enzymologist. Protein Science 3 528–532.

22 Baranowski, T., Illingworth, B., Brown, D.H. and Cori, C.F. (1957). The isolation of pyridoxal-5-phosphate from crystalline muscle phosphorylase. Biochim. Biophys. Acta 25 16–21.

23 Fischer, E.H., Kent, A.B., Snyder, E.R. and Krebs, E.G. (1958). The reaction of sodium borohydride with muscle phosphorylase. J. Am. Chem. Soc. 80 2906–2907.

24 Feldmann, K. and Hull, W.E. (1977). ^{31}P nuclear magnetic resonance studies of glycogen phosphorylase from rabbit skeletal muscle: Ionization states of pyridoxal 5'-phosphate. Proc. Natl. Acad. Sci. USA 74 856–860.

25 Feldmann, K., Hörl, M., Klein, H.W. and Helmreich, E.J.M. (1978). The role of pyridoxalphosphate in glycogen phosphorylases. In Regulatory Mechanisms of Carbohydrate Metabolism. (Esmann, V., Ed.). FEBS Fed. Eur. Biochem. Soc., Vol. 42, Symp. A1 pp. 205–218. Pergamon Press.

26   Helmreich, E.J.M. and Klein, H.W. (1980) The role of pyridoxal phos-
     phate in the catalysis of glycogen phosphorylases. Angew. Chem. Int.
     Ed. Engl. 19 441–455.
27   Klein, H.W., Schiltz, E. and Helmreich, E.J.M. (1981). A catalytic role
     of the dianionic 5′-phosphate of pyridoxal 5′-phosphate in glycogen
     phosphorylases: Formation of a covalent glucosyl intermediate. Cold
     Spring Harbor Conferences on Cell Proliferation, Vol. 8: Protein Phos-
     phorylation, pp. 305–320. Cold Spring Harbor Laboratory.
28   Withers, S.G., Madsen, N.B., Sykes, B.D., Takagi, M., Shimomura, S.
     and Fukui, T. (1981a). Evidence for direct phosphate-phosphate inter-
     action between pyridoxal phosphate and substrate in the glycogen
     phosphorylase catalytic mechanism. J. Biol. Chem. 256 10759–10762.
29   Withers, S.G., Madsen, N.B., Sprang, S.R. and Fletterick, R.J. (1982).
     Catalytic site of glycogen phosphorylase: Structural changes during ac-
     tivation and mechanistic implications. Biochemistry 21 5372–5382.
30   Withers, S.G., Madsen, N.B. and Sykes, B.D. (1981b). Active form of
     pyridoxal phosphate in glycogen phosphorylase. Phosphorus-31 nu-
     clear magnetic resonance investigation. Biochemistry 20 1748–1756.
31   Madsen, N.B. and Withers, S.G. (1986). Glycogen phosphorylase. In
     Vitamin B_6 Pyridoxal Phosphate. Part B (Dolphin, D., Paulson, R. and
     Avramovic, O., Eds.), pp. 355–389. J. Wiley, New York.
32   Klein, H.W., Im, M.J., Palm, D. and Helmreich, E.J.M. (1984). Does
     pyridoxal 5′-phosphate function in glycogen phosphorylase as an elec-
     trophilic or a general acid catalyst? Biochemistry 23 5853–5861.
33   Palm, D., Klein, H.W., Schinzel, R., Buehner, M. and Helmreich, E.J.M.
     (1990). The role of pyridoxal 5′-phosphate in glycogen phosphorylase
     catalysis. In Perspectives in Biochemistry, Vol. 2 (H. Neurath, Ed.), pp.
     132–140.
34   Hehre, E.J., Brewer, C.F., Uchiyama, T., Schlesselmann, P. and
     Lehmann, J. (1980). Scope and mechanism of carbohydrase action.
     Stereospecific hydration of 2,6-anhydro-1-deoxy-D-gluco-hept-1-enitol
     catalyzed by α- and β-glucosidases and an inverting exo-α-glucanase.
     Biochemistry 19 3557–3564.
34a Helmreich, E.J.M. (1992). How pyridoxal 5′-phosphate could function
     in glycogen phosphorylase catalysis. BioFactors 3 159–172.
35   McLaughlin, P.J., Stuart, D.I., Klein, H.W., Oikonomakos, N.G. and
     Johnson, L.N. (1984). Substrate-cofactor interactions for glycogen
     phosphorylase b: A binding study in the crystal with heptenitol and
     heptulose 2-phosphate. Biochemistry 23 5862–5873.
36   Hajdu, J., Acharya, K.R., Stuart, D.I., McLaughlin, P.J., Barford, D.,
     Oikonomakos, N.G., Klein, H. and Johnson, L.N. (1987). Catalysis in
     the crystal: synchotron radiation studies with glycogen phosphorylase
     b. EMBO J. 6 539–546.
37   Klein, H.W., Palm, D. and Helmreich, E.J.M. (1982). General acid-base
     catalysis of α-glucan phosphorylases: Stereospecific glucosyl transfer

from D-glucal is a pyridoxal 5'-phosphate and orthophosphate (arsenate) dependent reaction. Biochemistry 21 6675–6684.

38 Barford, D., Schwabe, J.W.R., Oikonomakos, N.G., Acharya, K.R., Hajdu, J., Papageorgiou, A.C., Martin, J.L., Knott, J.C.A., Vasella, A. and Johnson, L.N. (1988). Channels at the catalytic site of glycogen phosphorylase *b*: Binding and kinetic studies with the β-glycosidase inhibitor D-gluconohydroximo-1,5-lactone *N*-phenylurethane. Biochemistry 27 6733–6741.

39 Schinzel, R. and Palm, D. (1990). *Escherichia coli* maltodextrin phosphorylase: Contribution of active site residues glutamate-637 and tyrosine-538 to the phosphorolytic cleavage of α-glucans. Biochemistry 29 9956–9962.

40 Schinzel, R. (1991). Active site lysine promotes catalytic function of pyridoxal 5'-phosphate in α-glucan phosphorylases. J. Biol. Chem. 266 9428–9431.

41 Becker, S., Palm, D. and Schinzel, R. (1994). Dissecting differential binding in the forward and reverse reaction of *Escherichia coli* maltodextrin phosphorylase using 2-deoxyglucosyl-substrates. J. Biol. Chem. 269 2485–2490.

42 Pfeuffer, T., Ehrlich, J. and Helmreich, E. (1972). Role of pyridoxal 5'-phosphate in glycogen phosphorylase. II. Mode of binding of pyridoxal 5'-phosphate and analogs of pyridoxal 5'-phosphate to apophosphorylase *b* and the aggregation state of the reconstituted phosphorylase proteins. Biochemistry 11 2136–2145.

43 Barford, D. and Johnson, L.N. (1989). The allosteric transition of glycogen phosphorylase. Nature 340 609–616.

44 Johnson, L.N. and Barford, D. (1990). Glycogen phosphorylase. J. Biol. Chem. 265 2409–2412.

45 Johnson, L.N., Acharya, K.R., Jordan, M.D. and McLaughlin, P.J. (1990). Refined crystal structure of the phosphorylase-heptulose 2-phosphate-oligosaccharide-AMP complex. J. Mol. Biol. 211 645–661.

46 Perutz, M. (1990). Mechanisms of cooperativity and allosteric regulation in proteins. Cambridge University Press, Cambridge, U.K.

47 Pfeuffer, T. and Helmreich, E.J.M. (1975). Activation of pigeon erythrocyte membrane adenylate cyclase by guanylnucleotide analogues and separation of a nucleotide binding protein. J. Biol. Chem. 250 867–876.

48 Pfeuffer, T. (1977). GTP-binding proteins in membranes and the control of adenylate cyclase activity. J. Biol. Chem. 252 7224–7234.

49 Cassel, D. and Pfeuffer, T. (1978). Mechanism of cholera toxin action: Covalent modification of the guanyl nucleotide-binding protein of the adenylate cyclase system. Proc. Natl. Acad. Sci. USA 75 2669–2673.

50 Cassel, D. and Selinger, Z. (1977). Mechanism of adenylate cyclase activation by cholera toxin: Inhibition of GTP hydrolysis at the regulatory site. Proc. Natl. Acad. Sci. USA 74 3307–3311.

51 Pfeuffer, E., Mollner, S. and Pfeuffer, T. (1985). Adenylate cyclase from bovine brain cortex: purification and characterization of the catalytic unit. EMBO J. 4 3675–3679.

52 Burgermeister, W., Hekman, M. and Helmreich, E.J.M. (1982). Photoaffinity labeling of the β-adrenergic receptor with azide derivatives of iodocyanopindolol. J. Biol. Chem. 257 5306–5311.

53 May, D.C., Ross, E.M., Gilman, A.G. and Smigel, M.D. (1985). Reconstitution of catecholamine-stimulated adenylate cyclase activity using three purified proteins. J. Biol. Chem. 260 15829–15833.

54 Hekman, M., Feder, D., Keenan, A.K., Gal, A., Klein, H.W., Pfeuffer, T., Levitzki, A. and Helmreich, E.J.M. (1984). Reconstitution of β- adrenergic receptor with components of adenylate cyclase. EMBO J. 3 3339–3345.

55 Feder, D., Im, M.-J., Klein, H.W., Hekman, M., Holzhöfer, A., Dees, C., Levitzki, A., Helmreich, E.J.M. and Pfeuffer, T. (1986). Reconstitution of β_1-adrenoceptor-dependent adenylate cyclase from purified components. EMBO J. 5 1509–1514.

56 Im, M.-J., Holzhöfer, A., Böttinger, H., Pfeuffer, T. and Helmreich, E.J.M. (1988). Interactions of pure $\beta\gamma$-subunits of G-proteins with purified β_1-adrenoceptor. FEBS Lett. 227 225–229.

57 Hekman, M., Holzhöfer, A., Gierschik, P., Im, M.-J., Jakobs, K.-H., Pfeuffer, T. and Helmreich, E.J.M. (1987). Regulation of signal transfer from β_1-adrenoceptor to adenylate cyclase by $\beta\gamma$ subunits in a reconstituted system. Eur. J. Biochem. 169 431–439.

58 Kurstjens, N.P., Fröhlich, M., Dees, C., Cantrill, R.C., Hekman, M. and Helmreich, E.J.M. (1991). Binding of α- and β-subunits of Go to β_1-adrenoceptor in sealed unilamellar lipid vesicles. Eur. J. Biochem. 197 167–176.

59 Heithier, H., Fröhlich, M., Dees, C., Baumann, M., Häring, M., Gierschik, P., Schiltz, E., Vaz, W.L.C., Hekman, M. and Helmreich, E.J.M. (1992). Subunit interactions of GTP-binding proteins. Eur. J. Biochem. 204 1169–1181.

60 Kleuss, C., Scherübl, H., Hescheler, J., Schultz, G. and Wiltig, B. (1992). Different β-subunits determine G-protein interaction with transmembrane receptors. Nature 358 424–426.

61 Kleuss C., Scherübl, H., Hescheler, J., Schultz, G. and Wiltig, B. (1993). Selectivity in signal transduction determined by γ-subunits of heterotrimeric G-proteins. Science 259 832–834.

62 Tolkovsky, A.M. and Levitzki, A. (1978). Mode of coupling between the β-adrenergic receptor and adenylate cyclase in turkey erythrocytes. Biochemistry 17 3795–3810.

63 Henis, Y.I., Hekman, M., Elson, E.L. and Helmreich, E.J.M. (1982). Lateral motion of β receptors in membranes of cultured liver cells. Proc. Natl. Acad. Sci. USA 79 2907–2911.

64   Bakardjieva, A., Galla, H.J. and Helmreich, E.J.M. (1979). Modulation
     of the β-receptor adenylate cyclase interactions in cultured Chang liver
     cells by phospholipid enrichment. Biochemistry 18 3016–3023.
65   Helmreich, E.J.M. and Elson, E.L. (1984). Mobility of proteins and lip-
     ids in membranes. In Advances in Cyclic Nucleotide and Protein Phos-
     phorylation Research, Vol. 18 (Greengard, P. and Robison, G.A., Eds),
     pp. 1–62, Raven Press, New York.
66   Heithier, H., Hallmann, D., Boege, F., Reiländer, H., Dees, C., Jaeggi,
     K.A., Arndt-Jovin, D., Jovin, T.M. and Helmreich, E.J.M. (1994). Syn-
     thesis and properties of fluorescent β-adrenoceptor ligands. Biochemis-
     try (in press).

E.C. Slater, R. Jaenicke and G. Semenza (Eds.)
Selected Topics in the History of Biochemistry: Personal Recollections, IV
(Comprehensive Biochemistry Vol. 38) © 1995 Elsevier Science B.V.

Chapter 5

These are the Moments when we Live!
From Thunberg Tubes and Manometry to
Phone, Fax and Fedex*

HELMUT BEINERT

Department of Biochemistry and Biophysics Research Institute, Medical College of Wisconsin, Milwaukee, WI 53226 (U.S.A.)

Comments on being an experimental scientist

In this chapter I have in mind not to address primarily the few of my contemporaries who may want to read what I have to say, but rather to give members of the upcoming generation(s) an impression of life as a research scientist, the way I have experienced it. They may be struggling with the same doubts, uncertainties, and problems as I was; and it was the lack of such an impression that had me meandering some years before I settled for a career in science. I do not count myself among those who are predestined for a certain profession or vocation; if there are such people, they are, in my opinion, very few. Most of us have a variety of talents and interests, and it often depends on haphazard circumstances, which path we may decide to follow. In my teens I would have liked, among other things, to become a builder, an architect, and my older brother wanted to be an engineer; I ended up as a biochemist

---

* Fedex has become the generally used abbreviation for Federal Express.

194   H. BEINERT

The author in his laboratory (about 1970).

From left to right: George Feker, Richard H. Sands, Helmut Beinert and William H. Orme-Johnson during a meeting at Airlie House, Virginia (about 1965).

and my brother as a historian. It is my experience that for most people 'what' they are doing is not as characteristic for them, as 'how' they do what they are doing.

I have never regretted that I made the choice of profession that I did make, because, as I go along, I realize more and more how deficient my life would have been had I not gained the exciting insights into the mini-world within and the macroworld around us, and at the same time the 'attitude of humility' [1] that intensive occupation with the natural sciences conveys to us. Similar arguments may, of course, be made for other occupations, but it is hard to contest that nature's ingenuity, variety and depth far exceed those of any other field of endeavor based purely on human imagination. I am also aware that the insights I referred to above are a bare shadow of what others may already know or what eventually may be learned as time goes on, and that this, in turn, will always remain a crude approximation of what there really is.

In his recent book 'For The Love of Enzymes' Arthur Kornberg has a chapter entitled: 'Never a dull enzyme' [2]. He is probably right, because it is we who may simply fail to see in an enzyme the fascinating aspects, say in mechanism, structure or physiological significance; but I cannot promise that there will be 'never a dull moment' [3] or even some stretches of these. However, they are mostly not arising in the quest for understanding certain observations, but rather in the logistics around being able to do so, such as beating the bushes for money, filling out forms, writing requests or reports, arguing with referees and editors, or – being yourself an editor – arguing with authors, or sitting in poorly conducted committee meetings. The phase of research, when you are trying to find out something, to convince yourself that the ideas you have of a process or structure are correct, this phase is always exciting, while the next phase of research may become drudgery, namely when you have to convince the rest of your colleagues and potential, supercritical referees of your papers. The determination to weather these moments or stretches of drudgery, of bad luck and disappointments, must be fueled by the general satisfaction that occupation with the natural sciences entails, but above all, by the experience of 'the moments when we live', as my late colleague and benefactor, David E. Green,[1] used to call them, those moments when we realize that nature has let us in on one of her secrets, or, more prosaically expressed, when we happen to be clever enough to understand something that is a matter of course in nature. There are few of these moments and far between, often years or a lifetime for some of us. It has been my experience in observing young people's success or failure as scientists, that doggedness, namely the ability to survive the dry stretches, is one of the most important ingredients of success, outweighing brilliance of mind, if this is not combined with a certain measure of persistence and patience.

One of the great bioorganic chemists (if this designation can be applied to his period in history) of the beginning of this century, Richard Willstätter,[2] once spelled out, so I was told, what is necessary for success in the sciences, namely the four G's (in German): Geduld, Geschick, Geld und Glück. Patience, skill, money and luck. This recipe probably still stands today!

This saying brings me to the theme of my subtitle, namely, what one could call the three F's that play a prime role in research today (without making the four G's unnecessary!): namely phone, Fax and Fedex. It seems to me that I may be one of the first of my generation of scientists, who has reached the stage of writing reminiscences of his career and who witnessed and himself experienced the transition alluded to in the subtitle: from Thunberg tubes and manometry to the copious use of the indispensable three F's. In his article in one of these volumes [5] Nathan Kaplan recounts his moves from the

---

1 David E. Green (1910–1983), Institute for Enzyme Research, University of Wisconsin, Madison. The author's thoughts concerning this extraordinary personality and scientist have been summarized in reference 4.
2 Richard Willstätter (1872–1942), University of Munich, Germany.

McCollum-Pratt Institute at Baltimore to Brandeis University and then to the UCSD at LaJolla. He makes no secret of his nostalgia for the days of being at a relatively small institution such as McCollum-Pratt, where there was daily, hourly, interaction with other colleagues working on related problems, as compared to the large UCSD, where he felt much more distant from those who pursued similar goals. I have heard similar complaints from colleagues through the years: 'There is nobody to talk to'. I am sure that with the increased specialization that has unavoidably come into all fields, this development is irreversible; we can no longer expect to have colleagues in the same institution who have closely related interests. I found that out thirty years ago, when I ventured into metal biochemistry and, in connection with this, into electron paramagnetic resonance (EPR) spectroscopy at a university swarming with 'biochemists'. We must recognize the facts and make use of some of the good things that have come along, such as, in this case, the three F''s. The colleagues I may want to talk to are as close as the next phone or Fax station. Questions and answers can go back on the same day between continents. Along similar lines of thought: When I started in biochemistry in the forties, an investigator could still hope to have most of the required tools and instruments within his own laboratory, other than, maybe, something as extravagant as an ultracentrifuge. There just were not that many tools and instruments at that time. Today we are lucky, if they are in our department or even on our campus and if, what is there, is not an outdated model!

In 1951, during my first postdoctoral year with David Green at the Enzyme Institute at the University of Wisconsin at Madison, I had the good fortune to meet Hans Gaffron[3] and James Frank[4] at Chicago. It was the time when the chemical

---

3 Hans Gaffron (1902–1979), University of Chicago.
4 James Frank (1882–1964), University of Göttingen, later Chicago.

events in photosynthetic CO_2 fixation were hotly debated. I told Gaffron of our newly and, for that time, generously equipped institute at Madison, particularly our large and high speed centrifuges, which were a great help in tissue fractionation and enzyme purification, as practiced in those days. He said to me, with some expression of unease, that if he felt that he had to depend on such hardware and could not do experiments with simple means in his own laboratory, he would rather give up science altogether. I was impressed by this statement from a respected member of the generation before mine. Is there any way of stopping the wheels of history? The Romans had a saying that has stayed well in my memory from my school-days: 'Volentem fata ducunt, nolentem trahunt' (fate guides those who are willing to follow, it drags along those who resist). What has become of biochemistry? Not all has become bigger and heavier, as did the early centrifuges, but most has become more complicated, a lot faster, automated and heavily dependent on contemporary, ancillary technology, and has become often actually much smaller; and fortunately so. I may mention, in this context that, when I bought my first Cary spectrophotometer around 1960, I suggested to the Cary people that they should make it possible to use less solution than that required by the traditional light path which required 3 ml cuvettes. The answer was that this was not their business, but that of individual investigators who had such 'specialized' needs. This may sound amusing in our days of pico- and femtomol biochemistry.

The moments when...

Largely following the historical sequence, I shall mention below those moments referred to above, which have left an unforgettable impression in my mind and which had much influence on shaping my career and decisions. The reader should not expect that they always involved great advances in biochemistry as such; often they were only milestones in my own biochemistry in the laboratory, such as the overcoming of experimental hurdles or understanding what seemed previously mysterious to me.

Water, copper and yeast carboxylase[5]

In 1941, when during work for my doctor's degree at the Kaiser-Wilhelm Institute at Heidelberg, I was concerned with biological effects of quinones, my Doktor-Vater (major professor) Richard Kuhn[6] suggested to me preparing yeast pyruvate carboxylase to test the inhibitory action on that enzyme of a series of quinones and related compounds which were available from other work at the Institute [8]. My laboratory was in the wing of the institute that had previously been occupied by Otto Meyerhof's[7] group, and, in fact, not only was there much of the equipment which had been used by that group, but there were also several of the technicians, whom Meyerhof had brought with him on his move from Berlin about a decade earlier. Two, Walter Schulz and Walter Moehle, have been referred to in other historical accounts [9]. I learned much from them about laboratory manipulations, manometry and also about the Meyerhof laboratory and its occupants. They showed me all the advanced features of the institute which had only been built in 1928. Among these features was an elaborate system of shiny copper pipes and tanks which delivered distilled water to each laboratory. Of course, for the preparation of an

---

5 Pyruvate decarboxylase (EC 4.1.1.1).
6 Richard Kuhn (1900–1967), Max Planck (Kaiser-Wilhelm) Institute for Medical Research, Heidelberg, Germany. In an earlier volume of this series (6), Edgar Lederer comments on this outstanding scientist and an extensive description of his life and work is given by Otto Westphal in ref. 7. From my experience, I agree with these colleagues.
7 Otto Meyerhof (1884–1951), Kaiser Wilhelm Institute for Medical Research, Heidelberg, Germany, later University of Pennsylvania, Philadelphia.

enzyme such as carboxylase one must use distilled water, so I learned. I set out to prepare the enzyme with ammonium sulfate precipitations according to a paper by David Green [10] – unaware at that time that he would play a major role in my life years later. I obtained a nice yield of protein, but the enzymatic activity, assayed by manometry, was essentially zero. What went wrong? While going back to the literature and pondering about my failure, I recalled having read somewhere that carboxylase may be inhibited by copper ions. This and the shiny copper lines and tanks in the lab made me suspicious that the trouble may lie right there, and that I might have done better using Heidelberg's city water, which in those days probably still came from the surrounding mountains in the part of the city where we were. I set up a still, made my own distilled water and tried again. This time the manometer shot up in seconds. Guided by my previous disappointing experience, I had used much too much protein. When I then happily rinsed my glassware and got ready to go home after this victory, I saw a tiny speck of protein hanging on the top of one of the cylinders that I had used. I was curious and dissolved it in about 20 ml of water and added a little to a manometer vessel and ran it. There was plenty of activity. This manifestation of catalytic activity of a protein impressed me enormously; and this experience planted the seed in my mind that the then young field of enzymology may be the one I would like to get into after my thesis work. Sure, the whole story is trivial, certainly seen from today; but to someone who had never dealt with a protein before (other than as consumer, of course), let alone with an enzyme, this was an important event that eventually had a great influence on my future. One thing I have often wondered about is, how Meyerhof and his associates could have done all the pace-setting work they did in the same laboratory with the water that had killed my carboxylase. I can only come up with two suggestions, namely, that the enzymes they were dealing with were not as sensitive to copper as carboxylase, or that, when the water-delivery system was used heavily and continuously, the copper concentration could never have reached the levels that had piled up by the time I used it. Maybe, there was some of both.

Adventures in fatty acid oxidation

First, I should mention that we are now in the early fifties and in the meantime, the scenery has changed from the Kaiser-Wilhelm Institute at the shores of the Neckar River at Heidelberg to the Institute for Enzyme Research close to the shores of Lake Mendota in the Four-Lakes-City of Madison, Wisconsin.

With the identification of acetyl CoA as 'active acetate' by Feodor Lynen[8] in 1951 [11], the stage was set for an all-out attack on the problem of β-oxidation of fatty acids, and a number of laboratories became involved in attempts to arrive at soluble systems which would allow identifying individual enzymes of that pathway. The history of these events has been told many times from different viewpoints [12–14] and I have had a chance to give an account of my experience in this field [15]. A few of the great moments in this endeavor, however, deserve mention in the present context.

The quest for CoA

In the early 50's the knowledge of CoA and its properties far exceeded its actual availability. One had to go begging, hat in hand, to one of the microbiologists who would be willing to share a few micrograms of fairly pure material or one had to use crude concentrates, with which, however, one was likely to

---

8 Feodor Lynen (1911–1979), University of Munich and Max Planck Institute for Cell Chemistry; see Chapter 1 in this Volume.

add a number of unknown and undefined substances into the
system one was working with. After publishing my account of
the story of CoA production [15], my colleague of these days.
Dick von Korff[9], who was critically involved in this work, told
me that he remembered things somewhat differently. I am,
therefore, trying to take account of this and report the story
here with some amendments. What is sure: we were in dire
need of CoA for assaying the enzymes involved in β-oxidation
and preparing for this purpose the actual substrates for the
individual enzymes rather than using coupled systems, in
which such substrates were generated *in situ*. Dick van Korff
had worked out the most sensitive rapid assay for CoA then
available [16], which was fast enough to follow elution of a
column. One day, there was a discussion in the laboratory, in
which I worked, with David Green, Dick von Korff and me
being present. The question was raised, and I think it was
Dick von Korff who raised it, whether one could not make use
of the mercaptan nature of CoA for purification, by precipitating it with heavy metals such as Hg, Pb or Cu. It was, admit-
tedly, likely that a number of other things would also come
down under these conditions. Green recalled that, at Cambridge, F.G. Hopkins[10] used to precipitate and purify GSH
with mercury. There is, of course, a lot more GSH around than
there is CoA. I was charged with looking into this matter.
Luckily, a group in the Biochemistry Department at Madison,
headed by Frank Strong[11], had tried to produce CoA from
yeast extracts, but was, as was everybody else, lacking a decisive purification step to get beyond a modest enrichment of
CoA. However, they had plenty of yeast extract with reasonable CoA content and had a fair amount of experience with it. In
a joint effort, we decided to use this extract as starting material. After reading the relevant articles in the *Biochemical*

---

9 Richard W. von Korff.
10 Frederik G. Hopkins (1861–1947), University of Cambridge, England.
11 Frank M. Strong (1908–1993), University of Wisconsin, Madison.

Journal I tried my hand with Hg salts and yeast extract. Very little came down and whatever did did not contain much CoA to speak of, as Dick von Korff told me. During my reading I had come across a paper from Hopkins' laboratory, in which cuprous oxide was used to precipitate GSH [17]. It sounded like a clean procedure. I certainly preferred Cu_2O over Hg as a material to work with. I tried to repeat the precipitation of GSH and it gave a nice clean precipitate – of course I had used GSH out of a bottle! A minor success at least. When I tried to repeat the Cu_2O precipitation with yeast extract, the result was no more encouraging than it had been with Hg. Remembering the nice precipitate I had obtained with GSH, I tried adding GSH to yeast extract in the hope that I might be able to precipitate the CoA with the GSH. I did get a precipitate, i.e., the yeast did at least not spoil the precipitation of GSH. However, the crucial test was now up to Dick von Korff. I brought him some of the precipitate and retreated to my lab. I admit, I did not hold out great hope that there would be much CoA. However, it was not long that Dick came rushing in with the words: 'It's loaded with CoA!' Success! We had achieved an about 20-fold enrichment in CoA. From there on it was not a big problem to develop a practical method for preparing CoA from brewer's yeast and the skill and experience of our colleagues in the Biochemistry Department, Harvey Higgins, Bob Handschuhmacher, Don Buiske and Frank Strong was a great asset in these efforts [18]. The Pabst Laboratories became interested in this venture and soon CoA could be had in gram quantities. An interesting and amusing byproduct of our initial efforts was that Dick von Korff found that GSH out of the bottle actually was a fairly good source of CoA by itself, at least the commercial GSH of the 50's.

Flavoproteins of fatty acid β-oxidation

While I was busy making CoA and acyl derivatives of CoA, David Green, Henry Mahler[12] and Sanae Mii[13] had partially purified a green flavoprotein, which they called butyryl CoA dehydrogenase [19]. It was most active with butyryl CoA but could also oxidize hexanoyl CoA. They soon found out that it was contaminated with yet another dehydrogenase, which is specific for longer chain substrates. Since they were busy characterizing the green enzyme further, which Henry Mahler suspected to be a copper-flavoprotein, David Green entrusted me with going after the new enzyme. Fred Crane[14] joined me in this undertaking. After some purification steps, we knew that we had a yellow flavoprotein, which had maximum activity with octanoyl CoA [20]. We used the reduction of indophenol by octanoyl CoA as the assay for enzyme activity, which is easy to measure, can be carried out in the presence of O_2 and is fairly sensitive. As the enzyme became purer, we made the shortcut of using the direct reduction of the enzyme-bound flavin as our assay and also the ratio of flavin to protein absorbance (A_{445}/A_{280}). We noticed, however, that, as the enzyme became more and more yellow, the activities in the indophenol assay did not show a corresponding increase. So we started wondering what we were actually purifying and sought the help of Bob Bock[15], the physical biochemist in the biochemistry department and later our dean for many years. He was the expert in electrophoresis, which was boundary electrophoresis at that time. So, one day we handed him a beautifully yellow enzyme, which, we thought, was very pure, and curiously waited for the protein to migrate and prove our assumption. However, the first photos already showed that we had at least

12 Henry R. Mahler (1921–1983), University of Indiana, Bloomington.
13 Sanae Mii, Stella Maris College, Kobe, Japan.
14 Frederik L. Crane, Purdue University, West Lafayette, Indiana.
15 Robert M. Bock (1923–1991), University of Wisconsin, Madison.

three major protein components in our sample. Fortunately, they separated nicely within three hours and Bob Bock was able to fish out from the ascending limb of the apparatus the material showing the strong yellow peak, presumably the dehydrogenase, and from the descending limb the other major component, which was only pale yellow. I was the messenger and brought these fractions back to the Enzyme Institute, where Fred was sitting at the Beckman DU (manual, of course) ready to assay by both methods: direct flavin reduction and the catalytic assay with indophenol. I had promised to return to biochemistry to help Bob to clean up, and so I did not have the time to wait for the results of all the assays, but I learned before leaving that the yellow fraction was extensively reduced by substrate, better than we had ever seen before, but was absolutely dead in the catalytic assay. The other peak fraction showed neither activity. We looked at each other somewhat dumbfounded; what had we isolated? I left at this point and when I returned, Fred still sat at the Beckman, but with an entirely different look: he had a broad smile on his face, a rare occurrence with him at work. 'There is our enzyme', he said, showing me the rates he had observed with indophenol, when he combined the peak from the ascending limb, the dehydrogenase, with the peak that we had withdrawn from the descending limb. So we needed two enzymes [21]. Obviously one was the dehydrogenase and the other the link to indophenol. This second protein turned out also to be a flavoprotein, which meant, we had two flavoproteins acting in series, a completely new observation! According to the then existing picture of the mitochondrial electron transport system, we first thought that this new flavoprotein, which we called electron-transferring flavoprotein (ETF), was providing the link to cytochrome c, but we soon found that with ETF of higher purity, the rates with cytochrome c were so poor, that we gave up this idea. But what was ETF interacting with? This answer came some 20 years later, luckily also from our lab; however, this story fits better into a subsequent chapter (see p. 230), when some of the developments that made the discovery of the link to the respiratory chain feasible, have been discussed. It is clear that without these developments we could not possibly have solved this problem in the 50's, without running our careers aground in such attempts. It was also true that the acyl CoA dehydrogenases, of which, at the time, two were known, offered us more promising avenues of research than the sticky ETF problem.

Intermediates in fatty acid dehydrogenation; semiquinones?

We had noticed that, when substrate is added to the dehydrogenases, not only is the flavin color partly bleached, but new absorption arises at longer wavelengths [22], such as to make the protein solutions look anywhere from green to violet and almost black at high concentrations. Having been in Richard Kuhn's laboratory, I was, of course, aware of the colored flavin-semiquinones that had been crystallized in his laboratory [23] before I arrived there. Were we generating flavin semiquinones? Fred Crane and I did not think so, because we could make these colored species by adding product, e.g. octenoyl CoA, to the enzyme reduced by dithionite, so that little flavin absorption of the oxidized form remained. We thought that we rather had a complex of reduced enzyme and product. As we know today, there are states of interacting components of a $2e^-$ redox system other than oxidized, reduced and semiquinone, in which the equivalent of zero, one or two electrons are transferred between a two-electron redox couple. The concept of charge transfer complexes had not filtered down into biochemical circles at that time. I have discussed our search for answers to these questions at other occasions [15,24]; however, as this period showed a dearth of great moments other than those of disappointment, I will not enlarge on this here. It is worth emphasizing, however, that this search led me to EPR as a means of detecting semiquinones in a way independent of optical spectroscopy and of extinction coefficients, which are required for the quantitative interpretation of such spectra. My turn to EPR also brought me into contact with Dick Sands[16] from which developed our collaboration and friendship that has had an impact on the careers of both of us; in my case, I venture to call it a decisive impact.

There was some additional fallout from our work on the flavoproteins of fatty acid β-oxidation. While we were not able to clarify exactly what those colored intermediates were that we observed on addition of substrate to the enzymes, the phenomenon itself, namely to see the enzymes in action, was quite spectacular and lent itself to demonstrations. Fortunately, for this purpose, the reaction of the substrates to form the charge transfer complexes was relatively slow, so that it could be followed by simple means. At that time, the American Optical Company brought on the market a 'Rapid Scan' spectrophotometer, which was actually aimed at the dye or paint industry, where rapid scanning was really only a convenience rather than a necessity, as it is for studying enzymes. The instrument worked with a rotating mirror, which made one revolution per second, which means one could study reactions that occurred in the range of seconds. The wavelength range was 400 to 700 nm, which was fine for my purposes. It required about one ml of solution, but its greatest drawback was the feature that it measured transmittance, not absorbance; the reason for this was that circuitry for sufficiently rapid conversion of numbers to their logarithms was simply not available at the time. A colleague at the institute had bought the Rapid Scan instrument with the hope of looking at the cytochromes in mitochondria or submitochondrial fractions, which were then under study in Green's laboratory. However, the last mentioned feature, namely the transmission mode of the instrument, made it useless for my colleague, because there is a

---

16 Richard H. Sands, University of Michigan, Ann Arbor.

high background absorption in the materials he wanted to look at, so that the features he intended to see were all cramped together in the low transmittance region, which did not even allow one to see whether cytochromes were reduced or oxidized. Since it was a single-beam instrument, there was no way of reducing the background. However, for me this was just fine, since I was at the opposite end of the transmittance scale, because at about 550 to 700 nm there is very little absorption in pure flavoproteins, so that the charge transfer bands appeared on top of almost no background, which made them look very much absorption like, once one got used to the upside down feature of their appearance as % transmittance. There was no recorder available for these rapid events; it was originally thought just for the eye. While you may get away with this in the dye industry, for scientific purposes there had to be a record; I could not rely only on witnesses as I did in the first few trials. So, I borrowed a movie camera and set it up in front of the screen. It was great: an enzyme in action, as rarely anyone had seen it before! Sure, something like this had been done at single wavelengths, but to see the flavin absorption disappearing simultaneously with the appearance of the broad charge transfer bands was quite a feat in these days – 'flavoproteins up and down' as my colleague Sanae Mii of Osaka said with her charming Japanese accent. I had to show my film on several occasions.

On one of these occasions, at the biochemistry department of the University of Chicago, there was in the audience a young man, technical assistant to Mary Buell[17] at the time, not really a biochemist, rather a home grown electronics fan with mechanical skill and an alert mind. He was so impressed with the way I had shown enzyme action, that he followed an invitation by David Green to move to Madison and build a spectrophotometer for him, with which similar things could be done,

17 Mary V. Buell (1893–1969), University of Chicago.

although the rapid scan feature was not considered at that point. The young man was Raymond Hansen, now president of Update Instrument Company at Madison, who joined my lab in 1962 after a short interlude with David Green and who worked with my group until my retirement from the Enzyme Institute in 1984.

Additional dehydrogenases and 3D structure

I should not close this chapter on the flavoproteins of β-oxidation without mentioning that, shortly after we discovered ETF, we found yet a third dehydrogenase, specific for long chain acyl CoAs [25]. Jens Hauge[18] of Norway was mainly involved in this work. Today we know that there are three more dehydrogenases of a related kind in liver, all flavoproteins and all funneling into ETF [26]. Their specificity is toward branched chain fatty acids and glutarate. In addition, the flavoproteins sarcosine and dimethylglycine dehydrogenase also use ETF as a proximal link to the respiratory chain [27]. Deficiencies of all of these enzymes have been found in certain rare genetic diseases [28]. Finally, it is a source of great satisfaction to me now being at the institution and in the department, where my colleague, Jung-Ja Kim, has solved the crystal structure of the same enzyme on which Fred Crane and I labored years ago, the yellow medium-chain dehydrogenase [29]. I can now sit before Jung-Ja's graphics screen and look at the active site and see how the substrate squeezes itself into the site and wherefrom the chain-length specificity may be derived. I had often thought, thirty-five years ago, when we were working with those enzymes, that it would be a wonderful object for study having this series of homologous enzymes available with somewhat different specificity, so that one

---

18 Jens G. Hauge, University of Oslo, Norway.

might see how nature accomplishes this subtle specificity. We did not have the means to do this at the time. Today, we are 'almost' there: maybe we are, by the time this is printed and it will be exciting to see that day.

First ventures with EPR

Some of the background for my excursion into the field of EPR has been described elsewhere [15, 30, 31]. As one might have expected from what I said above, my first attempts to use EPR for the demonstration and determination of flavin semiquinones was at best a very modest success on top of much (duly unpublished) disappointment that went with it. However, with my attempts to 'see' metal components in tissue fractions and enzymes and verify their oxidation-reduction, I struck 'gold', albeit in the guise of iron, copper and molybdenum. As we were interested in metals most of this work was done at low temperature (80–180K). It was indeed a field day in the first few months to years after we got going, first at the physics laboratory at Ann Arbor with Dick Sands, and after 1960 in my laboratory with my own commercial instrument, which was, incidentally, quite primitive, as compared to today's spectrometers. Practically everything we looked at showed something interesting and often unsuspected, even if only impurities. There were indeed many of the great moments in those days. My colleagues at the Enzyme Institute almost became envious and the joke went around that I produced a paper a day, with the reprints coming right out of the back of my machine. I owe much to many of them for their support in generously furnishing materials and information about them, to David Green for his support in attracting most of the funds for the instrumentation and last, not least to Dick Sands for his advice and assistance and to my collaborator for many years, Ray Hansen, for keeping our instruments running and developing helpful auxiliary devices.

Copper in cytochrome *c* oxidase

One of the first feats was the independent confirmation by
EPR that cytochrome *c* oxidase (COX) of high purity did indeed
contain copper and the finding that this copper was reduced by
ascorbate in the presence of cytochrome *c* [32]. It was a great
moment, when Dick Sands and I, down in the stuffy second
basement of the physics department at Ann Arbor, saw the
recorder pen of the spectrometer go up tracing the copper signal, and when it did not do so in a second run, after addition of
cytochrome *c* to that sample. This could, however, not be repeated by my colleagues (and subsequently friends) A. Ehrenberg[19] and T. Yonetani[20] [33], who found at best negligible reduction of copper in their preparation of the oxidase and who
also reported distinct hyperfine structure in their spectrum,
which we had not observed. I should remark here that there
was a tendency at that time among the mitochondrial electron
(e⁻) transporters not to accept additional e⁻-carriers in mitochondria, such as Cu, non-heme iron or ubiquinone, because
'titrations' of the e⁻ capacity of cytochrome oxidase or the e⁻
transport system had been done [34] and the capacity so determined agreed well with the then known components of the
chain. Graham Palmer[21] and I could resolve the discrepancy
between the findings of Ehrenberg and Yonetani and ours,
when we showed that there were at least three different species of copper in COX [35, 36], namely the copper Dick Sands
and I had observed initially, and, what we called contaminating copper, which was not reducible by cytochrome *c* and had
the properties and hyperfine structure in the EPR signal like
the copper observed by Ehrenberg and Yonetani; and there
was, in addition, copper that became detectable by EPR only
after denaturation of the enzyme in the presence of a mercu-

---

19 Anders Ehrenberg, University of Stockholm.
20 Takashi Yonetani, University of Pennsylvania.
21 Graham Palmer, Rice University, Houston, Texas.

rial. The contaminating, so-called 'dirty' copper was observed to some degree in all the earlier preparations. It is absent or minimal in more recent ones [37]. The two species of copper detectable in the native enzyme could be readily differentiated by their very different saturation behavior with microwave power [36] and it gives me some satisfaction that in some of the most recent work, referred to below [38] – which is orders of magnitude more sophisticated than our old experiments – this same criterion is still found useful for telling the 'real' bound copper from the unwanted contamination.

Our chemical analyses always showed somewhat more than two coppers per two hemes, up to a ratio of 2.5:2.0. These ratios were based on the spectrophotometric determination of heme and on iron analysis. In view of the errors that could enter into such a determination, we did not pay much attention to this inequality of numbers and explained the excess of copper over heme by the presence of the dirty copper, which was not reducible by cytochrome c and was, therefore, not of particular interest to us. We attributed the copper that became detectable only after denaturation to a species of copper that was reducible by cytochrome c, but remained undetectable in the native enzyme, possibly because of interaction with another species. An important observation, that was repeatedly pointed out in the literature, was that the EPR signal of the EPR detectable copper, now called Cu_A, was unusual in its shape, g-value, lack of hyperfine structure typical for Cu, and temperature and power dependence. Dick Sands, therefore, had suggested as early as 1962 [39] that there may be two interacting copper ions represented in the signal. He suggested cupric- cuprous interaction, but stated that we could, at this point, not exclude cupric-cupric interaction. The finding that Cu appeared on denaturation could have supported the notion of cupric-cuprous interaction and titration experiments with ferricyanide as oxidant, which I did somewhat later [40], actually showed that the amount of oxidant required for the reappearance of the Cu EPR signal after reduction in the presence of CO, could not account for two cupric ions being involved in the process. Thus, with the chemical analyses showing a copper to heme ratio of ≤ 2.5, with one Cu obviously involved with cytochrome a_3 (see below) and with some contaminating Cu present, there was really not much incentive for postulating another Cu, a cuprous ion; I had a sufficiently hard time to have two coppers accepted by the community.

This problem lay dormant for some time, until we showed in collaboration with Jim Hyde[22], Wojciech Froncisz[23] and Charles Scholes[24] [41] that at lower microwave frequency (S-band) there was distinct, rather narrow hyperfine structure observable in the CuA signal at $g_\perp$. This had been seen at higher frequency (X-band), but was poorly resolved under these conditions. Moreover, at both frequencies, the EPR signal of cytochrome (cyt) a obscured the low field part of the spectrum through overlap of lines. Thus, we could not come up with a convincing interpretation of this hyperfine structure on the basis of the then accepted model of COX 'as isolated', namely a protein containing all metal components in the oxidized state (Cu_A, cyt a, Cu_B, cyt a_3), with the last two components spin-coupled, which would account for their lack of EPR in the native state. Again, the problem lay dormant for some years, when two new developments started shaking the ground once more. First, the preparative and analytical methods as well as the protein chemistry of COX made great strides forward and consensus developed that there was more Cu in the enzyme than one per heme, even if no dirty copper could be demonstrated. Ratios were found as high as 1.5:1.0 [37, 42, 43]. However, no EPR of this Cu was identified and it was also not titratable with reductants, indicating that it was most likely Cu(I). The second development was much less foreseeable than the gradual improvement of methods that took place in

---

22 James S. Hyde, Medical College of Wisconsin, Milwaukee.
23 Wojciech Froncisz, Jagellonian University of Krakow, Poland.
24 Charles P. Scholes, State University of New York, Albany.

the course of the years, as just described; it is in fact a typical example, how work that initially appears to be entirely unrelated, can bring about a solution for an old unyielding problem. The enzyme nitrous oxide reductase (NOR) was purified from Pseudomonas and turned out to be a copper protein with multiple Cu atoms per molecule. It can be obtained in several different forms according to its optical and EPR spectra. In one of these forms it shows an EPR signal very similar to the Cu_A signal of COX [44]. Since there is no heme present in NOR, the spectrum is much better resolved on the low field side. The hyperfine structure can be interpreted with reasonable probability as stemming from a Cu(I)-Cu(II) pair. W.E. Antholine, W. Zumft and P. Kroneck [45] and their associates drew attention to this and proposed that the unusual behavior of the Cu_A signal in COX may be explained by the presence of a mixed valence pair of Cu atoms. So, we were back at the interacting pair of Cu atoms, a mixed valence cupric-cuprous pair! This idea does not interfere with other findings on the enzyme, such as the number of Cu atoms now found by analysis and the values for total reducible components found in reductive titrations. The proposal by Kroneck and his colleagues first met with considerable skepticism [46] – I was among the skeptics, although among the less outspoken ones. I have known Peter Kroneck since his student days with Peter Hemmerich[25] at Konstanz, and Bill Antholine is right here in Biophysics at the Medical College where my office is. We had many discussions on the subject, however, when the proposal was first made [44], the stoichiometry of the number of Cu's per molecule was not settled and the EPR spectra shown and their analysis was preliminary. To all the old-timers in the field it seemed, and still seems, inconceivable that Cu_B, which, according to all evidence, interacts with cyt a_3, could at the same time be available for interaction with Cu_A. Also, the estimated distance sepa-

_____

25 Peter Hemmerich (1929–1981), University of Konstanz.

rating these metal atoms [47] would make this very unlikely. I found it, therefore, hazardous to rely only on an EPR spectrum, and for that matter only on the high-field part of it, as the low-field spectrum is obscured by overlap with a signal from cyt a. However, when a striking similarity in parts of the amino acid sequences of NOR and subunit II of COX was found [cf. 45], I gave the proposal a better chance of being right. There is fair evidence that Cu_A is in subunit II of COX, where there is a Cu binding site, as found in other Cu proteins [38, 48]. The dilemma finally boiled down to the question whether this Cu binding site can accommodate two Cu atoms. This has since been pursued with a combination of clever molecular biological approaches and it seems at this time that the answer may well be: 'yes' [38]. The demonstration of a plausible second Cu binding site in the vicinity of the site that was considered to bind Cu_A would, in my opinion, put the final piece into the jig-saw puzzle and a 30-year-old story will then have come to an end. It has long been forgotten that Dick Sands in 1962 suggested the cupric-couprous model from the EPR spectrum [39]. Unfortunately, we, the biochemists, had failed to come up with sufficiently hard data to support it at that time in a climate hostile even to a single significant Cu! I have discussed this example in some detail, because few people may have followed this development with as much attention and involvement as I have. It undoubtedly is an instructive example, as to how science progresses and what the ingredients are in such progress. Chance and luck are certainly among them. The details of the various experiments and arguments will have to be taken from the literature cited.

High spin heme in cytochrome c oxidase

From the standpoint of function, Cu_A may be to many the least interesting component of COX. It was, however, the easiest one to look at by EPR, if we ignore the repercussions of its detection just told above. A meaningful study of the heme components had to await the introduction of liquid helium technology, which we did not have in our laboratory until early 1969. In these studies I had the dedicated collaboration of Bob van Gelder[26], Charles R. Hartzell[27] and later his student Robert W. Shaw[28]. I will only recount the most exciting moments in this work. Bob van Gelder visited twice from Amsterdam: the first time we dealt mainly with Cu_A and at the end of his second stay we finally had everything going for the titration of the heme components with NADH and phenazine methosulfate although still only with liquid nitrogen cooling. The day when Bob had to go to the airport at noon, we started a titration and, as usual, Cu_A became reduced, i.e., its signal decreased, and so did that of cyt a (signal at g = 3), but to our complete surprise a huge signal at low field (g = 6) started to rise at the same time. This type of signal is typical for a high spin heme and cyt a_3 has always been thought to be a high spin heme. Was this then the first time we could detect this heme by EPR? Bob had to leave at this point, before we could find out what would happen to the signal on further titration. We sent Bob off to the airport, asking him to call us back when he was out there, as we would probably know the fate of the signal by then. He called and we did know: the signal disappeared on further titration and was practically gone when four equivalents of reductant per two hemes had been added. A plausible interpretation was that, as reduction proceeded, Cu_B, presumed to be the interacting partner of cyt a_3, became reduced before cyt a_3, the interaction between the heme and Cu was thus broken, and the high spin signal of cyt a_3 appeared temporarily until cyt a_3 also became reduced. This observation was one of the early clues to what is happening in COX on reduction and it supported the idea that there was a fairly

---

26 Bob F. van Gelder, University of Amsterdam.
27 Charles R. Hartzell, Alfred I. du Pont Institute, Wilmington, Delaware.
28 Robert W. Shaw, Texas Technical University, Lubbock, Texas.

strong interaction between Cu_B and cyt a_3, as suggested by the absence of signals for these components in the resting enzyme [49]. There was, however, an alternative explanation for the appearance of the high spin signal during reduction, towards which we became lured by the following circumstances. Despite the fact that the high spin signal at g = 6 was the largest signal by far, on quantitative evaluation it at most accounted only for one third of the total heme in the enzyme, while, simultaneously, the low spin heme signal disappeared. Whereas, in the interpretation suggested above, we had taken this disappearance as meaning reduction of the low spin heme, it might as well have meant conversion of this heme into another form, e.g., a high spin form, such as that represented by the g = 6 signal. Thus, with one signal coming and the other disappearing, we never observed simultaneously more than the equivalent of at most one heme. The question was, therefore: is the low spin heme converted to a high spin form, while reduction of some high potential component(s) in the enzyme proceeds? There was some additional evidence produced by D.F. Wilson and coworkers [50] which made us uncertain about our original interpretation. Namely, they concluded from their spectrophotometric observations that there is a high and a low potential heme in COX, and it was generally thought that cyt a_3 was the high potential one; and that should, of course, become reduced first. So, Chuck Hartzell and I proposed the second interpretation in a short paper [51]. However, as time went on, I started having my doubts whether this was actually correct, as we found one condition, namely rapid anaerobic reoxidation of reduced COX with our rapid freeze-quench technique [35], when we observed more than the equivalent of one heme in the spectra. However, the integration of the g = 6 signal and a quantitative evaluation were everything but trivial, so that we had to count on a considerable error in this procedure. Therefore, we had to find a hefty excess over and above a single heme in the spectra to be really sure that we saw both cyt a and a_3. I tried to improve the quantitation procedure, which was based on metmyoglobin as a standard, which also shows a g = 6 signal and the same temperature dependence as the COX signal [52]. In the meantime our friends Tore Vänngård[29] and Roland Aasa[29] at Göteborg had come up with a useful integration procedure for g = 6 signals [53] and communicated this to us, so that the stage was set to attack the problem anew. There was one problem yet to be solved, namely the choice of the oxidant; and there is not much choice, when you have to stay around pH 7. It had to be a high potential oxidant, and the old standby ferricyanide would have been suitable from that standpoint. However, at the concentrations required for rapid oxidation, ferricyanide produces a broad EPR signal around g = 3, which makes a reasonably accurate quantitation of the low spin heme impossible; and, of course, we had to have an accurate measurement of the concentration of both hemes. My time at the Kaiser Wilhelm Institute at Heidelberg had given me much valuable general experience, but what I had learned there about specific chemical procedures was mostly sufficiently off my now chosen track that it rarely had helped me; but in this case it did. I recalled the two compounds, of which Richard Kuhn had spoken with great affection many years ago, namely porphyrindin and porphyrexide. Kuhn loved free radicals, and they both were just that. They both had unusually high redox potentials (560 and 730 mV, respectively, at pH 7), higher than ferricyanide; and so it happened that porphyrexide was commercially available in the US at that time (I was recently told that, unfortunately, it is no longer). The EPR absorption of the porphyrexide radical, as one would expect, is right at g = 2, sufficiently far from that of cyt a that it did not interfere. With porphyrexide as an oxidant, Bob Shaw and I were then able to produce huge g = 6 signals in addition to a sizeable g = 3 signal and thus showed unambiguously that the signals at g = 3

---

29 Tore Vänngård and Roland Aasa, University of Göteborg, Sweden.

and at g = 6 must be due to different cytochromes. We found absorption equivalent to as much as 70–90% of the heme of COX! [54]

Non-heme iron: iron-sulfur proteins

By far my greatest effort from the late fifties on was directed toward non-heme iron in tissues and proteins, and more specifically toward iron-sulfur proteins. It started with the finding in David Green's laboratory that there was iron in mitochondria and membrane fragments prepared from them, significantly in excess of what could be accounted for in terms of hemeproteins [55]. I became very interested in this, because I saw the possibility here that some so far unknown 'factors', as such things were called at that time, might be involved. However, with iron being a ubiquitous metal and, therefore, a likely contaminant in almost everything, I wanted to be sure, before I embarked on a major project, that the butcher's knife and all the machinery involved in the preparation of mitochondria and their fragments or the reagents used had not furnished the detected non-heme iron. Together with Ernest Page[30], I prepared, therefore, some superclean submitochondrial particles and added ^{59}Fe to an aliquot as early in the preparation as feasible and then measured the radioactivity before and after dialysis against EDTA. I also determined iron chemically with a microprocedure that I had worked out [56]. I found that ^{59}Fe was indeed adsorbed or otherwise taken up by the particles, but most of this iron was readily removed by EDTA, while chemical analysis showed undiminished non-heme iron [57]. It was at about the same time that iron was detected by chemical analysis in NADH dehydrogenase [58], xanthine oxidase [59] and succinate dehydrogenase [60] and, luckily, it

---

30 Ernest Page, University of Chicago.

was also the time when EPR had just come on the scene. So, Dick Sands and I set out to look for EPR signals which might originate from this iron. Dick's thesis work had dealt with EPR of Fe(III) in glasses and he was, therefore, one of the experts, as far as expertise in EPR of iron went at that time; and that was not very far, as what was known was essentially the signal for high spin iron in a rhombic environment (g = 4.3), and this was far from being understood [61–63]. It was then no surprise that we found this signal in almost every protein we looked at, and it even disappeared on addition of some substrates such as NADH, which we interpreted as being due to reduction [64]. We soon found out, however, that this iron was prominent in material that had been dialyzed against EDTA, paradoxically, whereas undialyzed proteins usually showed less of it. Obviously, this iron was a minor and ubiquitous impurity in almost any protein and it may even have been attached as the EDTA complex after dialysis. This may seem contradictory to the results of the isotope experiment described above, but we were probably dealing here with small quantities of iron from endogenous sources, quantities which would certainly fall within the error of the chemical analysis. Dick Sands kept improving his home-built spectrometer following the state of the art; he installed 100 kHz modulation, which raised the signal to noise ratio several-fold. On scanning over the whole EPR spectrum, even in regions where iron was not known to have signals, we found in a number of different preparations, such as whole mitochondria, submitochondrial particles, succinate-ubiquinone oxidoreductase, and NADH dehydrogenase signals centered around g = 1.94 [65, 66]. In the more purified materials, such as the last two mentioned, these signals appeared only after addition of substrate and would disappear again on addition of an oxidant. This now seemed to be more like what we had been looking for and we both were very excited about these observations. At

about the same time Bob Bray[31] with Bo Malmström[32] and
Tore Vänngård found a similar signal when xanthine oxidase
was reduced with substrate [67]. Dick had no physical explanation for iron signals in this region of the spectrum, with an
average g value below 2.00; however, he mentioned the possibility of an interacting pair of iron atoms [68]. A physicist suggested to me seriously that, according to the g value, it could
be titanium what we were looking at! We had much fun in the
lab joking about the titanio-flavoproteins succinate and
NADH dehydrogenases and xanthine oxidase. If it was Fe
what we saw, the signal was obviously produced by a reduced
form of iron. Low spin Fe(II) is diamagnetic and should, therefore, be EPR-silent and high spin Fe(II), as a non-Kramers ion,
was at that time not expected to have a detectable signal; it is
now known that this is possible [69]. The situation was addi-
tionally complicated by the fact that there was much more
non-heme iron in the materials we had looked at than could be
accounted for by a quantitative evaluation of the spectra even
on the basis of the fictitious assumption that one Fe ion has
one unpaired electron; if it had more, this would have made
things even worse. Were we again looking at only a small fraction of the non-heme iron as with the g = 4.3 signal? I published some of this work and pointed out the dilemma, but
nevertheless stuck with the suggestion that we were dealing
with some reduced form of iron [70]. I received a hand-written
letter by John S. Griffith[33] of 'Theory of Transition Metal Ions'
[71] fame, saying that he was very interested in our findings,
but that it was absolutely impossible that the signal originated from ferrous iron; period. We met somewhat later during his visit to the U.S. and then again in Europe and became
friends. John even helped us barricade my first-floor laboratory windows in Madison during the days of student riots. The

---

31 Robert C. Bray, University of Sussex, Brighton, England.
32 Bo G. Malmström, University of Göteberg, Sweden.
33 John S. Griffith (1929–1972), University of Cambridge, England.

adjacent building housed the Naval ROTC and Madison was a hot spot of protests in these days. Unfortunately, not long thereafter John, the brilliant man, succumbed to cancer in his early forties.

I do not want to retell here the story of the discovery of iron sulfur proteins; however, as it finally turned out, both John and I were right. There was reduced iron, but only one of two or three of four atoms in these materials. It took some years and another clever mind, namely John F. Gibson[34] of Imperial College, to come up with the answer for the 2Fe case [72]. John Griffith was still alive when this happened and when I told him, he said: 'You did not tell me that there were two Fe atoms per molecule; of course, with two interacting Fe's your signal can be explained.' I had missed a chance; I was too conservative to mention that we suspected that there may be two iron atoms per molecule, but we were not sure. Our iron analyses were easy and accurate, but we had no reliable molecular weight for the protein on which we had concentrated our efforts at that time. This was a 2Fe ferredoxin from *Azotobacter vinelandii* [73], which, thanks to the collaboration of Perry Wilson[35] and Yogi Shethna[36], we had in quantity and high purity and which gave the typical g = 1.94 EPR signal. We chose a bacterial protein, because of the possibility of incorporating isotopes, which could eventually give us proof that the EPR signal was due to iron. Unfortunately, this protein was not very stable, so that the possibility of having breakdown products always existed, which could falsify analyses. Ultracentrifuge runs were not decisive. As so often in such cases, we found about 1.5 Fe per (what we thought was) the molecular weight. Less than 1 or more than 2 would have been all right, but 1.5 was hard to interpret. This was before the days, when so many now available aids for the practicing biochemist

---

34 John F. Gibson, Imperial College, London, England.
35 Perry W. Wilson (1902–1982), University of Wisconsin, Madison.
36 Yogi I. Shetna, University of Bombay.

were available. There was then no very practical way of concentrating protein samples. We used dialysis against polyvinylpyrrolidone (PVP). However, despite predialysis of the PVP used, there were apparently always smaller fragments, which could pass through the dialysis tubing and would then contaminate our protein with intractable goo. Amino acid analysis as well as attempted determinations of dry weight were spoiled in this way. Of course, it was after the fact, that we found this out. Nevertheless, with the Azotobacter protein labeled with ^{57}Fe that Shethna had prepared, we could unambiguously show by hyperfine broadening that iron was part of the compound from which the signal originated [74] and Bill Orme-Johnson[37] and I determined, by our titration method with solid-diluted dithionite [75], that one electron was taken up for two iron atoms with plant type ferredoxins we titrated and one for four iron atoms in what we now know as 4Fe ferredoxin.

I might mention here a typical experience that I had, which shows the problems we were up against with our work in these days. I mentioned above our attempts to prove by incorporation of ^{57}Fe that the signal observed at g = 1.94 was indeed related to the presence of iron. Before we had the labeled Azotobacter ferredoxin, we tried to do the experiment with whole Azotobacter cells. We obtained broadening, but it was not very impressive [68], at least to biochemists not familiar with EPR. When I presented this work in 1963 at a Gordon Conference on Metals in Biology, three of our physicist colleagues present stuck their heads together and did a back-of-the-envelope calculation, according to which even the little broadening of the signal that I had observed was too large in view of the low magnetic moment of ^{57}Fe! So, for the biochemists it was too small, for the physicists too large; where do you

37 William H. Orme-Johnson, Massachusetts Institute of Technology, Cambridge, Massachusetts.

go? I heard no comments from these colleagues after we showed our spectra from the isolated Azotobacter ferredoxin, which showed much more pronounced and unquestionable broadening. Obviously, in the first experiment with the whole cells, we had not succeeded to get rid of enough endogenous iron, so that the [57]Fe was more diluted. One of these physicists was my friend, the late Bill Blumberg[38], and when I once reminded him of that incident, he silently put on his engaging smile, so familiar to those who knew him.

As we went along in our attempts to define the nature of Fe-S proteins, our experiments, understandably, became more professional and sophisticated, but maybe the greatest moments of surprise and excitement were in those earliest years, of which I have given a glimpse above, the years when we were utterly ignorant of what we were dealing with. Nevertheless, I would like to report a few more episodes. Fortunately, we abandoned our unstable Azotobacter protein, when we had the opportunity to work on experimentally more tractable materials. They were an Fe-S protein from *Pseudomonas putida*, putida-redoxin, and the so-called adrenodoxin. We had a very successful collaboration with I.C. Gunsalus, generally known as 'Gunny'[39], and his group at Urbana who provided the Pseudomonas protein, and the adrenal ferredoxin was from our own production. We recognized that these proteins showed lower spin relaxation than other proteins we had dealt with and therefore had a narrower linewidth. Thus we could hope to resolve the hyperfine structure, so that we could determine the number of interacting Fe atoms represented in the signal. We also planned to do an analogous replacement with [33]S. With the Azotobacter protein it had not been possible to resolve even the [57]Fe hyperfine structure in the protein, we merely observed broadening. With [33]S one could not hope for

---

38 William E. Blumberg (1930–1989), Bell Telephone Laboratories, Murray Hill, N.J.
39 Irvin C. Gunsalus, University of Illinois, Urbana.

resolution, because this isotope was only available in < 60% enrichment in addition to being very expensive. By that time we had acquired a CAT, computer of average transients, with which we could average signals, to improve the signal to noise ratio. We needed this particularly for the ^{33}S experiments. However, the way our spectrometer was set up, there was no way for us precisely to synchronize the sweep of the CAT with the sweep of the magnet, which is required for superposition of the spectra, otherwise one would create as much noise as one tried to get rid of. Ray Hansen, inventive as always, threw some gadget together, which would come close to doing this. With ^{57}Fe we obtained nice spectra [76] in an hour or less; with ^{33}S, however, we might have to run all night; it was also necessary to keep a constant low temperature during that time. As expected, the broadening with ^{33}S was inferior to that seen with ^{57}Fe and was, of course, not resolved. After our experience with our biochemical community and audiences, among which there were always sceptics of the then new technique and approach, Bill Orme-Johnson and I decided that we could only convince everyone that the broadening was meaningful, if we did the reverse experiment, namely converting the ^{33}S labeled protein back to the ^{32}S form and see the broadening disappear. This we did, but as everyone in the business of Fe-S proteins knows, one always loses something in such operations, and there had not been much to start with. So we needed a long run and optimal signal to noise. The big centrifuges on our floor were the main culprits as far as noise is concerned. So, it was decided to run overnight from 7 p.m. to 7 in the morning in two shifts with Bill and Ray alternating for which they, incidentally, volunteered! It speaks for our mutually infective enthusiasm about this experiment, that Bill and Ray decided to bring a cot in the laboratory, spend part of the night there and keep an eye on the instruments and the temperature. The result was a marvelous spectrum with all the signal to noise that we could have hoped for. This spectrum – it is sad to say – never appeared in the literature, because at the time we got

around to writing it all down, John Tsibris[40] of the Urbana
group came up with the [77]Se substituted protein, which unambiguously showed that in these so-called 2Fe ferredoxins
there were 2Se – and by inference then also 2S – in the natural
protein involved in the structure producing the signal [77].
The great advantage of Se over S is that [77]Se has a spin of $\frac{1}{2}$ as
does [57]Fe, which produces a simple hyperfine structure pattern, whereas [33]S with spin $\frac{3}{2}$ produces a more complex pattern, which is practically impossible to resolve at only 60%
enrichment. [77]Se also was available in high enrichment and in
addition was less expensive. In view of the beautiful spectra
we obtained with the [77]Se substituted protein, our heroic experiment with [33]S became redundant and barely remained a
monument of enthusiasm and dedication in our memories (a
few decades earlier, it might not even have qualified for that
when such performance by collaborators was taken for
granted.)

Iron-sulfur proteins of the mitochondrial respiratory chain

After all these explorations we returned to where we started
from, namely to the Fe-S proteins of the mitochondrial, so-
called, respiratory chain, which in fact is not really a chain,
but a complicated three-dimensional network. In the meantime, as mentioned above, Bill Orme-Johnson had joined us.
The increased availability of liquid helium and the developing
technology of handling and applying it to spectroscopy added
a new dimension to this work, whose impact and extent we
had not expected. We stared in awe and surprise at our first
10K spectra of beef heart submitochondrial particles (SMP).
They were nothing like those we were used to from liquid nitrogen work; it seemed like a whole new world, and it indeed
turned out to be just that. There was a host of unknown sig-

---

40 John C.M. Tsibris, University of South Florida, Tampa, Florida.

nals, which even masked those we thought we knew. So when
Nan Orme-Johnson[41] also joined us to make her Pd.D., I let
her struggle with some of these problems, viz. the components
of NADH and succinate dehydrogenases and of the cytochrome
bc_1 particle [78, 79]. Joe Hatefi[42], Tom Singer[43], Edna Kearney[43] and Brian Ackrell[43] joined us in some of this work [80,
81]. We now could see several different Fe-S centers in these
materials and the spectral resolution was such that, by vary-
ing temperature and microwave power and by titration with
substrate we could identify the various components. It was
great fun for all of us and, what we concluded from our findings, has formed the basis for further developments in this
field. It is indeed a feat to see the EPR spectrum of a piece of
rat liver or of fly mitochondria and recognize in it so many
components of the respiratory chain and other components of
these tissues [82]. The respiratory chain experienced the
greatest inflation in its history, to the point that the expression 'chain' seemed to become more and more a misnomer.
What has happened in just about 20 years is shown in Figure
1.

   After the first successful assaults in the early seventies
there remained some puzzling holdovers and it became the lot
of Frank Ruzicka[44], when he joined me in 1973, to identify the
remaining unknowns; and so he did, quietly and steadily
working away. These unknowns were: (1) what was generally
known as the 'cubic iron' which appeared in oxidized particles
and Complex II in a nearly isotropic, highly temperature sensitive signal at g = 2.01; (2) a resonance at g = 2.08, whose
origin was unknown and which, for a while, was going under

41 Nanette R. Orme-Johnson, Tufts Univ. School of Medicine, Boston, Massachusetts.
42 Youssef Hatefi, Scripps Research Institute, LaJolla, California.
43 Thomas P. Singer, Edna B. Kearney and Brian A.C. Ackrell, Veterans Administration Medical Center, San Francisco, California.
44 Frank J. Ruzicka, Institute for Enzyme Research, University of Wisconsin, Madison.

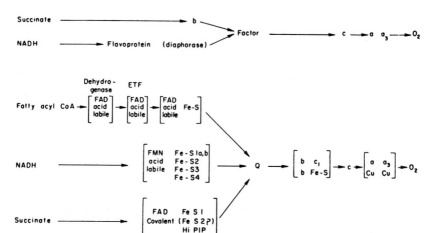

Fig. 1. Components of the respiratory chain of heart mitochondria as depicted in 1950 (adapted from Slater, ref. 84) and in 1975 (from Beinert, ref. 83, with permission from Plenum Press).

the name of 'center 5' [85]; and (3) a strange rather temperature-dependent signal in the $g \sim 2$ region [86, 87] which appeared on partial reduction and disappeared again on full reduction, as if it were due to the one-electron reduced state of a two-electron acceptor.

Concerning the first of these: in our attempts to identify the EPR signals found in mitochondria and submitochondrial particles, we had encountered a fairly strong signal centered at $g = 2.01$ that was characteristic of the oxidized state and was somewhat asymmetric, showing an extended tail toward high field. We soon realized that, on sonic treatment of mitochondria, some of the material producing this signal appeared in the soluble fraction, while part of it stayed with particles. We also noticed that the signal that followed the soluble fraction had a lower spin relaxation rate than that associated with particles, which means that the signals showed different temperature sensitivity. On further fractionation of mitochondria, the signal remaining with the particles was concentrated in

the purified succinate ubiquinone reductase [88]. We decided
to deal first with the more tractable, soluble fraction containing the signal, hoping that we might get some clues as to the
nature of the material which gave rise to the unknown signal.
Following purification by EPR, Frank then came up in 1974
with a reddish-brown protein of high purity, in which the signal was concentrated [89]. Iron analysis was positive and to
our surprise also determination of labile sulfide. So it was an
Fe-S protein that showed an EPR signal in the oxidized state!
The only Fe-S proteins then known that showed EPR signals
in the oxidized state were the high-potential Fe-S proteins
from a few bacteria, so-called HiPIPs [90]. Since the EPR signal readily disappeared with dithionite at neutral pH our protein certainly was not a very low potential Fe-S protein, as
some ferredoxins are. Thus, we called it provisionally the soluble mitochondrial HiPIP. Of course, we assumed that, as an
Fe-S protein, this protein would have some function in oxidation-reduction. However, whatever we tried, we found no reasonable reducing substrate for it. Eventually, we gave up and
pursued the identification of other so far unidentified signals
in mitochondria, as will be described in what follows.

   Number two, the g = 2.08 peak, 'center 5' was purified following this EPR resonance. It turned out to be a new Fe-S
flavoprotein, containing one FAD and one [4Fe-4S] cluster [91,
92]. Signs of the existence of a protein in mitochondria and
SMP containing acid dissociable FAD had previously emerged.
It must have been this protein, which follows NADH dehydrogenase in some purification procedures, and contributed the
FAD consistently found in these preparations and thus insti-
gated the controversy as to whether NADH dehydrogenase actually contained FMN or FAD [93]. We were now faced with
the task of finding a function for this FAD-protein. It certainly
is easier to purify an enzymatic activity and then establish its
properties than to purify a protein and then find out what it
does. This is a needle in a haystack proposition. It was our
good luck that the acyl CoA dehydrogenases and ETF were

contaminants and actually byproducts of the purification procedures which we applied to mitochondria. However, it did not occur to us that these proteins were 'meaningful' impurities of our new flavoprotein, until, one day, Edna Kearney, in a telephone conversation about our joint work on succinate dehydrogenase, dropped this remark concerning the new flavoprotein: 'Why do you look for a low M_r substrate for this protein? You yourself have discovered ETF; maybe the new protein also has a protein as substrate!' Edna was not more specific, but she did not have to be for the idea to spring up in my mind that, maybe, we finally had found the link of fatty-acid β-oxidation to the respiratory chain. The next day, we threw into EPR tubes various combinations of butyryl CoA, its dehydrogenase and ETF with our Fe-S flavoprotein and looked for appearance of the Fe-S signal of the reduced protein. Indeed it became apparent that this protein was an e⁻ acceptor for reduced ETF, not the dehydrogenase. The great moment came, when we then determined, by rapid freeze-quench EPR, the rate of reduction of the new enzyme by ETF, prereduced with butyryl CoA and catalytic amounts of acyl CoA dehydrogenase. The half-time of reduction was ~ 5 msec [91]! The question remained: what is the electron acceptor? By that time, of course, the logical choice was ubiquinone and we were not wrong. We could demonstrate the rapid reoxidation of reduced ETF by ubiquinone-1, catalyzed by the new flavoprotein [92]. It all went, as we now had thought it should! We called the new enzyme ETF-ubiquinone oxido-reductase. Thus, it took 20 years from the time we purified the enzymes of fatty acid oxidation and in addition much progress in many areas of biochemistry to make it possible for us to find the link to the respiratory chain! I am glad we did not try harder to find it 20 years earlier. Our chances of doing so, without EPR and without ubiquinone being known, would have been close to nil. Most people were still thinking about electron transfer directly to cytochromes at that time.

The most puzzling EPR signal of the lot was that of the

apparent two-electron acceptor in mitochondria. In titrations it behaved like a semiquinone but from the EPR standpoint it behaved like a transition metal ion, that is, in its temperature dependence and saturation properties. 'Why might it not be some of both?' I argued and guessed at a semiquinone-HiPIP interaction, because the signal was only seen in materials where the bound mitochondrial HiPIP (see above) was present and at oxidation states when some of this HiPIP signal was seen. The semiquinone could originate from a flavin or ubiquinone (Q) but since the signal had never been seen in soluble Fe-S flavoproteins but only in particles, our guess was at ubiquinone. The moment of truth came when Frank extracted Q from the particles and we looked at them with EPR during stepwise reduction. The signal did not appear with extracted particles, but reappeared again after restoration of Q to the particles [86]. Dick Sands, Dick Dunham[45] and Ken Schepler at Ann Arbor were then able to simulate EPR spectra coming very close to those observed, when it was assumed that there was interaction between two ubisemiquinones rather than semiquinone and HiPIP [86], but it is likely that the neighborhood of the HiPIP cluster has a relaxing influence on the semiquinone [94]. Since we had seen the signal typical of the semiquinone pair in rapidly frozen whole tissue we became convinced that this signal was no artefact but that ubisemiquinone could be stabilized in this paired species *in vivo* in a rather specific way.

On fractionation of SMP into Complexes I to III, the EPR signal of the paired species was not seen in all fractions that contain ample Q, such as Complexes I or III, but only in fractions that also showed the signal at $g = 2.01$, typical of the particle-bound Fe-S center, which is now known to belong to Fe-S center 3 of Complex II or succinate dehydrogenase. Thus, there seemed to be a relationship of the paired ubisemiqui-

---

45 W. Richard Dunham, University of Michigan, Ann Arbor.

none species to succinate dehydrogenase. However, Complex II contains little Q and the signal of the semiquinone pair had never been seen with preparations of Complex II. We then found, to our surprise, that the signal does indeed appear on reoxidation of reduced complex II with a soluble Q analog [95]! We still do not quite understand what this observation is telling us about the disposition of Q or its analogs with respect to the active site of succinate dehydrogenase.

Iron-sulfur clusters in non-oxidative enzymes; aconitase

Around 1972 Walter Lovenberg asked me to write the introductory chapter for the first volume of Iron-Sulfur Proteins [96]. Among speculations, what the future of this field might hold, I mentioned that, for instance, a long known, non-oxidative enzyme, such as aconitase, had been found to contain iron and that we may have to expect more surprises in this field. I never found that statement in the final text as printed, and I can only conclude that, at some point, I must have decided to take it out of the manuscript as irrelevant talk. Little did I know at that time that, while I may have been able to remove this statement out of the text, instead, aconitase and the surprises connected with this enzyme would enter my life with full force years later.

When Frank Ruzicka and I gave up trying to find a substrate or function for our soluble mitochondrial HiPIP, as mentioned above, we hoped, of course, that, as time went on, we might eventually get some clue from our work on other proteins of the respiratory chain, as to what our HiPIP could be or do. This clue, however, did not come from our, but from other colleagues' work or writing. R. Switzer published a paper in 1977 on glutamine phosphoribosyl-pyrophosphate amidotransferase, in which he stated that it was obviously an Fe-S protein, but that there was no evidence that the Fe-S cluster was involved in any oxidation-reduction in the func-

tion of this enzyme [97]. He referred to the finding in 1972 of
Fe and labile sulfide in aconitase [98]. We knew, of course,
that Fe and S^{2-} had been found in aconitase, but nobody paid
much attention to this at the time, including ourselves. When
I had read it in *Biochem. Biophys. Res. Commun.*, I called up
Joe Villafranca, who had done extensive work on aconitase
and asked him what he thought of this. He shrugged his shoulders (as far as I could notice on the phone). The authors of that
paper themselves apparently did not pursue this finding any
further. I told Frank Ruzicka about Switzer's paper and his
reference to aconitase. A few days later Frank came to my
office and asked me whether I thought that our HiPIP could be
aconitase; he said that the isoelectric point mentioned for aconitase, namely 8.5, was what we found for our HiPIP (8.6) and
this is not a very common value among proteins. I was surprised, but did not find Frank's suggestion outlandish. So we
immediately set out to collect everything for an aconitase
assay, which, at the Enzyme Institute, was not difficult; and
that same afternoon we had the answer, namely that our
HiPIP had strong aconitase activity, when we 'activated' it
with iron and thiol, as the assay protocol requires [99]. These
really were exciting moments, and, in a way, I am still living
today on the consequences of this event. The next step then
was to make aconitase by the method published in the paper
which had reported on the labile sulfide in aconitase [98] and
then compare that protein with our HiPIP as to its activity,
optical and EPR spectra. Data for activity and the light absorption spectrum of aconitase had been published; however,
the way the spectrum had been plotted, there was a huge peak
at 280 nm and the long wavelength absorption was very indistinct. Had I known aconitase was a brown protein, I would
have gone after it a long time ago! The comparison of the data
showed little, if any difference to those from our HiPIP. An
EPR spectrum only at liquid nitrogen temperature had been
published, which showed the ubiquitous g = 4.3 high spin ferric signal. This signal could be expected to be, as in most cases,

from an impurity. When the great day came, on which we had
the purified, authentic aconitase ready to be run by EPR, we
sat there before the recorder of the spectrometer, curiously
waiting for the pen to trace the g = 2.01 signal which we had
seen in our HiPIP: and indeed it did so faithfully. We knew, we
had it in the bag!

There followed three years of explorations in which we were
joined by Jean-Luc Dreyer[46] of Switzerland, Claire Kennedy[47],
who had discovered the labile sulfide in aconitase [98], and by
Mark Emptage[48] who had been at Minneapolis and established our link with Eckard Münck[49]. We learned isolated bits
and pieces about properties of aconitase, but why iron was
necessary for activity and why reduction could partially activate the enzyme and what the actual active species was, we
did not find out so quickly. Again, to our help came apparently
unrelated work in other laboratories and this was the time
when telephone, then airfreight and later Federal Express
and Fax started making increasing and significant contributions. David Stout[50] was coming close to completing the X-ray
structure of ferredoxin from Azotobacter and Eckard Münck
was looking with Mössbauer spectroscopy at this protein and
a ferredoxin (FdII) from *Desulfovibrio gigas*. Stout first published a note proposing that the Azotobacter protein had a
[4Fe-4S] and a [2Fe-2S] cluster [100]. Eckard Münck, however, concluded from his Mössbauer spectra of the oxidized
forms of Fd I of Azotobacter and of Fd II of *D. gigas* that he was
dealing with a system of spin coupled ferric ions and he
pointed out that two ferric ions in the two iron site, as proposed by Stout for Fd I, cannot be spin-coupled in such a way
that one spin is left over, as the EPR spectra demanded [101].

---

46 Jean-Luc Dreyer, University of Fribourg, Switzerland.
47 Mary Claire Kennedy, Medical College of Wisconsin, Milwaukee, Wisconsin.
48 Mark H. Emptage, Dupont de Nemours and Company, Wilmington, Delaware.
49 Eckard Münck, Carnegie-Mellon University, Pittsburgh, Pennsylvania.
50 C. David Stout, Scripps Research Institute, LaJolla, California.

Eckard postulated that there must be a third spin to produce the known EPR spectrum of these proteins in the oxidized state (g = 2.01), which, most likely, was also due to an iron atom. He concluded, therefore, that the Azotobacter as well as the Desulfovibrio ferredoxin must have a 3Fe cluster. As the same kind of signal was observed in aconitase and our HiPIP (because of their identity we will from now on use only the name aconitase for this protein), Münck also proposed that aconitase has a 3Fe cluster. Dave Stout was quick to pick this up and found indeed a better fit for a [3Fe-3S] than for a [2Fe-2S] cluster (note that there is in addition a traditional [4Fe-4S] cluster in this protein which had been correctly identified) [102]. Eckard then looked at an aconitase sample from us[51] and found for the 'as isolated' state the same Mössbauer spectrum as with Fd II of *D. gigas*. We had no reason not to accept his interpretation, as our Fe and S^{2-} stoichiometries with respect to the M_r had always been odd. Eckard's arguments were so convincing that the Fe-S community almost eagerly embraced the new interpretation. We knew that the form of aconitase which gave the EPR signal at g = 2.01 was the inactive form of the enzyme. What was then the active form? We needed iron for full activation, but reduction alone gave about 70% activation. Was iron then necessary? We had Eckard look at a spectrum of reduced aconitase [103]. It looked indeed like the spectrum of the reduced 3Fe cluster of Fd II, but Eckard discovered a minor, superimposed Mössbauer spectrum, which had all the characteristics of that from a [4Fe-4S] cluster in the 1+ oxidation state. He suggested therefore that we reduce the material extensively and look for an EPR signal. He predicted there had to be the typical spectrum of a reduced [4Fe-4S]$^+$ Fd. We did that and indeed, to our great joy, there it was [104]. We learned then, using EPR as a monitor, how to improve the yield of the reduced form, which had an unexpect-

---

51 *Us* in this case refers to: J.-L. Dreyer, M.H. Emptage, M.C. Kennedy and myself.

edly low oxidation-reduction potential. The recipe was to go to pH 8.5, use an about 10 fold molar excess of dithionite under anaerobic conditions, and methylviologen as a mediator. This way we could get up to 85% yields of the reduced $[4Fe-4S]^{1+}$ form. As we had not been aware of the importance of observing all of these conditions simultaneously, we never had seen a significant signal for the reduced 4Fe form of aconitase. We mainly had been working at neutral pH and, as compared to the relatively narrow signal of the 3Fe form, the spectrum of the reduced form is sufficiently spread out that one has to use about ten times more material to get a similar signal to noise ratio as for the signal of the 3Fe form. In fact, Frank Ruzicka and I had repeatedly seen weak signals of the kind which we now know are due to $[4Fe-4S]^{1+}$ aconitase, when we looked at dithionite-reduced samples during the preparative attempts with our HiPIP, but were never able to enrich the species responsible for the signal, but rather lost it. We know now that the 4Fe cluster of aconitase is unstable when exposed to dithionite in the presence of oxygen.

The discovery of a [4Fe-4S] species was crucial for all further developments, because from our activity measurements and the Mössbauer and EPR observations it now became clear that the $[4Fe-4S]^{2+}$ form was the active form and the 3Fe form presumably was only a breakdown product arising during isolation. It also became obvious that the iron was needed for 'activation' to build up the 4Fe cluster from the 3Fe form; and the activation on mere reduction resulted from transfer of iron from decaying 3Fe clusters to other 3Fe clusters to complete the $[4Fe-4S]^{1+}$ form, which is more stable under reducing conditions. We also could now make use of radioactive iron isotopes such as ^{55}Fe and ^{59}Fe and of isotopes with nuclear spin such as ^{57}Fe to investigate the iron sites further. ^{57}Fe is not only useful as an ENDOR isotope because of its nuclear spin, but is also the isotope required for Mössbauer spectroscopy of iron compounds. ^{57}Fe constitutes 2.2% of iron in nature. Thus, we just had enough sensitivity to observe Mössbauer spectra of the 3Fe cluster at very high concentrations. If, however, we converted the 3Fe into the 4Fe cluster with > 90% ^{57}Fe and observed at a low sample concentration, we could only observe the incorporated fourth iron, which is critical for activity. Vice versa, by rebuilding the 4Fe cluster with pure ^{56}Fe and observing at a high concentration, we only saw the three original Fe sites stemming from the 3Fe cluster.

The question, where the sulfide was coming from in the build-up of the 4Fe from the 3Fe cluster was still unanswered. We had to add iron for full activation but never sulfide! This is where accurate analyses for iron and sulfide were needed and I spent some months to get these to a level of precision and accuracy in small scale applications, so that we could draw convincing conclusions from them [105]. It turned out that the putative [3Fe-3S] clusters were in fact [3Fe-4S] clusters, that is, they were cubane clusters with just one Fe missing in one corner of the cube [106]. Such clusters have so far been found to be stable only in a protein environment. There was now one thing left that made little sense, namely the notion of a [3Fe-3S] cluster, as Stout had deduced from his X-ray diffraction experiments on Azotobacter Fd I. This putative form of cluster had much larger Fe-S and Fe-Fe distances than observed with any other Fe-S cluster. There were no crystal structures available yet for FdII of D. gigas or for aconitase. An EXAFS study of FdII from D. gigas showed, however, that the distances in FdII were, in fact, the normal ones observed previously with other Fe-S clusters [107]. This was confirmed in our EXAFS study on aconitase in collaboration with Bob Scott[52] and J. Penner-Hahn[53] [106]. Reevaluation of the crystal structure of Azotobacter FdI [108] then showed that, indeed, there is a [3Fe-4S] cluster, as it must be in Fd II of D. gigas and in aconitase according to analyses and EXAFS spectroscopy (this has

---

52 Robert A. Scott, University of Georgia, Athens.
53 James E. Penner-Hahn, University of Michigan, Ann Arbor.

now been confirmed by X-ray crystallography on both of these proteins [109, 110]). There was still a problem to be solved. We had the ratio of Fe to S^{2-} for aconitase, but we were not entirely sure of the stoichiometry with respect to the protein, because we did not have a reliable molecular weight. Gel filtration gave a low M_r, dry weight determination indicated a considerably higher one and SDS PAGE gave a value in between[54]. We suspected that we may have an appreciable amount of apo-aconitase in our preparations. By the expert collaboration of Lars Rydén[55] at Uppsala, we obtained determinations by three methods: SDS PAGE, ultracentrifuge and amino acid analysis, which together indicated an M_r of 81,000 [111]. When we received these data from Lars, we were delighted and greatly relieved, because they gave us the assurance that what we isolated was > 95% holoprotein with a [3Fe-4S] cluster.

Before I proceed with the story of aconitase: in the context of the identification of the metal cluster in aconitase as a 3Fe-4S cluster, it may be of interest to briefly consider the fate of its cousin, namely the particle-bound, presumed HiPIP, Fe-S center 3 of Complex II and succinate dehydrogenase. As Complex II (in which center 3 is stable as opposed to the situation with succinate dehydrogenase) contains in addition a heme component and two FeS clusters and is also not easily labeled with ^{57}Fe, prospects for a successful application of Mössbauer spectroscopy are very poor. However, in the meantime Jack Peisach[56] had shown that all the known Fe-S cluster types, viz., 2Fe, 3Fe, and 4Fe in Fds and HiPIPs have different and characteristic patterns of the Linear Electric-Field Effect (LEFE)

<hr/>

54 It is pertinent to mention here that, 10 years earlier in her thesis work at Duquesne University, Claire Kennedy had faced a similar problem, when she had first found the unusual Fe-S ratio of 2:3 and tried to relate this to the M_r [98]. At that time, knowledge of Fe-S clusters was not sufficiently advanced that these findings could be interpreted.
55 Lars Rydén, University of Upsala, Sweden.
56 Jack Peisach, Albert Einstein College of Medicine, New York.

in EPR, which should make it possible to distinguish them. So, I joined forces with Brian Ackrell and Edna Kearney, who provided the Complex II preparation, and with Jack Peisach to do the spectroscopy, and, indeed, the verdict was that the LEFE of the g = 2.01 signal in Complex II was that typical of [3Fe-4S] clusters [112]. This assignment was later independently verified by magnetic circular dichroism.

In the years following this we consolidated our knowledge of the EPR and Mössbauer spectra of aconitase in its various forms with and without substrate present. Substrate had a profound effect on EPR and Mössbauer spectra of the 4Fe form. One of the exciting findings from Mössbauer was that the unique Fe in aconitase furnishes at least one of the binding sites for substrate [113]. This Fe, called Fe_a, is the one that gets lost on purification and has to be incorporated for enzymatic activity to be restored [114]. This led us to ENDOR experiments, a technique ideally suited for finding out which groups in the substrate were actually bound to Fe_a. After Claire Kennedy and I had moved to the Medical College of Wisconsin at Milwaukee in 1985[57], we initiated an extensive ENDOR study with Brian Hoffman[58] and Melanie Werst and arrived at a model for binding of the various substrates [115]. In this model one carboxyl and the hydroxyl of citrate or isocitrate are bound to Fe_a in addition to a water molecule. From ^1H and ^2H ENDOR spectroscopy we concluded that in the absence of substrate the fourth ligand to Fe_a is a hydroxyl. This was the first example of a cubane cluster in a protein with a ligand at one of the irons that is not derived from protein. This example provided a stimulus for the synthesis of models for 'subsite-differentiated' cubane clusters [116].

The 3D structure of aconitase with substrate bound was still not known at that time. There actually was some uncertainty,

---

57 On invitation by James S. Hyde and Bettie Sue Masters.
58 Brian M. Hoffman, Northwestern University, Evanston, Illinois.

whether the determination of the 3D structure of the enzyme
with substrate bound would be successful, because it was very
likely that there would be a mixture of the three substrates
present on the enzyme, which would almost certainly blur the
picture. Mössbauer spectra had shown such a mixture, when
any one of the substrates had been added [117]. For this case,
however, we could have had something up our sleeve: we could
add one of the nitro-inhibitors, which had given us Mössbauer
spectra of a single species! Nevertheless, we tried substrate,
and, to everybody's surprise, there was no evidence from the
X-ray data that more than one species of substrate was present in the active site, and this species, according to Dave
Stout's analysis of the X-ray data, was beyond doubt isocitrate. Of course, we were now anxious to see the Mössbauer
spectra, and so we sent some ^{57}Fe labeled protein to Dave
Stout for crystallization, filtered off the crystals, and packed
them into a Mössbauer cell. When Eckard Münck called us,
there was great joy in our lab and Dave Stout's: Mössbauer
has shown the isocitrate spectrum! Obviously, at the pH and
(or) the composition of the crystallization medium the enzyme
must have selected isocitrate. What luck! And it was again one
of the great moments, when we then received from Dave Stout
the pictures of his X-ray structure of the protein with isocitrate in the active site. It was exactly as our ENDOR studies
had suggested [118]. Of course, it is true that Dave had the
results of our ENDOR studies available to guide him in the
construction of his 3D model, but it is rather far-fetched to
think that all his distances and intensities could have been
fitted, were the structures not compatible (we must consider
here that ENDOR does not provide a 3D structure). Similarly
the structures we deduced from ENDOR for aconitase with
citrate, nitroisocitrate and fluorocitrate bound agreed with
the X-ray results. Something we could not have guessed from
spectroscopy was: What is the base that abstracts the proton
from citrate or isocitrate, that is not readily exchanged with
other protons of the medium and can thus be added to the

product of the reaction? There had been all sorts of guesses: aspartate or glutamate, for instance. According to the 3D model, however, to many enzymologists' surprise, it turned out to be a serine!

During Dave Stout's work on the 3D structure, Jim Howard[59] and David Plank at Minneapolis determined the sequences of nine cysteinyl-peptides obtained from our beef heart aconitase preparation [119]. Stout was able to locate these peptides in his structure of the enzyme by the increased electron density at the sulfur atoms. This was a great help in tracing the peptide chain. However, shortly thereafter, Howard Zalkin[60] of Purdue kindly communicated to David Stout and to us the complete amino acid sequence of the pig heart enzyme, as deduced from the cDNA [120]. In a total of 754 residues there was only an eight residue difference between the pig and the beef heart aconitase sequences. The latter was first determined partially from our cysteinyl-peptides and later completed from the final high resolution X-ray structure of the enzyme. I might mention here that, when in the course of these events, we submitted our final manuscript on the cysteinyl-peptides and the cysteine ligands to the Fe-S cluster [121], referees voiced the criticism that this was redundant now that the X-ray structure was known. This was not a very encouraging experience. I had been the one who sponsored Dave Stout's paper on the structure for the Proceedings of the National Academy of Sciences to provide for rapid publication and without the availability of the sequences of these peptides, communicated to Stout a good year earlier, the final structure determination would have had to await the complete sequence determination from Howard Zalkin's lab! Knowing the aconitase sequence was a great step forward and was a boost to our morale, if such a boost was still needed at this

---

59 James B. Howard, University of Minnesota, Minneapolis.
60 Howard Zalkin, Purdue University, West Lafayette, Indiana.

stage. However, the excitement was even greater, when shortly thereafter the primary structure of yeast mitochondrial aconitase was published [122]. It would have been little help, if it had been, say, the beef or horse heart aconitase, because there would have been too little difference to tell the essential from the non-essential elements in the structure. However, fortunately, the yeast aconitase sequence differed just enough (63% identity) that it was possible to recognize the essential features. It turned out that all the amino acids in the active site that have contact with the substrate are conserved as are the Fe-S cluster ligands. Limin Zheng in Howard Zalkin's laboratory was then able to express the pig heart enzyme in *E. coli* and sent us proteins with mutations in most of the active-site residues [123]. All of these mutants showed major changes in the enzyme's behavior in various tests, in line with the expectations derived from the X-ray structure.

The Janus-faced protein

The next unexpected, major upheaval in the pursuit of our research came again entirely from the outside. I must digress here somewhat and describe first some developments in the field of iron metabolism. Iron is one of the most universally used elements in living creatures. Various elaborate mechanisms have been developed to accomplish and control its uptake, transport, storage, utilization and disposal. Control is exerted at several levels. At the level of individual cells, two substances play a pivotal role, namely, transferrin receptors for entry of iron and ferritin for storage. Both are controlled at the stage of messenger RNA translation. mRNAs for the synthesis of both transferrin receptor and for the heavy subunit of ferritin contain iron sensing structures, called iron-responsive elements (IREs), which are RNA stem-loop structures, located in the untranslated regions of their respective mRNAs [124–126]. The structures of these IREs, although basically similar, differ in some detail among each other. The IREs, in turn, are controlled by a cytosolic protein that binds to them tightly when the intracellular iron level is low, and binds loosely when iron is available in sufficient amounts. Thus, this protein, called IRE-binding protein (IRE-BP), apparently is the ultimate sensor of iron availability. When it binds tightly to IREs, the transferrin receptor mRNA is stabilized and translation of the ferritin mRNA is blocked. I had no occasion yet to hear about this, when one day in late 1990, while we were still engaged in work on mutations directed toward active site residues in aconitase, we received a phone call that electrified us. Richard Klausner[61] from the NIH, unknown to me at that time, told us that there is a protein in mammalian cells, called 'Iron-responsive element binding protein' (IRE-BP) – as just explained in the foregoing paragraph – and that the closest relative to this protein known from comparisons of amino acid sequences is pig heart mitochondrial (m)-aconitase (30% identity)! [127, 128]. He had already communicated with Dave Stout, and according to the 3D model of aconitase, all active site residues are conserved and the cysteines that are cluster ligands are the only conserved cysteines in both structures. Yet another surprise with aconitase!

The IRE-BP, however, is a cytosolic protein and m-aconitase therefore can not serve as IRE-BP in the cytosol. Moreover, each one of these proteins is coded for in a different chromosome. There is, however, a cytosolic aconitase known, which had so far not received much attention [129, 130], and it is actually not clear what its physiological function is. It is coded for in the same chromosome as is IRE-BP. Are cytosolic aconitase and IRE-BP identical? Klausner wondered whether the 3Fe to 4Fe conversion could be the iron sensing mechanism. We thought that, for a sensitive iron response, this conversion in m-aconitase was almost too slow, and, from all we knew,

---

61 Richard D. Klausner, National Institutes of Health, Bethesda, Maryland.

this conversion was even more sluggish with the cytosolic(c)-aconitase. However, our experience with c-aconitase was very limited. We had, on two occasions, produced small quantities of fairly pure c-aconitase, so that we knew its EPR spectrum and knew about the requirement for activation and its enzymatic activity, however, as this protein is much harder to purify in quantity than its mitochondrial counterpart, I had little enthusiasm for this enzyme; it seemed to me like studying the methyl ester of a compound, after you have studied the ethyl ester. However, in the case at hand, this now seemed to be different; namely m-aconitase is not an IRE-BP. Curiously, it turns out that, in its RNA binding form, c-aconitase binds to the IRE of m-aconitase mRNA and may thus control the synthesis of m-aconitase [131]. So, we set out to purify the c-aconitase. It was fortunate that we could mobilize George Blondin's[62] enthusiasm in this venture; he had been kind enough to prepare for us m-aconitase with the facilities of the Enzyme Institute at Madison. Within a few months he came up with some clean c-aconitase and we soon found out that c-aconitase was a beast quite different from m-aconitase in a number of ways, which had a bearing on its presumed function as IRE- BP. The 4Fe cluster is even more stable than we thought; we regularly obtained about 80% holoprotein of which at least three-fourths was in the [4Fe-4S] form, unheard of for m-aconitase. It also binds substrate very tightly, so that it is difficult to free it completely from substrate, even when going from the 4Fe back to 3Fe state. We furnished some of this material in the form of apoenzyme, 3Fe, and 4Fe enzyme to Klausner and his group and all three forms could be induced to serve as IRE-BP by exposure to high levels of thiols, but only the so treated apoenzyme retained RNA binding activity, when the activating thiol had been removed [132]. No iron was required for the apoenzyme to display RNA binding activity.

---

62 George A. Blondin, University of Wisconsin, Madison, Wisconsin.

Other experiments [133] that will not be detailed here all pointed to the apo-c-aconitase being closest if not identical to the IRE-BP which has been isolated in ng to μg amounts from human liver and placenta [134, 135]. The M_r of both proteins, c-aconitase and IRE-BP, was around 98,000; however, while the amino acid sequences of the IRE-BP's from human, rat, and mouse liver had been deduced from their respective c-DNAs, the sequence of c-aconitase was not known. Thus, we considered it incumbent on us to fill in some of this gap. In collaboration with our protein-chemist colleague, Liane Mende-Müller[63], we obtained six peptides by cyanogen bromide digestion of c- aconitase, which together constituted 10% of the 889 residues present in the enzyme [136]. As we know now, all four domains of the protein were represented in these peptides. While these analyses were going on, we were of course very apprehensive what we might learn. At one point, Liane called us up and said she thought she had a 17-residue sequence of a large peptide, but now had her doubts, whether something might have gone wrong and this was actually a contaminant, because she could not find this peptide in the sequence of human IRE-BP, which she had called up on her computer from the data bank. However, this is where phone and Fax come in handy: we had already received an amended sequence from our colleagues at the NIH, and it took just a few minutes to find the 17-residue sequence in the amended part of the sequence! Of the 89 residues in our 6 peptides there were only two residue-differences, and, for that matter, conservative ones, between beef liver c-aconitase and human IRE-BP from liver and placenta. According to our experience with various m-aconitase sequences, such a small difference can easily be accounted for by a difference between the IRE-BPs of related species such as beef and rodents or human. Yet the identity of our six peptides to the corresponding peptides in

---

63 Liane Mende-Müller, Medical College of Wisconsin, Milwaukee, Wisconsin.

beef heart m-aconitase was only about 30%. The conclusion from all this was: c-aconitase is identical with IRE-BP-except for one relatively small, but essential detail, namely the cubane Fe-S cluster: aconitase needs it, but for RNA binding it is not only not required but it is obviously an obstacle. How nature manages the reversible conversion between these two states and consequently the functions, we yet have to figure out.

As more and more protein sequences become available and relationships between apparently unrelated proteins thus become apparent, we will certainly find more cases of proteins that can 'wear two hats' as we have now experienced it with the aconitases. Concerning evolution of proteins it is interesting that the aconitase of *E. coli* is much closer in sequence and M_r to beef heart c-aconitase than to m-aconitase. One might, therefore, argue that the *E. coli* and mammalian c-aconitase are the older forms of the enzyme and that the m-aconitase has evolved from this form. One might also wonder, whether the IRE-BP function was not the original function of the molecule, which might have utilized a citrate binding site for capturing citrate-chelated Fe and that, such a site being present, the aconitase function evolved.

Reminder: Some moments when not...

Readers who have worked their way through this chapter up to this point may have come away with the impression that this field of biochemistry is all joy and fun. I should, therefore, remind them that, according to the title given to this chapter, I have only selected the events of joy and fun; these are the ones that have settled most vividly in my memory. Fortunately, we probably all have the ability to relegate to some of our more remote memory the less pleasant events – which are, most of the time, actually nonevents – the many hours, days and weeks of often futile struggles and frustrations. If I sum

up the most exciting 'moments' in my research, recounted
above, and spread them over the forty-odd years in which they
occurred, it is easy to see that the density is low. There were,
of course, enjoyable events of lesser significance, often leading
up to those most memorable ones I reported. There are, however, also the big disppointments, 'the moments when there
did not...' happen what one has hoped and worked for. For the
sake of contrast, I may mention one of our major near misses.
In the mid-sixties we did not yet have our liquid helium system ready. At that time such systems were not commercially
available and it was also cumbersome to get liquid helium. It
had to be flown in from Amarillo, Texas to Chicago and from
there brought up by truck. By the time a 30 l container arrived, there certainly was no longer 30 l in it. We were anxious
to look at a plant ferredoxin at very low temperature so that
we could see whether we might not find an EPR signal. By that
time it was known that there were many similarities between
ferredoxins and the Fe-S proteins of the respiratory chain, and
particularly to Fe-S proteins of low M_r, such as those we had
purified from Azotobacter. We had noticed that the spin relaxation properties could differ significantly between various
Fe-S proteins, which means the EPR resonances of some of them
were broader than those of others. Those with broader lines
could only be detected at lower temperatures. We suspected
that some of the ferredoxins might have even faster spin relaxation (i.e., broader lines) than those we had so far been dealing
with. So, I got some spinach ferredoxin from Tony San Pietro[64]
of the Kettering Laboratories, and Dick Sands had promised to
come over to Madison and bring his helium Dewar and EPR
cavity, and we were to contribute the helium. And, indeed, one
day everything was together: ferredoxin, helium, Dick Sands
and his hard- and quartz-ware, and the decisive experiment
was to be run. The question was: should we right away run the

---

64 Anthony San Pietro, Indiana University, Bloomington, Indiana.

reduced sample, because with all Fe-S proteins other than HiPIPs, EPR signals appeared only in the reduced state. However, we did not dare to split the sample into a reduced and an oxidized one, because there was not all that much of it and Dick's cavity required 0.5 ml; we were worried about signal to noise! If we reduced the ferredoxin right away, there would have been no chance to run the oxidized, i.e., as is, control. The decision was mine, and I proposed 'quickly' to run the oxidized control and then reduce the sample. The running of the oxidized sample, which showed nothing, as expected, was, unfortunately, not all that quick with the make-shift assembly, which we had put together. Then we removed the sample, reduced it, put everything together again and were ready to fill the Dewar with liquid helium. Unfortunately, while we were transferring helium with great expectations, there was, at one point, a final puff from the helium-tank and that was the end of the helium. The instrument did not become stable with the amount we had been able to transfer, and that was then also the end of the experiment. Inexperienced as I was with liquid helium, I had made the wrong decision! We warmed up everything, dismantled, and Dick left for Ann Arbor. The next helium would have been some days and some dollars away, and Dick had to be back home to attend to his duties. The next time, when again all ingredients were together, about a year later, in 1966, the experiment was done and it worked and showed a typical Fe-S spectrum for the reduced ferredoxin! The only flaw, as far as we were concerned: this crucial experiment was done at Ann Arbor by Dick Sands and Graham Palmer [137] and not by us. We had missed the boat; but we took comfort in recognizing the fact that, at least, it remained within the family and, above all, it meant progress for science, and that this actually had been our goal. A similar experiment was done at the Imperial College in London [138] at about the same time, and even without liquid helium, rather with brute force, namely with a very concentrated solution of ferredoxin, which we never had. The detection of an EPR signal for spinach ferredoxin was an important step forward in our knowledge of Fe-S proteins, as it 'unified', so to speak, the field, in that it removed the apparent difference between Fe-S proteins, that had EPR signals and the ferredoxins, for which none had been observed so far.

Another set of experiments, which, however were not of the 'no second chance' type, brought me to Ann Arbor some time around 1959/60. Dick Sands always was a busy man, in the department, with students, four kids at home, teaching Sunday school and having other obligations. Vital parts of Dick's home-made spectrometer were used by students in other experiments and so, whenever we wanted to run, it had to be reassembled from parts scavenged from other apparatus or taken from big tables in the lab, which were scattered with parts and looked somewhat like an EPR spectroscopist's flea market to me. So, whenever I arrived, Dick reassembled the spectrometer. One time then, it simply would not produce decent spectra, rather nothing but ghastly noise spikes. Dick exhausted all the knowledge of the man who had himself built more than one spectrometer, he exchanged parts, he pounded the chassis' of various critical components to check for loose connections (these were still the days of vacuum tubes!), but nothing showed up that would have indicated where the trouble might come from. After some hours of this futile search for the source of the problem, Dick decided that a thorough study of the problem was necessary, for which he did not have the time and leisure at this point and that the best thing was for me to return and come back when everything was working again. So, I packed up my goodies and retreated across the great lake to Madison, with my expectations at a low ebb. A few days later I had a phone call in the lab. There was Dick: 'Helmut, I found the trouble!', he said. 'A spider must have gotten into some of our waveguide and built some webs there, which were shaking in the breeze and producing our noise spikes!' I remembered the big tables with the flea market appearance, from which some of the waveguide must have been coming, and it sure sounded like a quite plausible explanation to me.[65]

Dedication

In closing, I should not miss to point to one of the greatest pleasures in research – and probably in all worthwhile human activity – namely to the joy, satisfaction and stimulation that can come from collaboration and exchange of thoughts with congenial and dedicated colleagues. I would also like here to apologize to those few of my collaborators, intra- or extramural, whose names do not surface in my stories. This does not mean that I do not appreciate their efforts and contributions, but luck – or rather lack of it – would have it that we did not experience any of those rare moments together. With these statements, I dedicate this chapter to all of those who have, through the years, shared with me goals, labor, disappointment and success, or who have in other ways supported our research in my or collaborating laboratories.

Addendum

In writing historical accounts one has the great advantage over writing on hot, current subjects that little or nothing changes with time. This applies to most of what I have written

---

65 I mailed a copy of this paragraph to Dick Sands and he provided the following details which I did not know but which may be of interest to EPR spectroscopists: 'If you will remember, the waveguide going down to the cavity (we were using an immersion cavity at the time) had a double bend at the top to permit a tuning rod to pass next to the waveguide to tune the iris (matching) of the cavity at low temperature. Because of the double bend, you could not see through the waveguide. I isolated the problem to that piece of waveguide, and in desperation tied a string to a piece of cloth and pulled the cloth through the bends in the waveguide to clean out anything that might have gotten in the waveguide. What came out was a dusty spiderweb, which, indeed, was the cause of the problem.'

above, but the two coherent stories that I have followed up to the time of writing, namely that on copper in cytochrome c oxidase and on the 'Janus-faced protein', have moved ahead considerably in the year that passed since I wrote the manuscript. I feel, therefore, that I owe it to the reader bringing these stories up to date. At the time of writing (September, 1993) all further experimentation, spectroscopic [139] and molecular biological [140] has nothing but supported the proposal that, what was called copper A, is in fact a mixed-valence, dinuclear copper site in subunit II of cytochrome oxidase, so that there remains little resistance to acceptance of this model [141, 142]. The all around most convincing evidence appeared in a poster by de Vries, van der Oost and Wachenfeldt of Holland [143], which showed X- and S-band EPR spectra of a soluble subunit II preparation of COX from *Bacillus subtilis*, containing copper, but no heme. At both frequencies the four low-field lines of the seven-line pattern expected at g_z were clearly resolved.

The story of the Janus-faced protein has developed at an even more rapid pace. With recombinant techniques, the iron-responsive element binding protein (IRE-BP) has been expressed in insect tissues infected with baculovirus in two laboratories [144, 145] and the identity of this material with cytosolic aconitase has been confirmed. The protein can be expressed with or without iron and sulfide in the molecule, depending on conditions: with iron and sulfide it is an enzymatically active aconitase, without iron and sulfide it is IRE-BP. The latter can be reconstituted with iron and sulfide to an active aconitase and, on oxidation with ferricyanide, its [4Fe-4S] cluster is converted to the [3Fe-4S] cluster of the inactive enzyme, as indicated by the EPR signal typical of the [3Fe-4S]$^+$ cluster [145]. In two laboratories it was shown that, in several lines of cultured cells, c-aconitase is converted to IRE-BP, when in these cells NO synthase activity is stimulated by various means [146, 147]. In our laboratory, we have shown that isolated c-aconitase can be inactivated by NO, depending on dose and time of exposure. Substrate protects the enzyme substantially; however, it seems entirely feasible, though not proven at this time, that, with continued exposure to NO, nature may indeed use this path to accomplish the conversion of c-aconitase to IRE-BP; note that it takes 10–15 hours in cells to inactivate aconitase and form IRE-BP! If iron and sulfide can be supplied by suitable carriers (or by an enzyme in the case of sulfide [148]), c-aconitase from IRE-BP should also be readily resynthesized *in vivo*. However, the regulation of uptake and storage of iron is probably yet more complicated. There have been reports of phosphorylation of IRE-BP [149] and for the requirement of interleukin-2 in this process [150].

REFERENCES

1   A. Einstein as quoted by G.H. Raner and L.S. Lerner (1992) Nature, 358, 102.
2   A. Kornberg (1989) in 'For the Love of Enzymes' Harvard University Press, Cambridge, Mass. 336 p.
3   H. Eisenberg (1990) Comprehensive Biochem. 37, 265–348.
4   H. Beinert and P.K. Stumpf (1983) Trends Biochem. Sci. 8, 434–436.
5   N.O. Kaplan (1986) Comprehensive Biochem. 36, 255–296.
6   E. Lederer (1986) Comprehensive Biochem. 36, 437–490.
7   O. Westphal (1968) Ang. Chem. Internat. Ed. 68, 489–506.
8   R. Kuhn and H. Beinert (1947) Chem. Ber. 80, 101–109.
9   S. Ochoa (1980) Ann. Rev. Biochem. 49, 1–30.
10 D.E. Green, D. Herbert and V. Subrahmanyan (1941) J. Biol. Chem. 138, 327–339.
11 F. Lynen, E. Reichert and L. Rueff (1951) Ann. Chem. 574, 1–32.
12 D.E. Green (1954) Biol. Rev. 29, 330–366.
13 F. Lynen (1955) Ann. Rev. Biochem. 24, 653–688.
14 G.D. Greville and H.B. Stewart (1953) Ann. Repts. Progr. Chem. (Chem. Soc. London) 50, 301–322.
15 H. Beinert (1990) in P.M. Coates and K. Tanaka Eds. Progress in Clinical and Biological Research. Vol. 321, Fatty Acid Oxidation: Clinical, Biochemical and Molecular Aspects. Alan R Liss, Inc. pp. 1–22.
16 R.W. Von Korff (1953) J. Biol. Chem. 200, 401–405.
17 N.W. Pirie (1930) Biochem. J. 24, 51–54.

18 H. Beinert, R.W. von Korff, D.E. Green, D.A. Buyske, R.E. Hand-
    schuhmacher, H. Higgins and F.M. Strong (1953) J. Biol. Chem. 200,
    385–400.
19 D.E. Green, S. Mii, H.R. Mahler and R.M. Bock (1954) J. Biol. Chem.
    206, 1–12.
20 F.L. Crane, S. Mii, J.G. Hauge, D.E. Green and H. Beinert (1956) J.
    Biol. Chem. 218, 701–716.
21 F.L. Crane and H. Beinert (1956) J. Biol. Chem. 218, 717–731.
22 H. Beinert (1957) J. Biol. Chem. 225, 465–478.
23 R. Kuhn and R. Ströbele (1937) Ber. dtsch. Chem. Ges. 70, 753–760.
24 H. Beinert and R.H. Sands (1961) in M.S. Blois Jr., H.W. Brown, R.M.
    Lemmon, R.O. Lindblom and M. Weissbluth Eds. Free radicals in bio-
    logical systems, Semiquinone formation of flavins and flavoproteins,
    Academic Press, New York, NY pp. 17–52.
25 J.G. Hauge, F.L. Crane and H. Beinert (1956) J. Biol. Chem. 219,
    727–733.
26 Y. Ikeda, C. Dabrowski and K. Tanaka (1983) J. Biol. Chem. 258,
    1066–1076.
27 H. Beinert and W.R. Frisell (1962) J. Biol. Chem. 237, 2988–2990.
28 P.M. Coates and K. Tanaka, Eds. (1990) Progress in Clinical and Bio-
    logical Research, Fatty acid oxidation, Vol. 321, Wiley and Sons, New
    York, 1990, pp. 273–427.
29 J.-J. Kim, M. Wang, S. Djordjevic and R. Paschke (1992) in Progress in
    Clinical and Biological Research, Vol. 375, P.M. Coates and K.
    Tanaka, eds., New developments in fatty acid oxidation. Wiley-Liss,
    New York (1992) pg. 111–126.
30 H. Beinert (1985) Biochem. Soc. Trans. 13, 542–547.
31 R. Cammack (1992) in Advances in Inorg. Chem. 38, IX–XI.
32 R.H. Sands and H. Beinert (1959) Biochem. Biophys. Res. Commun. 1,
    175–178.
33 A. Ehrenberg and T. Yonetani (1961) Acta Chem Scand. 15 1071–
    1080.
34 T. Yonetani (1960) Federation Proc. 19, 32.
35 H. Beinert and G. Palmer (1964) J. Biol. Chem. 239, 1221–1227.
36 H. Beinert and G. Palmer (1965) in Oxidases and Related Redox Sys-
    tems, Vol. 2, T.E. King, H.S. Mason, M. Morrison Eds., On the Func-
    tion and Disposition of Copper in Cytochrome Oxidase, John Wiley &
    Sons, Inc. pp. 567–585.
37 M. Öblad, E. Selin, B. Malmström, L. Strid, R. Aasa and B.G. Malm-
    ström (1989) Biochim. Biophys. Acta 975, 267–270.
38 J. van der Oost, P. Lappalainen, A. Musacchio, A. Warne, L. Lemieux,
    J. Rumbley, R.B. Gennis, R. Aasa, T. Pascher, B.G. Malmström and M.
    Saraste (1992) EMBO. J., 3209-3217.
39 H. Beinert, D.E. Griffiths, D.C. Wharton and R.H. Sands (1962) J.
    Biol. Chem. 237, 2337–2346.

40 C.R. Hartzell and H. Beinert (1976) Biochim. Biophys. Acta 423, 323–338.

41 W. Froncisz, C.P. Scholes, J.S. Hyde, Y-H. Wei, T.E. King, R.W. Shaw and H. Beinert (1979) J. Biol. Chem. 254, 7482–7484.

42 G.C.M. Steffens, R. Biewald and G. Buse (1987) Eur. J. Biochem. 164, 295–300.

43 E. Bombelka, F.-W. Richter, A. Stroh and B. Kadenbach (1986) Biochem. Biophys. Res. Commun. 140, 1007–1014.

44 P.M.H. Kroneck, W.E. Antholine, J. Riester and W.G. Zumft (1988) Febs Lett. 242, 70–74.

45 W.E. Antholine, D.H.W. Kastrau, G.C.M. Steffens, G. Buse, W.G. Zumft and P.M.H. Kroneck (1992) Eur. J. Biochem. 209, 875–881.

46 P.M. Li, B.G. Malmström and S.I. Chan (1989) Febs Lett. 248, 210–211.

47 G.W. Brudvig, D.F. Blair and S.I. Chan (1984) J. Biol. Chem. 259, 11001–11009.

48 V. Chepuri, L. Lemieux, D.C.-T. Au and R.B. Gennis (1990) J. Biol. Chem. 265, 11185–11192.

49 B.F. Van Gelder and H. Beinert (1969) Biochim. Biophys. Acta 189, 1–24.

50 D.F. Wilson, J.G. Lindsay and E.S. Brocklehurst (1972) Biochim. Biophys. Acta 256, 277–286.

51 C.R. Hartzell, R.E. Hansen and H. Beinert (1973) Proc. Natl. Acad. Sci. USA 70, 2477–2481.

52 C.R. Hartzell and H. Beinert (1974) Biochim. Biophys. Acta 368, 318–338.

53 R. Aasa, S.P.J. Albracht, K.E. Falk, B. Lanne and T. Vänngård (1976) Biochim. Biophys. Acta 422, 260–272.

54 H. Beinert and R.W. Shaw (1977) Biochim. Biophys. Acta 462, 121–130.

55 D.E. Green (1959) in Advances in Enzymol., F.F. Nord ed. Interscience Publishers, Inc., New York vol. 31, pp. 73–129.

56 H. Beinert (1978) Methods Enzymol. 54, 435–445.

57 H. Beinert, in A. San Pietro, Ed. Non-Heme Iron Proteins: Role in Energy Conversion, The Antioch Press, Yellow Springs, Ohio, 1965, pp. 23–42.

58 H.R. Mahler and E. Elowe (1954) J. Biol. Chem. 210, 165–179.

59 D.A. Richert and W.W. Westerfeld (1954) J. Biol. Chem. 209, 179–189.

60 E.B. Kearney and T.P. Singer (1955) Biochim. Biophys. Acta 17, 596–597.

61 H.H. Wickman, M.P. Klein and D.A. Shirley (1965) J. Chem. Phys. 42, 2113–2117.

62 W.E. Blumberg (1967) in A. Ehrenberg, B.G. Malmström and T. Vänngård Eds., Magnetic Resonance in Biological Systems, Pergamon Press, Oxford, pp. 119–133.

63 R. Aasa (1970) J. Chem. Phys. 52, 3919–3930.

64 H. Beinert and R.H. Sands (1959) Biochem. Biophys. Res. Commun. 1, 175–178.

65 H. Beinert and R.H. Sands (1960) Biochem. Biophys. Res. Commun. 3, 41–46.

66 R.H. Sands and H. Beinert (1960) Biochem. Biophys. Res. Commun. 3, 47–52.

67 R.C. Bray (1961) Biochem. J. 81, 196–199.

68 H. Beinert, W. Heinen and G. Palmer (1962) in Enzyme Models and Enzyme Structure, Brookhaven Symposia in Biology: No. 15 229–265.

69 M.P. Hendrich and P.G. Debrunner (1989) Biophys. J. 56, 489–506.

70 H. Beinert and W. Lee (1961) Biochem. Biophys. Res. Commun. 5, 40–45.

71 J.S. Griffith (1961) The Theory of Transition Metal Ions, Cambridge University Press.

72 J.F. Gibson, D.O. Hall, J.H.M. Thornley and F.R. Whatley (1966) Proc. Natl. Acad. Sci. U.S. 56, 987–990.

73 Y.I. Shethna, P.W. Wilson and H. Beinert (1966) Biochim. Biophys. Acta 113, 225–234.

74 Y.I. Shethna, P.W. Wilson, R.E. Hansen and H. Beinert (1964) Proc. Natl. Acad. Sci. USA 52, 1263–1271.

75 W.H. Orme-Johnson and H. Beinert (1969) J. Biol. Chem. 244, 6143–6148.

76 J.C.M. Tsibris, R.L. Tsai, I.C. Gunsalus, W.H. Orme-Johnson, R.E. Hansen and H. Beinert (1968) Proc. Natl. Acad. Sci. USA 59, 959–965.

77 W.H. Orme-Johnson, R.E. Hansen, H. Beinert, J.C.M. Tsibris, R.C. Bartholomaus and I.C. Gunsalus (1968) Proc. Natl. Acad. Sci. USA 60, 368–372.

78 N.R. Orme-Johnson, R.E. Hansen and H. Beinert (1974) J. Biol. Chem. 249, 1922–1927.

79 N.R. Orme-Johnson, R.E. Hansen and H. Beinert (1974) J. Biol. Chem. 249, 1928–1939.

80 N.R. Orme-Johnson, W.H. Orme-Johnson, R.E. Hansen, H. Beinert and Y. Hatefi (1971) Biochem. Biophys. Res. Commun. 44, 446–452.

81 H. Beinert, B.A.C. Ackrell, E.B. Kearney and T.P. Singer (1975) Eur. J. Biochem. 54, 185–194.

82 H. Beinert (1982) in A.N. Martonosi Ed., Membranes and Transport, Plenum Press, New York, pp. 389–396.

83 H. Beinert (1976) Adv. Exp. Med. Biol. 74, 137–149.

84 E.C. Slater (1949) Biochem. J. 44, XLVIII–XLIX.

85 T. Ohnishi, D.F. Wilson, T. Asakura and B. Chance (1972) Biochem. Biophys. Res. Commun. 46, 1631–1638.

86 F.J. Ruzicka, H. Beinert, K.L. Schepler, W.R. Dunham and R.H. Sands (1975) Proc. Natl. Acad. Sci. USA 72, 2886–2890.

87 W.J. Ingledew and T. Ohnishi (1975) FEBS Lett. 54, 167–171.

88 H. Beinert, B.A.C. Ackrell, E.B. Kearney and T.P. Singer (1974) Biochem. Biophys. Res. Commun. 58, 564–572.

89 F.J. Ruzicka and H. Beinert (1974) Biochem. Biophys. Res. Commun.
    58, 556–563.
90 K. Dus, H. DeKlerk, K. Sletten and R.G. Bartsch (1967) Biochim. Bio-
    phys. Acta 140, 291–311.
91 F.J. Ruzicka and H. Beinert (1975) Biochem. Biophys. res. Commun.
    66, 622–631.
92 F.J. Ruzicka and H. Beinert (1977) J. Biol. Chem. 252, 8440–8445.
93 T.P. Singer (1966) Comprehensive Biochem. 14, 127–198.
94 W.J. Ingledew, J.C. Salerno and T. Ohnishi (1976) Arch. Biochem.
    Biophys. 177, 176–184.
95 B.A.C. Ackrell, E.B. Kearney, C.J. Coles, T.P. Singer, H. Beinert, Y.-P.
    Wan and K. Folkers (1977) Arch. Biochem. Biophys. 182, 107–117.
96 H. Beinert (1973) in Iron-Sulfur Proteins Vol. 1 W. Lovenberg, ed.
    Academic Press, Inc., New York and London pp. 1–30.
97 J.Y. Wong, E. Meyer and R.L. Switzer (1977) J. Biol. Chem. 252, 7424–
    7426.
98 Sr.C. Kennedy, R. Rauner and O. Gawron (1972) Biochem. Biophys.
    Res. Commun. 47, 740–745.
99 F.J. Ruzicka and H. Beinert (1978) J. Biol. Chem. 253, 2514–2517.
100 C.D. Stout (1979) Nature 279, 83–84.
101 M.H. Emptage, T.A. Kent, B.H. Huynh, J. Rawlings, W.H. Orme-
    Johnson and E. Münck (1980) J. Biol. Chem. 255, 1793–1796.
102 C.D. Stout, D. Ghosh, V. Pattabhi and A.H. Robbins (1980) J. Biol.
    Chem. 255, 1797–1800.
103 T.A. Kent, J.-L. Dreyer, M.C. Kennedy, B.H. Huynh, M.H. Emptage,
    H. Beinert and E. Münck (1982) Proc. Natl. Acad. Sci. USA 79, 1096–
    1100.
104 M.H. Emptage, J.-L. Dreyer, M.C. Kennedy and H. Beinert (1983) J.
    Biol. Chem. 258, 11106–11111.
105 H. Beinert (1983) Anal. Biochem. 131, 373–378.
106 H. Beinert, M.H. Emptage, J.-L. Dreyer, R.A. Scott, J.E. Hahn, K.O.
    Hodgson and A.J. Thomson (1983) Proc. Natl. Acad. Sci. USA 80, 393–
    396.
107 M.R. Antonio, B.A. Averill, I. Moura, J.J.G. Moura, W.H. Orme-
    Johnson, B.-K. Teo and A.V. Xavier (1982) J. Biol. Chem. 257, 6646–
    6649.
108 C.D. Stout (1988) J. Biol. Chem. 263, 9256–9260.
109 A.H. Robbins and C.D. Stout (1985) J. Biol. Chem. 260, 2328–2333.
110 C.R. Kissinger, L.C. Sieker, E.T. Adman and L.H. Jensen (1991) J.
    Mol. Biol. 219, 693–715.
111 L. Rydén, L.-G. Öfverstedt, H. Beinert, M.H. Emptage and M.C. Ken-
    nedy (1984) J. Biol. Chem. 259, 3141–3144.
112 B.A.C. Ackrell, E.B. Kearney, W.B. Mims, J. Peisach and H. Beinert
    (1984) J. Biol. Chem. 259, 4015–4018.
113 M.H. Emptage, T.A. Kent, M.C. Kennedy, H. Beinert and E. Münck
    (1983) Proc. Natl. Acad. Sci. USA 80, 4674–4678.

114 M.C. Kennedy, M.H. Emptage, J.-L. Dreyer and H. Beinert (1983) J. Biol. Chem. 258, 11098–11005.
115 M.M. Werst, M.C. Kennedy, A.L.P. Houseman, H. Beinert and B.M. Hoffman (1990) Biochemistry 29, 10533–10540.
116 T.D.P. Stack and R.H. Holm (1988) J. Am. Chem. Soc. 110, 2484–2494.
117 T.A. Kent, M.H. Emptage, H. Merkle, M.C. Kennedy, H. Beinert and E. Münck (1985) J. Biol. Chem. 260, 6871–6881.
118 H. Lauble, M.C. Kennedy, H. Beinert and C.D. Stout (1992) Biochemistry 31, 2735–2748.
119 D.W. Plank and J.B. Howard (1988) J. Biol. Chem. 263, 8184–8189.
120 L. Zheng, P.C. Andrews, M.A. Hermodson, J.E. Dixon and H. Zalkin (1990) J. Biol. Chem. 265, 2814–2821.
121 D.W. Plank, M.C. Kennedy, H. Beinert and J.B. Howard (1989) J. Biol. Chem. 264, 20385–20393.
122 S.P. Gangloff, D. Marguet and G.J.-M. Lauquin (1990) Mol. and Cell. Biol. 10, 3551–3561.
123 L. Zheng, M.C. Kennedy, H. Beinert and H. Zalkin (1992) J. Biol. Chem. 267, 7895–7903.
124 R.D. Klausner and J.B. Harford (1989) Science 246, 870–872.
125 E.C. Theil (1990) J. Biol. Chem. 265, 4771–4774.
126 E.A. Leibold and B. Guo (1992) Ann. Rev. Nutr. 12, 325–348.
127 T.A. Rouault, C.D. Stout, S. Kaptain, J.B. Harford and R.D. Klausner (1991) Cell 64, 881–883.
128 M.W. Hentze and P. Argos (1991) Nucleic Acids Res. 19, 1739–1740.
129 V. Guarriero-Bobyleva, M.A. Volpi-Becchi and A. Masini (1973) Eur. J. Biochem. 34, 455–458.
130 R.Z. Eanes and E. Kun (1973) Mol. Pharm. 10, 130–139.
131 L. Zheng, M.C. Kennedy, G.A. Blondin, H. Beinert and H. Zalkin (1992) Arch. Biochem. Biophys., 299, 356–360.
132 D.J. Haile, T.A. Rouault, C.K. Tang, J. Chin, J.B. Harford and R.D. Klausner (1992) Proc. Natl. Acad. Sci. USA 89, 7536–7540.
133 D.J. Haile, T.A. Rouault, J.B. Harford, M.C. Kennedy, G.A. Blondin, H. Beinert and R.D. Klausner (1992) Proc. Natl. Acad. Sci. USA, 89, 11735–11739.
134 T.A. Rouault, C.K. Tang, S. Kaptain, W.H. Burgess, D.J. Haile, F. Samaniego, O.W. McBride, J.B. Harford and R.D. Klausner (1990) Proc. Natl. Acad. Sci. USA 87, 7958–7962.
135 H. Hirling, A. Emery-Goodman, N. Thompson, B. Neupert, C. Seiser and L.C. Kühn (1992) Nucleic Acids Res. 20, 33–39.
136 M.C. Kennedy, L. Mende-Mueller, G.A. Blondin and H. Beinert, Proc. Natl. Acad. Sci. USA, 89, 11730–11734.
137 G. Palmer and R.H. Sands (1966) J. Biol. Chem. 241, 253.
138 D.O. Hall, J.F. Gibson and F.R. Whatley (1966) Biochem. Biophys. Res. Commun. 23, 81–84.
139 P.M.H. Kroneck, W.E. Antholine, H. Koteich, D.H.W. Kastrau, F. Neese and W.G. Zumft (1993) in Bioinorganic Chemistry of Copper

(K.D. Karlin and Z. Tyeklár, eds.) Chapman and Hall, N.Y., (1993) pg. 419–426.
140 M. Kelly, P. Lappalainen, G. Talbo, T. Haltia, J. van der Oost and M. Saraste (1993) J. Biol. Chem. 268, 16781–16787.
141 G. Palmer (1993) J. Bioenerg. Biomem. 25, 145–151.
142 B.G. Malmström and R. Aasa (1993) FEBS Letters 325, 49–52.
143 C. von Wachenfeldt, S. de Vries and J. van der Oost (1994) FEBS Lett. 340, 109–113.
144 E. Emery-Goodman, H. Hirling, L. Scarpellino, B. Henderson and L.C. Kühn (1993) Nucl. Acids Res. 21, 1437–1461.
145 M.C. Kennedy, H. Beinert, T.A. Rouault and R.D. Klausner (1993) J. Inorg. Biochem. 51, 442.
146 J.-C. Drapier, H. Hirling, J. Wietzerbin, P. Kaldy and L.C. Kühn (1993) EMBO J. 12, 3643–3649.
147 G. Weiss, B. Goosen, W. Doppler, D. Fuchs, K. Pantopoulos, G. Werner-Felmayer, H. Wachter and M.W. Hentze (1993) EMBO J. 12, 3651–3657.
148 L. Zheng, R.H. White, V.L. Cash, R.F. Jack and D.R. Dean (1993) Proc. Natl. Acad. Sci. USA, 90, 2754–2758.
149 K.L. Schalinske, P.T. Tuazon, S.A. Anderson, J.A. Traugh and R.S. Eisenstein (1993) FASEB J. 7, A1081.
150 C. Seiser, S. Teixeira and L.C. Kühn (1993) J. Biol. Chem. 268, 13074–13080.

E.C. Slater, R. Jaenicke and G. Semenza (Eds.)
Selected Topics in the History of Biochemistry: Personal Recollections, IV
(Comprehensive Biochemistry Vol. 38) © 1995 Elsevier Science B.V.

Chapter 6

Chorismic Acid and Beyond

FRANK GIBSON

The John Curtin School of Medical Research The Australian National
University, Canberra (Australia)

There was nothing in my family background or early schooling
which hinted at an academic career. I was born in 1923, in
Melbourne, Victoria, as a third generation Australian of Irish
and Scottish extraction. My father, after returning from the
1st World War was employed by the Adelaide Steamship Company and eventually became a foreman stevedore, working on
coal ships. My schooling started in public primary schools and
I showed no great enthusiasm for home work or other scholas-
tic activities. My childhood, with two sisters, was a very happy
one and our parents were not only supportive but somehow
managed to shield us from the worst effects of the Great Depression. We did not have much in the way of luxuries but I do
not recall feeling deprived. In fact my only clear memories of
'hardships' were eating 'bread and dripping' and that the card-
board that I wore inside my shoes, when the soles developed
holes, was not very effective in the wet weather not unknown
in Melbourne. After primary school I was encouraged by my
parents to attend the Collingwood Technical School with the
aim of becoming a draftsman. The two years spent at this
school were sufficient to show that I was not very good at doing
woodwork or mechanical work and that drafting was not my
thing (apart from everything else the sheets of paper always
finished up far from white). It was not surprising therefore

that at the earliest permissible legal age of 14 years I was looking for a job.

Scientific beginnings

The choice of my first job was fortuitous, but was a happy one directly influencing my future. The Bacteriology Department in the University of Melbourne needed to recruit junior laboratory technicians every year and the Head of the Department, Professor H.A. Woodruff, was in the habit of approaching the local technical college rather than, as might be expected, the high school. I suspect that he felt that in this way he was able to assist some 'socially deprived' youths (to the best of my recollection there were no female technicians at that time) to get 'a leg up in the world'.

Probably my only exposure to things scientific at secondary school was some rudimentary physics and there was no outstanding teacher to enthuse the pupils. The only memorable teacher was one who administered excessive corporal punishment and today would have been considered to scar the tender psyches of his pupils permanently! However I do remember having a cheap microscope and avidly reading about this time 'The Science of Life', a monumental work edited by H.G. Wells et al. [1] which was published in what seemed like an infinity of weekly parts. I was so impressed that I had the parts bound but unfortunately the massive volume was eventually lost.

I applied for the laboratory technicians' job a couple of months after I turned 14, possibly stimulated by one of my friends who had started work in the Department earlier in the year. I started work for 15 shillings a week in September, 1937.

The next three years, spent as a technician in the Melbourne Department were very happy ones. While there was a distinctly 'upstairs/downstairs' atmosphere which separated the technical and academic staff, it was certainly not oppressive and it was a relaxed place in which to work. The technical staff were recruited young and started life by washing glassware and running errands such as collecting lunches and specimens in the city for the Public Health Laboratory. When a technician at the top rung of the promotion ladder left the department, everyone moved up a rung. From 'washing up' the progression was through 'plugging' (every test tube and flask was individually plugged with cotton wool) to media making. From the media room the movement was either to the classroom area to prepare material for the practical classes or to assist in one of the research laboratories.

Like the other technicians I was encouraged, in part by having the fees paid, to take classes at the Royal Melbourne Technical College, the usual course being six years of the evening Chemistry Diploma Course, involving about 3 hours for each of 4 or 5 nights a week. There was little or no biology in this course, the emphasis being on chemistry, physics and mathematics, but it did include a short course in Scientific German and it was compulsory to take Leaving English, a subject also a prerequisite for University matriculation. My examination results were not outstanding and I recall failing miserably in Chemistry II, both theory and practical, largely due to the fact that I had smashed a number of my front teeth following an excursion over the handlebars of my bicycle when riding home one night. During the long series of attendances at the Dental Hospital, pipetting was rather a painful experience. As will be seen, the subjects taken in the early years of this course later proved invaluable as a route to a university degree.

My routine was then established; I travelled the 4 miles to the University by bicycle and, as there were no time clocks, the arrival time was somewhat variable, although invariably late. On the other hand, as a group of us were going to 'tech', we never worried about working late if there were jobs to be done and, following a standard meal of fish and chips, we went to our classes. The time between work and 'tech' was spent in the Department and our amusements during this period were unsophisticated and would not be encouraged by a modern Health and Safety Committee. One youthful prank that comes to mind is locking someone in the 37° room and squirting pyridine under the door. Unlike the modern cold or warm room doors there was no way of opening the door from inside and the occupant was usually rather distressed by the time he was released.

By 1940 I had passed through the early stages of my 'apprenticeship' and had graduated to the research laboratory of Dr. (later Professor) Sydney D. Rubbo who was working in collaboration with Adrien Albert on the mode of action of acridines. One of the chores associated with this promotion was setting the coal fire in the laboratory in the winter. I was then encouraged by both H.A. Woodruff and S.D. Rubbo to consider taking a University course but it seemed beyond my reach. However the opportunity arose when bacteriology was being introduced as a new subject in the University of Queensland in the Department of Pathology headed by Professor J.V. Duhig. The intention was to provide courses for medical, dental and science students and the courses were to be run by Dr. (later Professor) D.F. Gray, who was based in the Veterinary School, and a technician. I applied for the technical position and shifted to Queensland where I formed the junior half of the embryonic department which was eventually to become one of the largest in Australia. A major attraction for me was that the University of Queensland would pay my fees and it was possible to attend evening courses, in at least the subjects of the first two years of the three-year science course.

The first problem was the hurdle of matriculation. After much cogitation the University of Queensland pronounced that, if I passed chemistry and physics at matriculation level, they would accept my Melbourne Leaving English and the Chemistry Diploma mathematics together with the (one term of) Scientific German as fulfilling the language and mathematical requirements for matriculation[1]. After a year at night school I fulfilled the necessary requirements and started a science course, the first two full years of which, actually taking 4 years at night, consisted of Chemistry I and II, Physics I, Biology I, Physiology and Biochemistry I, Pure Mathematics I and Statistical and Actuarial Mathematics. The last subject, I have to confess, I never really understood, but I managed to obtain a bare pass mark by committing enough material to memory. I cannot claim that I was a model student; with 4 nights, at least, of classes and full time work, I spent most weekends either surfing or walking around the mountains of south eastern Queensland. For a couple of weeks before the annual examinations I 'swotted' enough to obtain a pass. My lifestyle would not have been possible under the modern regime of continuous assessment. The lecturer who stands out as the most stimulating was Professor 'Ernie' Goddard who, in first year biology, spiced the botany and zoology with lectures on comparative biochemistry. It was C.H. Williams, a lecturer in Pharmacy, who eventually persuaded me that a pass of 51% was not quite good enough and I did obtain Honours in Chemistry II.

My course was interrupted for a year by a period in the Army. My military career was a rather bizarre one. Being a laboratory technician, I was in a reserved occupation. I joined the VDC (Volunteer Defence Force) and was assigned to the 'intelligence section'. In January 1943, just when I had reached the dizzy heights of Lance-Corporal (Acting) despite never mastering the Morse code, there was some administrative mistake and I was called up for the regular army. On reporting to the Brisbane showground, I promptly sent a telegram to my father for permission to volunteer for the Austra-

---

1 It would never have done for the University of Melbourne! In fact the University of Queensland originally decreed that I had to take matriculation mathematics, but later waived this requirement. This was much to my relief because maths was never my strong point.

lian Imperial Forces (AIF) and service abroad. This permission was necessary because I was still a minor and it was immediately given. After undergoing aptitude tests and taking into account that I was a bacteriological laboratory technician, I was placed in the Medical Corps as a stretcher bearer! This was not my idea of military service as my two best friends were in the Air Force and the Commandos, respectively. My enrolment in the AIF led to complications because the University was trying to get me back and now that I was a volunteer, rather than a conscript, it seemed more difficult to get me released, especially as in my youthful enthusiasm I was determined to stay in the Army. My resistance must have considerably hindered the war effort because I was eventually hauled up before the officer commanding the Medical Corps in Queensland to state my case. Apparently it was not too strong as I was sent back to work in the Medical School on leave and after some months discharged from the Army. My discharge certificate states that I served 268 days active service in Australia!

By the end of 1946 I had completed two of the three years (full time) of the Science degree. Having decided that I wished to major in Biochemistry and Bacteriology I made application to be admitted to the University of Melbourne so that I could work in the Bacteriology Department again while completing the degree. Despite my peculiar route to matriculation I was admitted *ad eundum statum* and started my course again in 1947 while working in the 'Bugs School' as a research assistant. I took Biochemistry during the first year and was then given permission to take Bacteriology I and II simultaneously to complete the degree which I did in 1948.

First steps in research

I was then employed as a demonstrator (later Temporary Lecturer) and, having developed an interest in chemotherapy, managed to carry out some research using manometry testing the effects of acridines on oxidase reactions. This was prompted by the notion that the acridines could be metabolic analogues of the flavin coenzymes[2]. This work led to my first scientific publication [2] although some years earlier, being an enthusiastic amateur photographer, I had a note on intensifying under-exposed negatives published in a photographic journal.

In 1949 I married Margaret Burvill who was working with S.D. Rubbo and A. Albert on the mode of action of acridines and 8-hydroxyquinoline and together we carried out a small project on the basis of 'low level' bacterial resistance to streptomycin, the competing claims of those believing in adaptation and in mutation to drug resistance being topical.

The aim now was to obtain postgraduate research experience and at that time there was little chance of doing this in Australia. However the Australian National University had recently been established and while they were soon to take Ph.D. students in Canberra, in the interim they were offering generously funded postgraduate scholarships to be taken in Britain. I applied for a scholarship although, as I did not have an Honours degree, my chances of success were thin. It was therefore disappointing, but no great surprise, when I received a letter of rejection. However, shortly afterwards, I received a further letter from the ANU reversing the previous decision and letting me know that I had been awarded a scholarship and that I was to work for my D. Phil. in Oxford with D.D. Woods who had elucidated the mode of action of the sulphonamides.

Around December 1949 we sailed for England, my sailing orders being to report to Professor Howard Florey in the Sir

---

2 My interest in chemotherapy and biochemistry was no doubt stimulated by the work on acridines in the Department, but also by three outstanding books published about that time. They were 'The Bacterial Cell' by Rene Dubos, 'Dynamic Aspects of Biochemistry' by Ernest Baldwin and 'The Basis of Chemotherapy' by Thomas and Elizabeth Work.

William Dunn School of Pathology in Oxford. Florey alerted me to the fact that I had to be attached to a college before I could be admitted to the University and asked if I had a preference for a particular college. Taken aback, and knowing nothing about the Oxford system or any of the colleges I mumbled that I would rather like to join 'an old one'! Florey's reply was the 'My College is Lincoln, 1429, will that do?' I agreed that Lincoln was suitably mature and Florey arranged for my admittance. He conducted me to the Microbiology Unit of the Biochemistry Department, introduced me to 'D.D.' as he was known to the occupants of the 'temporary' hut which comprised the Microbiology Unit[3]. The Microbiology Unit was like a home from home with June Lascelles and Bill Murrell from Sydney, Dick Batt from New Zealand and Ken Back from Brisbane. D.D. Woods was a meticulous research supervisor and a fount of knowledge on microbial metabolism. My topic of research was to study the metabolism of methionine with particular reference to the source of the methyl group and the role of p-aminobenzoic acid (and folic acid) and vitamin B_{12} in transmethylation.

This period was one of changing emphasis in the study of microbial metabolism from the study of degradative reactions to the study of synthetic reactions, largely brought about by detailed study of the nutrition of various bacterial species and strains. The pioneering work of Beadle and Tatum which showed that mutants of *Neurospora crassa* could be isolated which, as a result of a single gene change, appeared to have lost a single enzymic activity, stimulated the search for nutritional mutants in bacteria. These, when isolated, generally proved more convenient tools for studies on specific reactions and for microbiological assays of metabolites.

In order to study methionine metabolism I was presented

---

3 D.D. was Dr. Woods to his face, although in all official University notices he was 'Mr. D.D. Woods'. After all, his Ph.D. had been awarded at 'the other place'!

with various bacterial strains such as lactobacilli for microbiological assays of methionine and folic acid, and mutants of *Escherichia coli* requiring 4-aminobenzoic acid or vitamin B_{12} (or methionine), or serine (or glycine). My project gradually developed into a search for the source of the methyl group of methionine using washed cell suspensions of *E. coli* mutants. Progress was not fast by today's standards as the range of compounds available commercially was limited and it was often necessary to synthesize biochemicals. For example, at one stage I needed cystathionine to test some now forgotten hypothesis and set about making it both synthetically and by accumulation in a *Neurospora* mutant. These operations involved the unusual (at least in those days) feat of working simultaneously in more than one Oxford Department. At one stage I was growing up large volume *Neurospora* cultures in the Pathology School, carrying out a catalytic hyrogenation reaction in the Dyson Perrin Laboratory and reading the barometer in the Physical Chemistry Laboratory as well as reducing cystine with sodium in liquid ammonia in the Microbiology Unit. I eventually obtained cystathionine by both methods after working night and day, but not before I had decided that the original hypothesis was not tenable, and the cystathionine went into the store of 'rare' chemicals. Such is research! Those chemicals that were commercially available were not always purified to standards necessary for microbiological assay. On one memorable occasion I was testing whether adenine would replace methionine for the growth of a bacterial strain. To my delight it did, but further investigation, by paper chromatography, showed that the 'adenine' contained about 20% methionine!

It was not unusual in the lab, when using nutritional tests, to test the effect of gas mixtures on growth. I noticed that 20% CO_2 inhibited the growth of *E. coli* and that the addition of serine to the growth medium reversed the inhibition. Although this was not understood, I decided to clutch at straws and tested the effect of serine on methionine formation from homocysteine. Using microbiological assays and the appropri- ate mutants it was possible to show that when the methionine was formed from homocysteine and serine, glycine was formed in equimolar amounts indicating that the serine was probably the source of the 1-carbon atom for methionine biosynthesis. While I had the excitement of finding this out, it did not make much of an impact elsewhere as, except for one very brief ab- stract from a meeting, the work was not published [3] until about 7 years later, by which time it was generally accepted that serine was the source of the methyl group of methionine.

Early in 1953 I faced my oral examination by Professor Sir Rudolph Peters, head of the Biochemistry Department in Ox- ford and Professor H.A. Krebs, at that time from the Biochem- istry Department in the University of Sheffield. I remember the latter saying, as he handed back a copy of my thesis 'All this without manometry?' My thesis was passed and we sailed for home. Margaret also had completed a D. Phil., although she was awaiting the result of her oral examination. She had obtained a grant to read for her degree under Professor Sir Cyril Hinshelwood, head of the Physical Chemistry Labora- tory. This was the era of conflicting views on the roles of adap- tation and mutation in the development of drug resistance and Hinshelwood was the main proponent of the adaptation hy- pothesis. The opportunity to work in the 'PCL' was therefore an exciting prospect. In the event, however, it has to be said that Margaret's stay in the PCL was not a particularly happy one. While she did get enough work done to obtain her degree the limited range of techniques she could apply to her problem (mainly measurement of growth curves) left her dissatisfied with the results and she never attempted to publish them.

Before we left Oxford I had offers of two academic positions. One was a Fellowship in the Microbiology Department of the John Curtin School of Medical Research in the Australian Na- tional University in Canberra and the other was a Senior Lec- tureship in the Bacteriology Department of Melbourne Uni- versity where S.D. Rubbo was now Head. The former was a research only position while the latter was a teaching position involving Science, Medical, Dental and Agricultural students with research to be fitted in whenever possible. I had always assumed that I would have a teaching career and also the position in Canberra was to work in the field of virology about which I knew very little. I chose Melbourne. Although the Melbourne Department was a large one, the teaching load was relatively light and Rubbo was keen that the Department should become noted for its research effort. I was to be given a free hand in the choice of a research field.

Aromatic biosynthesis

Thus I settled in to the routine of preparing and delivering lectures and practical courses and starting on an independent research career. It was obvious that the use of mutants for solving biochemical problems was a field about to explode and the choice finally narrowed down to one of two problems. The first was the question of nitrogen fixation which seemed (and later proved) ripe for the use of mutants and the other was to follow up an observation I had made as a 'Saturday morning experiment' just before I left Oxford. This derived from my interest in one-carbon transfers. It seemed that in one of the few biosynthetic pathways postulated at the time the conversion of anthranilic acid to indole in the tryptophan pathway could be such a reaction. I therefore took a mutant strain of *E. coli* which responded to tryptophan but not indole and was presumably blocked in the metabolism of indole, and incubated a cell suspension with, among other things, anthranilic acid and serine. The serine stimulated the formation of indole somewhat, even in the presence of glucose.

Repeating this experiment formed the basis of my future research activities and it was one that I should not have carried out! One of my life-long failings has been to rush to the laboratory bench when I should have rushed to the library.

This may relate back to a school excursion, when I was about 13 years old, to a munitions factory in Bendigo, Victoria. We were shown a laboratory and above the door was a large sign which read 'ONE EXPERIMENT IS WORTH A THOUSAND EXPERT OPINIONS'; perhaps not such good advice when dealing with explosives! Had I read the literature on the conversion of anthranilic acid to indole I would have known that Yanofsky [4] had shown that the carboxyl group of anthranilic acid was lost during the conversion and that probably a two-carbon fragment derived from ribose was utilized.

However, as is not uncommon in my experience, an ill-considered experiment is not always without reward. In this case the cell suspension in the control reaction mixture containing glucose and ammonium ions alone formed almost as much indole as the various test mixtures. This suggested that the entire biosynthetic pathway leading to indole could be investigated using cell suspensions. My interest in the mode of anti-bacterials led to experiments on the effects of antibiotics on the pathways [5–7]. These in turn led to the study of the biochemistry of the pathway, as there were obviously new intermediates to be found although the pathways of aromatic biosynthesis were being actively studied by B.D. Davis and D.B. Sprinson and their colleagues. The first new compound we positively identified was 1-o-carboxyphenyl-amino1-deoxyribulose (Fig. 1) [8, 9], the phosphorylated form of which is an intermediate between anthranilic acid and indoleglycerol phosphate.

During a period of study leave in Stanford I was introduced to the use of cell-free preparations for studies of enzymic reactions and enzyme purification [10, 11] while in Melbourne the

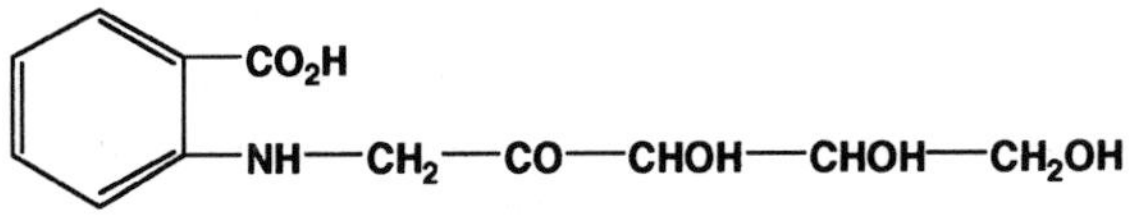

Fig. 1. Structure of 1-o-carboxyphenyl-amino1-deoxyribulose.

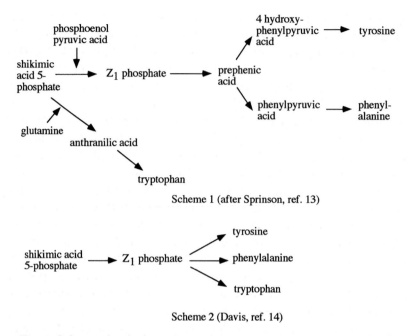

Scheme 1 (after Sprinson, ref. 13)

Scheme 2 (Davis, ref. 14)

Fig. 2. Schemes for the branching of the aromatic pathway. From [15].

work with washed cell suspensions continued. Colin Doy and
Jim Pittard showed that a variety of phenolic compounds were
formed by strains blocked in the early stages of tryptophan
biosynthesis [12]. The significance of these compounds was not
clear and this line of work was dropped for some time in order
to tackle the problem of the position and nature of the branch
point(s) leading from the 'common' aromatic pathway to the
individual pathways for the various amino acids and vitamins.

It was apparent that the common pathway branched at
some point(s) leading to the specific pathways to phenylalanine, tyrosine and tryptophan, as well as the aromatic vitamins
p-aminobenzoic acid (and thence to folic acid) and *p*-hy-
droxybenzoic acid, a growth factor of unknown function. However, the points of branching of the common aromatic pathway

was far from clear. Sprinson [13] suggested that the tryptophan pathway branched at shikimate-5-phosphate, while B.D.
Davis [14] suggested that the pathways to the three aromatic
amino acids diverged at 3-enolpyruvyl-shikimate-5-phosphate
(or Z_1-phosphate as it was then known) (Fig. 2).

The results of experiments in which cell extracts of mutants
blocked in the last known reactions of the common pathway
and a survey of the available evidence convinced us that the
pathways could be visualized as in Fig. 3 [15], in which there
was an undiscovered compound (compound X) which formed
the true branch point leading to the aromatic amino acids and
possibly 4-aminobenzoate and 4-hydroxybenzoate.

It is one thing to postulate the existence of a compound and
another to confirm its existence. It was decided to isolate a
mutant which, as a result of three mutations, required tryptophan, tyrosine and phenylalanine for growth in the hope that
some evidence could be obtained for the accumulation of compound X. This appeared to be a long shot since, as can be seen
from Fig. 3, such mutants should accumulate prephenate or

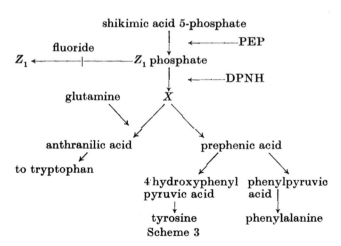

Fig. 3. Proposed scheme for the branching of the aromatic pathway. From
[15].

the phenylpyruvic acids. The starting strain was an auxotroph of *A. aerogenes* unable to convert anthranilate to tryptophan[4]. This strain was irradiated with ultraviolet light and a double auxotroph isolated, following penicillin selection, which required tryptophan plus tyrosine. This mutant was further mutated to give the triple mutant *A. aerogenes* 62-1, requiring tryptophan, tyrosine plus phenylalanine.

The first experiments [16] with strain 62-1 were carried out by Margaret Gibson and the strain grown on limiting tryptophan medium gave exciting results. Cell extracts incubated with shikimate and glutamine metabolized the shikimate to anthranilate. However, if glutamine was omitted from the reaction mixture, a new compound was formed which, like anthranilate, could be extracted into ether but had a different ultraviolet spectrum (Fig. 4). It was also shown in preliminary experiments that the new compound could be formed from 3-enolpyruvyl-shikimate-5-phosphate, converted enzymically or non-enzymically to 4-hydroxybenzoate and phenylpyruvate, non-enzymically to prephenate and enzymically to 4-hydroxybenzoate [17]. It seemed that the elusive compound X, the branch point compound in the aromatic pathway, had been found[5].

The immediate problem was to isolate and determine the structure of compound X and its instability under mild conditions indicated that it might be difficult to isolate. On this assumption ether extracts of culture supernatants were applied to a wide variety of chromatographic columns. After

---

4 A strain able to form anthranilate was chosen in order to facilitate the assay of any compound X formed.

5 The preliminary experiments were published as a note in Biochimica Biophysica Acta and as a result I received a letter which would be considered very unusual today. This came very shortly after we had determined the structure of the new compound and was from Professor H. Plieninger of Heidelberg. He wrote that he and his colleagues had discussed the properties of our compound as described in the preliminary note and predicted the structure, which he drew correctly. He then went on to request our permission for him to attempt to synthesize the hypothetical compound!

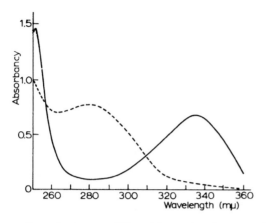

Fig. 4. The formation of compound X and its conversion to anthranilic acid. The spectra of ethyl acetate extracts of: ...... the conversion of shikimate to compound X; —— the further conversion of compound X to anthranilic acid. From [16].

much trial and error it transpired that simply passing culture supernatants through an ion-exchange column, elution with NH_4Cl, addition of barium acetate and precipitation with ethanol gave the barium salt which could be further purified by additional precipitations [18].

The structure of the new compound was determined largely by spectroscopy. The ultraviolet spectrum indicated a cyclohexadiene carboxylic acid structure and the infrared spectrum of the barium salt showed that the enolpyruvyl group of 3-enolpyruvyl-shikimate-5-phosphate had been retained. The remaining problem was the position of the double bonds of the conjugated hexadiene. This was solved after consultation with L.M. Jackman an NMR spectroscopist who had recently been appointed as head of the Chemistry Department in Melbourne University and who had just set up the first NMR spectrometer in the Department. Examination of the barium salt of compound X immediately determined the position of the double bonds and confirmed the other structural features [19] (Fig. 5).

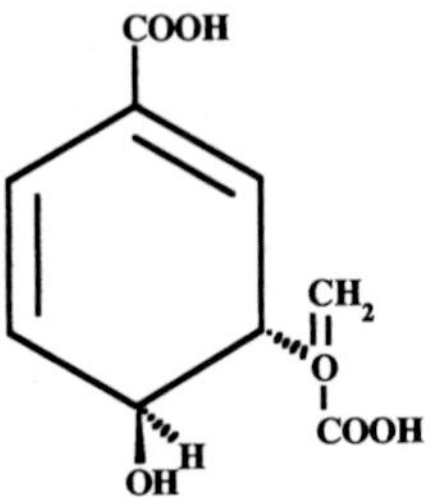

Fig. 5. Structure of chorismic acid.

Compound X was now respectable enough to be given a trivial name and Canon W. Burvill of Brisbane, after having been
given the concept of a branching metabolic pathway, suggested that 'chorismic acid' would be appropriate. His letter of
Oct. 25, 1962, reads, in part:

> I understand what you really do require is a 'trivial' name for 'Compound
> X' in which the sequence diverges – like a place on the 'family tree' –
> towards anthranilic acid in one direction and prephenic acid in another.
> Assuming that I am right so far, I would suggest that – Apo-chorismic –
> might do. It is an adjective one would be quite justified in forming from the
> Greek verb 'apo-chorizo' found perhaps not in strictly classical Greek, but
> in the Septuagint, i.e. the Greek Old Testament and meaning to 'sepa-
> rate', 'part asunder' and much later, of course, in the New Testament, Acts
> 15 39 telling of the quarrel between St. Paul and Barnabas over the young
> (St.) Mark. The English translation runs: – 'And there arose a sharp con-
> tention, so that *they parted asunder* one from the other...'. There is an-
> other 'Classical Greek' word, where we don't have the apo which conveys
> the idea of 'away from' (amongst many other shades of meaning). The
> word is 'chorismic', my first suggestion without the 'apo'. It means *inter
> alia*: – a being separated, separations, partings, *etc.*'

This name of chorismic acid was adopted and the stage was
now set for a detailed examination of the branching of the
aromatic pathway. Over the next few years this problem more
than fully occupied the lab. Around this time the group was
fortunate enough to have attracted a number of gifted graduate students including C.H. Doy, A.J. Pittard, G.B. Cox, I.G.

Young and R.G.H. Cotton, but tragically lost the participation of Margaret Gibson through an illness which prevented her from working again.

Studies were soon made on the conversions of chorismate to anthranilate [20], 4-aminobenzoate [21] and prephenate [22]. The last proved of particular interest as it provided the explanation as to why strain 62-1 accumulated chorismate rather than prephenate, as might have been expected. Prephenate synthase was being purified on chromatographic columns and was found to be eluted in two peaks. The opportunity was taken to assay the fractions for the ability to convert prephenate to phenylpyruvate and 4-hydroxyphenylpyruvate. To our surprise, it was found that the peaks of the latter activities coincided with the two peaks observed for prephenate synthase. It seemed, and was later confirmed [22–28], that we had two multifunctional enzymes, chorismate mutase-prephenate dehydratase and chorismate mutase-prephenate dehydrogenase. These findings explained why the triple auxotroph *A. aerogenes* 62-1 accumulated chorismate rather than prephenate. This was another stroke of luck as not all bacteria have these multifunctional enzymes.

A change of direction

Soon after the discovery of chorismic acid several significant events occurred. I was appointed to a Personal Chair in Chemical Microbiology in the University of Melbourne. Our work was being supported by grants from the Australian National Health and Medical Research Council and the Robertson Committee which was the precursor of the Australian Research Grants Committee but I decided to apply for a grant for equipment and running expenses from the U.S. Public Health Service. This was a period when the funding for research in the United States was generous to the extent that applications from abroad were accepted and often funded. The application was successful and the grant was extremely generous by Australian standards and included funds for a 40 litre fermenter which is in use to this day.

Early in 1966 I was summoned to breakfast at the Menzies Hotel in Melbourne by Sir Hugh Ennor who was Dean of the John Curtin School of Medical Research at the Australian National University in Canberra. I duly attended wondering what was about to befall me and was surprised to be asked if I would be interested in becoming head of the Department of Biochemistry in the John Curtin School. This was a shock as I had always considered myself a microbiologist rather than a biochemist, and knew nothing about the Department. However, as the advantages of the position were outlined, its attractiveness grew. The Department was small with three groups led by R.L. Blakley, J.F. Morrison and H. Rosenberg. The School taught postgraduate students only, there was very little administration and the School was amply funded by direct grants from the Commonwealth Government. Staff members, therefore, did not have to spend time making out research grant applications. Furthermore, if any of the existing group in Melbourne wished to go to Canberra, this could be arranged. With some trepidation I conveyed the news to members of the lab and to my surprise and delight they all expressed interest in moving. Those who had finished their PhD work were made Research Fellows and graduate students transferred to the Australian National University. So seven of us together with families moved to Canberra and to the best of my knowledge no one ever regretted the move. We found that Canberra, with its small population and excellent working, schooling and sporting facilities was an ideal place to live and bring up our two daughters. In the ensuing quarter of a century the population has more than quadrupled from the 80,000 when we arrived but the living conditions still have many advantages over those in the big cities. The migration of almost the whole group to Canberra meant that there was little break in the continuity of the work, although conditions were rather crowded in the Department until Ray Blakley eventually migrated to the United States.

Menaquinone and ubiquinone: biosynthesis and function

Other pathways leading from chorismate began to emerge. ^{14}C-shikimate was shown to be incorporated into menaquinone (vitamin K) and ubiquinone in *E. coli* [29, 30]. The incorporation of radioactivity into ubiquinone was suppressed by the addition of unlabelled 4-hydroxybenzoate confirming the role of the latter compound in ubiquinone biosynthesis. The incorporation of the label into menaquinone indicated the route of biosynthesis of the naphthoquinone nucleus for the first time.

About this time we stopped working with *A. aerogenes*, which had originally been chosen as an organism which grew vigorously in synthetic media and shifted to derivatives of *E. coli* K-12. Jim Pittard, who had spent several years working with E. Adelberg on bacterial genetics, had returned to the Department and it was obvious that bacterial genetics was to play an important role in biochemical investigations.

With the aim of studying the pathway of ubiquinone biosynthesis mutants lacking ubiquinone were sought. There was no direct method of selecting for ubiquinone-deficient mutants. On the assumption that ubiquinone was essential for oxidative metabolism it was decided to isolate mutants which could grow fermentatively on glucose but unable to grow with a non-fermentable carbohydrate (malate) as sole bulk source of carbon. Graeme Cox set about the tedious task of mutagenizing *E. coli*, isolating such mutants and then extracting the quinones from 1 litre cultures. In the original experiment 60 mutants were examined [31] and one strain was found to be devoid of ubiquinone. As a bonus, one other strain was found to be deficient in menaquinone biosynthesis. The genes affected

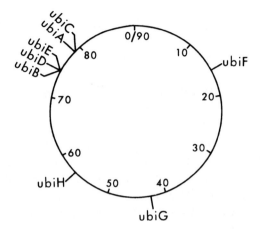

Fig. 6. Genetic map and intermediates in the pathway of ubiquinone biosynthesis established by using ubiquinone-deficient mutants of E. coli. From
[36].

in the mutants were mapped by interrupted mating and transduction.

The pathway of biosynthesis of ubiquinone was to occupy the attention of the group, particularly I.G. Young, for several years. The intermediates in the pathway were postulated by others following extraction of quinones from the photosynthetic bacterium *Rhodospirillum rubrum*, but the work with *E. coli* allowed the pathway to be defined more clearly, the relevant genes mapped [32–38] (Fig. 6) and the accumulation of sufficient quantities of the intermediates for studies on the enzymic reactions of the pathway.

The availability of ubiquinone mutants allowed a new approach to the study of ubiquinone function which had previously relied on extraction of quinones from membranes and reconstitution of the extracted membranes, usually with short chain analogues of ubiquinone. With mutants it was possible to obtain membrane preparations lacking ubiquinone, with or without non-functional intermediates. Such studies led to the conclusion that ubisemiquinone functioned in at least two positions in the electron transport sequence and that ubisemiquinone played an important role [39–41] (Fig. 7). This scheme

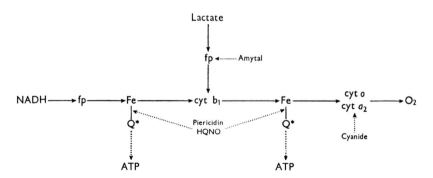

Fig. 7. Scheme for ubiquinone function in E. coli. *Abbreviations: fp, flavoprotein; cyt, cytochrome. From [39].*

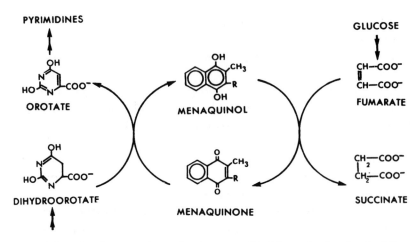

Fig. 8. The role of menaquinone in the anaerobic biosynthesis of pyrimidine. From [43].

served as a precursor of the hypothesis of the protonmotive ubiquinone cycle advanced by Mitchell [42].

Similarly, a menaquinone mutant was used to study the function of this quinone [43]. It appeared that menaquinone was only essential in *E. coli* when the organism was growing under anaerobic conditions. It was then required for the oxidation of dihydroorotate to orotate during pyrimidine biosynthesis (Fig. 8).

Phenolic compounds and iron transport

As mentioned earlier, phenolic compounds are formed by cell suspensions of *A. aerogenes* [12]. These included 4-hydroxybenzoate, 2,3-dihydroxybenzoate and 3,4-dihydroxybenzoate but some years elapsed before the latter compounds were investigated in detail. The last compound was shown to be formed enzymically from chorismate and suspected to be involved in iron metabolism since the enzyme(s) concerned in

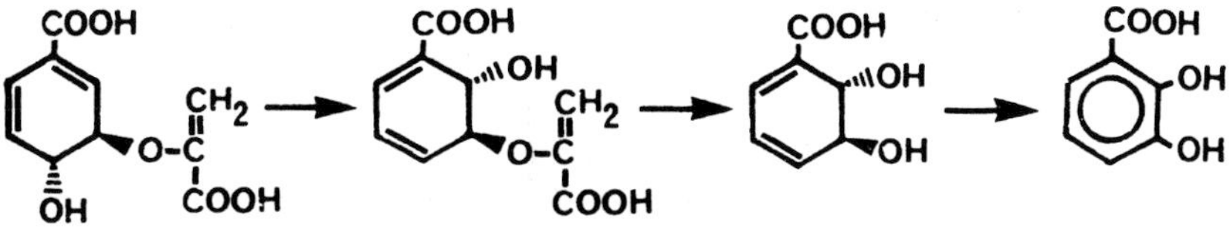

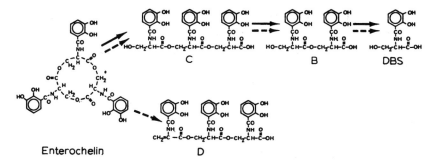

Fig. 9. The biosynthesis of 2,3-dihydroxybenzoic acid from chorismic acid. From [49].

its formation was repressed by iron, but no other evidence for function was shown [44]. However it had been shown that 2,3-dihydroxybenzoate was an important metabolite in that it would replace the requirement for Fe shown by some multiple aromatic auxotrophs of *E. coli* [45] and was formed in three steps from chorismate [46–50] (Fig. 9).

When cells of *E. coli* were grown on iron-deficient media, five phenolic compounds could be isolated from culture supernatants [51, 52]. The relationships between these compounds were determined (Fig. 10) and three clustered genes shown to be concerned with the conversion of 2.3-dihydroxybenzoate into the cyclic trimer of 2,3-dihydroxybenzoylserine (enteroch-

Fig. 10. The conversions of enterochelin and related compounds. Compounds B, C, D and 2,3-dihydroxy-N-benzoylserine (DBS) were identified after formation from enterochelin. The solid arrows indicate enzymic conversions and the broken arrows indicate spontaneous breakdown products. From [51].

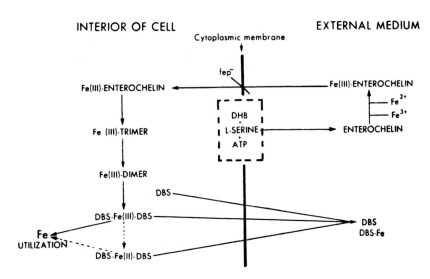

Fig. 11. Proposed scheme for the function of enterochelin and its hydrolysis products in iron transport in E. coli. *Following the synthesis of the avid chelator, enterochelin, from 2,3-dihydroxybenzoate (DHB), serine and ATP, iron is sequestered from the external medium and the iron-enterchelin complex transported into the cell where enzymic degradation of the enterochelin to less avid complexes (DBS, 2,3-dihydroxy-N-benzoylserine) allows utilization of the iron. From [53].*

elin). Enterochelin was shown to be an avid iron-binding compound and further experiments [53, 54] allowed a scheme to be proposed for the function of enterochelin and the related compounds in iron transport (Fig. 11).

The observation was made during the above work that, while multiple aromatic auxotrophs of *E. coli* had an absolute requirement for 2,3-dihdyroxybenzoate, there was no such requirement in the case of *A. aerogenes* 62-1. This suggested the possibility that the latter strain produced an alternate iron-binding compound and examination of culture supernatants showed the presence of a hydroxamate which was isolated and its structure determined (Fig. 12). The compound was given the trivial name aerobactin [55].

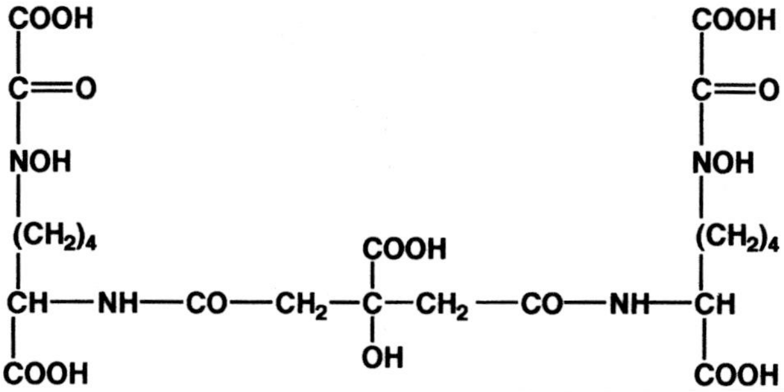

Fig. 12. Structure of aerobactin.

ATPase and oxidative phosphorylation

During the work on the function of ubiquinone we realized that among the discarded mutants which would not grow on non-fermentable carbohydrates (Suc⁻ mutants) there might be some affected in oxidative phosphorylation. At the time there was very little work on ATPases or oxidative phosphorylation in *E. coli*. The P/O ratios obtained with cell free preparations were very low and there was no evidence for the F_0F_1 structure known in mitochondria. The apparent lack of interest in *E. coli* was puzzling considering the rapid advances that had been made before 1970 in fields such as protein synthesis and DNA synthesis using a combination of microbial genetics and biochemistry. It seems probable that most workers in the bioenergetics field had backgrounds which did not lead them to appreciate the power of what is now called molecular biology[6].

A graduate student, Janet Butlin, was given the task of ex-

6 This was brought home to me when in 1974 I attended a Gordon Conference on Energy Coupling and during a discussion on a paper on ubiquinone function in yeast I made the point that the same approach as used with *E. coli*, namely the study of Suc⁻ mutants, could be used to advantage. This comment elicited no reaction at the time, but during the coffee break a young participant came up to me and said 'We of the underground salute you' and promptly melted back into the crowd.

amining the ATPase activities of Ubi^+, Suc^- mutants. She identified two strains in which the membranes lacked ATPase activity. One of these was examined in detail and found to have normal levels of electron transport but no detectable levels of oxidative phosphorylation. The gene concerned was mapped at about minute 74 on the *E. coli* chromosome and designated *unc*A. In the 1971 paper describing these findings [56] a statement was made which proved to be prophetic:

> 'The use of bacteria with their simpler cellular organization than eucaryotic cells, and of *Escherichia coli* in particular, with its amenability to genetic manipulations, seems a promising experimental system for a combined genetic and biochemical approach to the problem of coupling of phosphorylation to electron transport.'

The potential significance of these observations was obvious and the ramifications occupied an increasing proportion of the activities of our group until the present time with Graeme Cox playing a prominent role.

Another early observation which proved to be invaluable in the search for further *unc* mutants was that the aerobic growth yield (the final yield of cells grown on a limiting level of glucose under conditions of aeration) of the *unc*A mutant was intermediate between that shown by a normal strain grown under aerobic or anaerobic conditions. This was to prove a very simple and rapid test for the initial screening of Suc^- strains for possible *unc* strains.

A second mutant with an 'uncoupled' growth yield and which could not carry out oxidative phosphorylation was found to have normal ATPase activities and the gene affected (*unc*B) was shown to map in the same region as the *unc*A gene [57]. Utilizing a known method of stripping the Mg,Ca-ATPase from bacterial membranes by low-ionic-strength buffer it was possible to fractionate the cell membranes from the *unc*A and *unc*B strains. Reconstitution experiments showed that the former strain was affected in the ATPase which could be stripped off the membranes while the *unc*B strain was affected in an 'intrinsic' membrane protein. Eventually the nomenclature used by workers with mitochondria was adopted and the two strains would be described as affected in the F_1-ATPase and F_0 portions of the F_0F_1-ATPase or ATP synthase respectively.

The experiments with membrane preparations were assisted by the chance observation while investigating the effect of protease inhibitors. It was found that the addition of 4-aminobenzamidine to the low ionic strength buffer used to wash the membranes resulted in the F_1-ATPase being retained on the membrane preparations while other proteins were washed off. Thus the specific activity of the membrane ATPase was increased and subsequent washes without 4-aminobenzamidine yielded more active F_1 preparations [58]. The cause of this effect is not known, but is not due to protease inhibition [59].

The search then began for more unc mutants and a number of other groups were now working in the same area. The techniques for the isolation of such mutants was now clear and the problems were to identify and characterise mutants in new unc genes. Genetic complementation tests were used after developing the technique of putting unc genes on plasmids [60] and the relationship between the various mutant unc alleles and the known F_0F_1-ATPase genes was investigated, initially by 2-dimensional electrophoresis [61,62]. Two further unc genes C and D were identified [63,64] and all the known unc genes appeared to be closely linked; using Mu-phage induced polarity mutants it was possible to show that the genes were part of an operon and that the order of the known genes was uncBADC [65]. To these was added the uncE gene coding for a component of the F_0 [66] and then the uncF [67] and uncG [68] genes.

The order of the genes comprising unc operon and identification of the corresponding proteins was now well advanced (Table 1). The obvious absentee from the table is the gene coding for the δ-subunit of the F_1 and mutants in which the δ-subunit was affected were finally described by other workers

TABLE I
unc Genes and relation to subunits (1980)

Gene	Subunit location	Subunit
uncB	F_0	a-subunit
uncE	F_0	c-subunit (DCCD-binding protein)
uncF	F_0	b-subunit
uncA	F_1	α-subunit
uncG	F_1	γ-subunit
uncD	F_1	β-subunit
uncC	F_1	ε-subunit

several years later [69]. In the meantime a major advance had been made by the sequencing of the *unc* operon. Soon after we had started DNA sequencing, the sequence of the entire operon was reported from other laboratories (see Ref. [70]). This revealed the sequence of the gene (*uncH*) coding for the δ-subunit and an unsuspected gene (*uncI*) at the beginning of the operon which does not code for a subunit of the F_0F_1-ATPase and the function of which is still unknown.

Our attention had also turned to the route of assembly of the F_0F_1-ATPase complex and an observation was made which was difficult to reconcile with the simplistic view that the F_0 was inserted into the plasma membrane and a preformed F_1 from the cytoplasm attached. The finding was that if a mutant strain formed no normal or altered protein for the β-subunit of the F_1-ATPase then the b-subunit of the F_0 was not inserted into the membrane, although its gene was normal, and the resulting membranes were impermeable to protons. Other mutants were found to contain abnormal amounts or range of subunits in their membranes and consideration of all the available data allowed the formulation of scheme for the assembly of the complex in normal membranes [71] (Fig. 13).

The details of the mechanism of action of ATP synthase has been an intractable problem; the nature of the proton channel and the mechanism by which proton translocation drives ATP

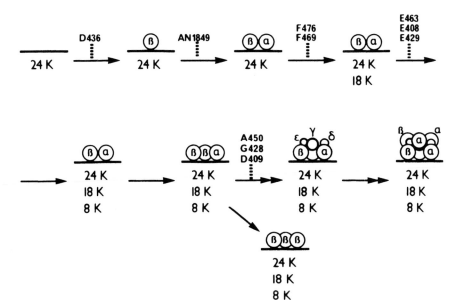

Fig. 13. Proposed sequence for the assembly of the F_oF_1-ATPase complex. The scheme is based on the presence or absence of particular subunits of the complex in various mutant strains of E. coli and the other properties of the membranes. The 24 K subunit is now designated the a-subunit, the 18 K the b-subunit and the 8 K the c-subunit. From [71].

synthesis have only recently been amenable to investigation at the molecular level. A model for the structure and function of ATP synthase was proposed by Graeme Cox [72] who took into account kinetic studies, mutations affecting the F_0 sector of ATP synthase and possible analogy with flagellar motion. This hypothesis was refined after comparison of the predicted secondary and tertiary structures of 'a-subunits' of *E. coli* and human, bovine and yeast mitochondria. This, together with site-directed mutagenesis experiments, led to incorporation of the α-subunit in the model [73]. The novel feature of the model is that there is a core consisting of the *a*-subunit and two copies of the *b*-subunit with the minor subunits of the F_1 (γ, ε and δ) attached. It is proposed that this core rotates within a ring of c-subunits and, with the asymmetry caused by the minor subunits, forces conformational changes resulting in release of the newly formed ATP from the active site on the F_1. A diagrammatic representation of the structure of the ATP synthase complex is given in Figure 14.

The energy required, and the rotation, is provided by proton passage through the F_0 involving charged residues on the α-subunit interacting sequentially with the aspartic acid-61 residues on the ring of c-subunits (Fig. 15).

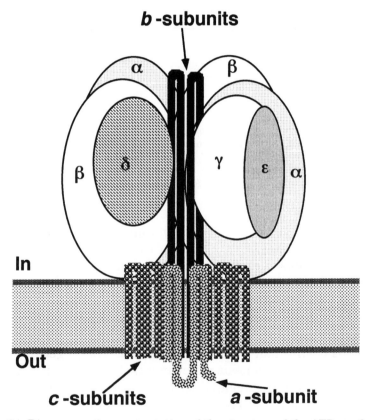

Fig. 14. Diagrammatic representation of the structure of the ATP synthase complex. From [85].

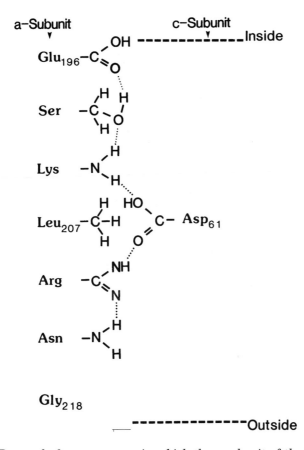

Fig. 15. Proposal of a proton pore in which the a-subunit of the E. coli F_0 interacts with the Asp-61 of the c-subunit. From [73].

Site-directed mutagenesis of the a-subunit refined the model and allowed definition of two of the charged residues (arginine-210 and glutamate-219) concerned in proton translocation [74,75] and the isolation of a mutant by Cain and Simoni [76], in which histidine-245 was altered, suggested that these three amino acids were the important residues concerned in proton translocation (Fig. 16).

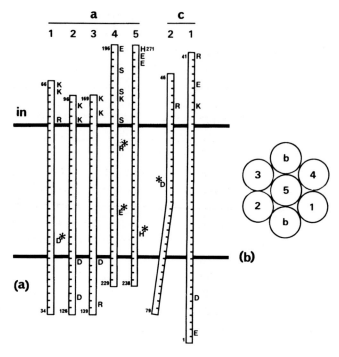

Fig. 16. Proposed transmembrane helices of the a- *and* b-*subunits of the* E. *coli ATP synthase complex. From [77].*

A further series of experiments in recent years [77–79] has provided more information about the positioning of the trans-membrane helices, their arrangement in the F_0 and the inter-action between different groups within the a-subunit. During the latter part of this time I have been an interested bystander and have noted that the accumulating evidence (see Ref. [80]) favours the model, which now seems to have been accepted by other workers [81,82].

A trail of research

I have been fortunate in that, once the initial decision was made to study tryptophan formation, there has never been a shortage of problems to study. The difficulties of research direction with a relatively small research group has only been one, particularly in the early days, of deciding which problems were soluble with the existing facilities and techniques and determining priorities. All my research interests for about 40 years can be related back to the discovery of chorismic acid and the course of the journey may be depicted as in Figure 17.

The isolation of chorismic acid opened a Pandora's box and the investigation of the pathways of biosynthesis of the aromatic amino acids was an obvious starting point for the study of its metabolism. This work led to the discovery of the multifunctional enzymes chorismate mutase-prephenate dehydrogenase and chorismate mutase–prephenate dehydratase. The study of the aromatic 'vitamins' was more challenging and the observation that labelled shikimate was converted to both

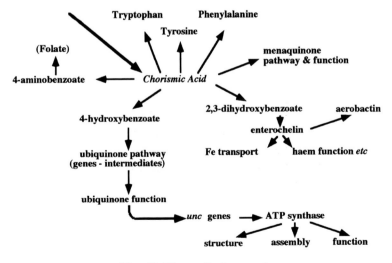

Fig. 17. The trail of research.

ubiquinone and to menaquinone stimulated work on their pathways of biosynthesis, particularly as the source of the latter compound was previously unknown.

Our interest had been aroused previously in the phenolic compounds formed by cell suspensions and when it was found that these were formed from chorismate, their biosynthesis, the genes concerned and eventually their function proved fruitful sources of study. Aerobactin was discovered, as described above, as a sideline to the investigation of the phenolic chelating compounds.

The availability of a variety of mutants unable to form ubiquinone led naturally to a study of their function and the realization that among our store of strains unable to grow on succinate but having normal levels of ubiquinone there were probably useful mutants for studying other aspects of bacterial respiration. The first strain examined in detail had a lowered aerobic growth yield and lacked Mg, Ca-ATPase activity and we found ourselves in the murky waters of oxidative phosphorylation. This was to remain my main pre-occupation for almost 20 years and the fact that the study of 'ox phos' in *E. coli* has now contributed so much to our still incomplete understanding of this intractable problem is a source of considerable satisfaction.

Miscellanea

One of the great attractions of an academic life is the opportunity to travel to conferences and to work for periods overseas. My first period of study leave was made possible by a Carnegie Foundation Travel Grant and I spent about 6 months in California. The first 10 weeks were spent at the Hopkins Marine Station at Pacific Grove where I sat in on the course in General Microbiology given by C.B. van Niel. This memorable course was based on the history of the subject and it is a measure of van Niel's abilities as a teacher that he could hold the attention of an audience for over 8 hours (with breaks for coffee and lunch), with a lecture on respiration. This was all the more remarkable in that the venue was a laboratory with the students perched on uncomfortable stools and at the end of the day he had just reached 1934!

The rest of the time was spent in the laboratory of Charles Yanofsky in Stanford where Margaret Gibson and I worked on indoleglycerol and tryptophan synthetases [10,11] and gained valuable experience in the preparation and use of cell-free preparations. Later, other short periods of study leave were spent in the Chemical Microbiology Unit in Cambridge with Ernest Gale and in the Rockefeller University in New York with Ed Reich. The longest period away from the Department was as a recipient of the Newton-Abraham Visiting Professorship at Oxford during 1981–1982 which, by coincidence, carried with it a Fellowship in Lincoln College to which I had been attached for my post-graduate work. I have to say that I relished the contrast of the two situations since, as a married graduate student spending long hours in the laboratory there was little contact with the College. The Newton-Abraham Visiting Professorship carried with it a residence in sharp contrast to that provided for another Visiting Professorship held by G.W. Beadle many years before [83]. During my year in Oxford I worked again in the Microbiology Unit and the Biochemistry Department. I had re-married and Robin and I, together with our infant son, lived in luxury. It was during this period that, for my 60th birthday, Robin presented me with a microcomputer and a new world was revealed.

Such periods of study leave were not only valuable scientifically but resulted in lasting friendships. The contacts with overseas workers have not only resulted from travel from Australia, but a number of workers from abroad, including F.L. Crane, R.K. Poole, E. Reich, A.E. Senior, S. Silver and C. Yanofsky have spent time working in our laboratory. My contacts on the international scene have been most enjoyable, although I have been aware of a couple of instances of plagiarism, the frequency of which seems to have increased by the highly competitive atmosphere in which we now work.

Despite years of practice lecturing to undergraduates, I have never considered myself an exciting lecturer. However I must have had something of interest to say as I have been invited to give a number of named lectures in Australia and the U.K. Probably the most memorable was the 1981 Leeuwenhoek Lecture of the Royal Society which I was asked to deliver three times – in London, Manchester and Durham. As the last two lectures were to be the first lectures of the Royal Society to be given in the United Kingdom outside London in the 321 years since its formation, they were seen to be something special. The topic was the biochemistry and genetics of oxidative phosphorylation and the lecture in London seemed to go well as did the Manchester lecture which was held in conjunction with the Manchester 'Lit and Phil'. The Durham lecture was different. The President of the R.S. came to Durham and the event seemed to be something of a social occasion with Faculty and their wives in attendance in a large hall more adapted to ceremonials than a scientific lecture as it could not be blacked out successfully for showing slides. Following afternoon tea I did my best and although everyone assured me that it was a great success, I have my doubts. Compensation came later in the form of a splendid dinner hosted by the Vice-Chancellor in the Senate room of Durham Castle.

I have always found it difficult to refuse appeals to my better nature to take on various administrative chores and as a result have found myself on a variety of committees over the years. I have to say that most of these have at least turned out to be interesting and given me an insight into the workings of Universities and government departments which I would not have otherwise gained. For more than three years I was Director of the John Curtin School with its total staff of over 300 including about 80 academics. This period taught me quite a lot about human behaviour. Luckily I was able to carry out some laboratory work during this time but was glad to be able to return to the Department full-time. At this period the School was still well funded and able non-tenured staff could find positions when their contract expired. The Director's job has since become much more onerous with the contractions of funding and the job market.

The half century since I started working in laboratories has seen the most extraordinary expansion of knowledge of biological processes. It is difficult to believe now that my first biochemical text [84] concluded that 'The pyrimidines – may possibly be of great biochemical importance. So little is known of their function of their fate, however, that no useful purpose is served by discussing them further'. The availability of more and more sophisticated apparatus and experimental techniques seemed to increase exponentially. New ones always seemed to emerge just in time to allow apparently faltering projects to be advanced. Thus paper chromatography, NMR, mass spectrometry and the various techniques of gene transfer and mapping were followed by the introduction of restriction enzymes, DNA sequencing and site-directed mutagenesis in the 1970s and then the extraordinary growth of information retrieval, data banks and computer graphics in the 80s. I consider it a great privilege to have been involved over this time in the biological sciences.

I was also fortunate in that I started my career at a time when tenured positions were not difficult to obtain and that I have always been allowed to pursue any line of research which engaged by curiosity rather than towards goals influenced by research managers. My retirement in 1988 came just before the Universities in Australia were thrown into turmoil by government initiated 'reforms' with a consequent flood of committee and paper work. Capitalizing on my relatively new-found experience with computers I have, in my current role of Visiting Fellow in the Membrane Biochemistry Group, been of some use, in setting up systems for the molecular biological work and to indulge in computer modelling of membrane proteins, I am divorced from the trials and tribulations of academic life today.

I could not have wished, throughout my career, for a more congenial and enthusiastic group of colleagues, technical staff, students and academics. Graeme Cox and I have now worked together harmoniously for more than 30 years and his intellect and good humour have been invaluable to such successes as our group may claim. I regret that it has not been possible to name all those who have passed through our research group, or even describe much of their work, but many of their names are to be found in the references cited.

Acknowledgements

I am grateful to the John Curtin School of Medical Research in the Australian National University for the support of a quarter of a century of 'curiosity-motivated' research. In recent years generous support has also been received from Professor Sir Rutherford Robertson, Sir John Proud, The Raymond E. Purves Foundation. The James N. Kirby Foundation, C.H. Warman and the Bruce and Joy Reid Foundation.

REFERENCES

1   H.G. Wells, J. Huxley and G.P. Wells, The Science of Life, 1938, Cassell, London.
2   F. Gibson, Aust. J. Exper. Biol. Med. Sc. 28 (1950) 459–463.
3   F. Gibson and D.D. Woods, Biochem. J. 74 (1960) 160–172.
4   C. Yanofsky, Biochim. Biophys. Acta 16 (1955) 594–595.
5   F. Gibson, B. McDougall, M.J. Jones and H. Teltscher, J. Gen. Microbiol. 15 (1956) 446–458.
6   B. McDougall and F. Gibson, Aust. J. Biol. Exper. Med. 36 (1958) 245–250.
7   F. Gibson and B. McDougall, Aust. J. Exper. Biol. Med. Sci., 39 (1961) 171–178.

8   F. Gibson, C.H. Doy and S.B. Segall, Nature 181 (1958) 549–550.
9   C.H. Doy and F. Gibson, Biochem. J. 72 (1959) 586–597.
10 F. Gibson and C. Yanofsky, Biochim. Biophys. Acta 43 (1960) 489–500.
11 F. Gibson, M.I. Gibson and C. Yanofsky, J. Gen. Microbiol. 24 (1961) 301–312.
12 A.J. Pittard, F. Gibson and C.H. Doy, Biochim. Biophys. Acta 49 (1961) 485–494.
13 D.B. Sprinson, Adv. Carbohydrate Chem. 15 (1960) 235.
14 B.D. Davis, Biochemists Handbook, pp. 595, Spon Ltd., 1961, London.
15 M.I. Gibson, F. Gibson, C.H. Doy and P. Morgan, Nature 195 (1962) 1173–1175.
16 M.I. Gibson and F. Gibson, Biochim. Biophys. Acta 65 (1962) 160–163.
17 M.I. Gibson and F. Gibson, Biochem. J. 90 (1964) 248–256.
18 F. Gibson, Biochem. J. 90 (1964) 256–261.
19 F. Gibson and L.M. Jackman, Nature 198 (1963) 388–389.
20 J.M. Edwards, F. Gibson, I.M. Jackman and J.S. Shannon, Biochim. Biophys. Acta 93 (1964) 78–84.
21 F. Gibson, M.I. Gibson and G.B. Cox, Biochim. Biophys. Acta 82 (1964) 637–638.
22 R.G. Cotton and F. Gibson, Biochim. Biophys. Acta 147 (1967) 222– 237.
23 R.G. Cotton and F. Gibson, Biochim. Biophys. Acta 160 (1968) 188– 195.
24 G.L. Koch, D.C. Shaw and F. Gibson, Biochim. Biophys. Acta 212 (1970) 375–386.
25 G.L. Koch, D.C. Shaw and F. Gibson, Biochim. Biophys. Acta 212 (1970) 387–395.
26 G.L. Koch, D.C. Shaw and F. Gibson, Biochim. Biophys. Acta 229 (1971) 805–812.
27 G.L. Koch, D.C. Shaw and F. Gibson, Biochim. Biophys. Acta 229 (1971) 795–804.
28 G.L. Koch, D.C. Shaw and F. Gibson, Biochim. Biophys. Acta 258 (1972) 719–730.
29 G.B. Cox and F. Gibson, Biochim. Biophys. Acta 93 (1964) 204–206.
30 G.B. Cox and F. Gibson, Biochem. J. 100 (1966) 1–6.
31 G.B. Cox, F. Gibson and J. Pittard, J. Bacteriol. 95 (1968) 1591–1598.
32 G.B. Cox, I.G. Young, L.M. McCann and F. Gibson, J. Bacteriol. 99 (1969) 450–458.
33 I.G. Young, L.M. McCann, P. Stroobant and F. Gibson, J. Bacteriol. 105, (1971) 769–778.
34 P. Stroobant, I.G. Young and F. Gibson, J. Bacteriol. 109 (1972) 134– 139.
35 I.G. Young, R.A. Leppik, J.A. Hamilton and F. Gibson, J. Bacteriol. 110 (1972) 18–25.
36 I.G. Young, P. Stroobant, C.G. Macdonald and F. Gibson, J. Bacteriol. 114 (1973) 42–52.
37 J. Lawrence, G.B. Cox and F. Gibson, J. Bacteriol. 118 (1974) 41–45.

38 R.A. Leppik, P. Stroobant, B. Shineberg, I.G. Young and F. Gibson, Biochim. Biophys. Acta 428 (1976) 146–156.
39 G.B. Cox, N.A. Newton, F. Gibson, A.M. Snoswell and J.A. Hamilton, Biochem. J. 117 (1970) 551–562.
40 J.A. Hamilton, G.B. Cox, F.D. Looney and F. Gibson, Biochem. J. 116 (1970) 319–320.
41 N.A. Newton, G.B. Cox and F. Gibson, J. Bacteriol. 109 (1972) 69–73.
42 P. Mitchell, Febs Lett. 56 (1975) 1–6.
43 N.A. Newton, G.B. Cox and F. Gibson, Biochim. Biophys. Acta 244 (1971) 155–166.
44 I.G. Young and F. Gibson, Biochim. Biophys. Acta 177 (1969) 182–183.
45 G.B. Cox and F. Gibson, J. Bacteriol. 93 (1967) 502–503.
46 I.G. Young, G.B. Cox and F. Gibson, Biochim. Biophys. Acta 141 (1967) 319–331.
47 I.G. Young, I.M. Jackman and F. Gibson, Biochim. Biophys. Acta 148 (1967) 313–315.
48 I.G. Young, T. Batterham and F. Gibson, Biochim. Biophys. Acta 165 (1968) 567–568.
49 I.G. Young, T.J. Batterham and F. Gibson, Biochim. Biophys. Acta 177 (1969) 389–400.
50 I.G. Young, L.M. Jackman and F. Gibson, Biochim. Biophys. Acta 177 (1969) 381–388.
51 I.G. O'Brien and F. Gibson, Biochim. Biophys. Acta 215 (1970) 393–402.
52 I.G. O'Brien, G.B. Cox and F. Gibson, Biochim. Biophys. Acta 201 (1970) 453–460.
53 I.G. O'Brien, G.B. Cox and F. Gibson, Biochim. Biophys. Acta 237 (1971) 537–549.
54 R.J. Porra, L. Langman, I.G. Young and F. Gibson, Arch. Biochem. Biophys. 153 (1972) 74–78.
55 F. Gibson and D.I. Magrath, Biochim. Biophys. Acta 192 (1969) 175–184.
56 J.D. Butlin, G.B. Cox and F. Gibson, Biochem. J. 124 (1971) 75–81.
57 J.D. Butlin, G.B. Cox and F. Gibson, Biochim. Biophys. Acta 292 (1973) 366–375.
58 G.B. Cox, J.A. Downie, D.R. Fayle, F. Gibson, and J. Radik, J. Bacteriol. 133 (1978) 287–292.
59 J.A. Downie, A.E. Senior, G.B. Cox and F. Gibson, J. Bacteriol. 138 (1979) 87–91.
60 F. Gibson, G.B. Cox, J.A. Downie and J. Radik, Biochem. J. 162 (1977) 665–670.
61 D.R. Fayle, J.A. Downie, G.B. Cox, F. Gibson and J. Radik, Biochem. J. 172 (1978) 523–531.
62 A.E. Senior, J.A. Downie, G.B. Cox, F. Gibson, L. Langman and D.R. Fayle, Biochem. J. 180 (1979) 103–109.

63   F. Gibson, G.B. Cox, J.A. Downie and J. Radik, Biochem. J. 164 (1977) 193–198.
64   G.B. Cox, J.A. Downie, F. Gibson and J. Radik, Biochem. J. 170 (1978) 593–598.
65   F. Gibson, J.A. Downie, G.B. Cox and J. Radik, J. Bacteriol. 134 (1978) 728–736.
66   J.A. Downie, A.E. Senior, F. Gibson and G.B. Cox, J. Bacteriol. 137 (1979) 711–718.
67   J.A. Downie, G.B. Cox, L. Langman, G. Ash, M. Becker and F. Gibson, J. Bacteriol. 145 (1981) 200–210.
68   J.A. Downie, L. Langman, G.B. Cox, C. Yanofsky and F. Gibson, J. Bacteriol. 143 (1980) 8–17.
69   R. Humbert, W.S.A. Brusilow, R.P. Gunsalus, D.J. Klionsky and R.D. Simoni, J. Bacteriol. 153 (1983) 416–422.
70   J.E. Walker, M. Saraste and N.J. Gay, Biochim. Biophys. Acta 768 (1984) 164–200.
71   G.B. Cox, J.A. Downie, L. Langman, A.E. Senior, G. Ash, D.R. Fayle and F. Gibson, J. Bacteriol. 148 (1981) 30–42.
72   G.B. Cox, D.A. Jans, A.L. Fimmel, F. Gibson and L. Hatch, Biochim. Biophys. Acta 768 (1984) 201–208.
73   G.B. Cox, A.L. Fimmel, F. Gibson and L. Hatch, Biochim. Biophys. Acta 849 (1986) 62–69.
74   R.N. Lightowlers, S.M. Howitt, L. Hatch, F. Gibson and G.B. Cox, Biochim. Biophys. Acta 894 (1987) 399–406.
75   R.N. Lightowlers, S.M. Howitt, L. Hatch, F. Gibson and G. Cox, Biochim. Biophys. Acta 933 (1988) 241–248.
76   B.D. Cain and R.D. Simoni, J. Biol. Chem. 261 (1986) 10043–10050.
77   S.M. Howitt, F. Gibson and G.B. Cox, Biochim. Biophys. Acta 936 (1988) 74–80.
78   S.M. Howitt, R.N. Lightowlers, F. Gibson and G.B. Cox, Biochim. Biophys. Acta 1015 (1990) 264–268.
79   S.M. Howitt and G.B. Cox, Proc. Natl. Acad. Sci. USA 89 (1992) 9799–9803.
80   G.B. Cox, R.J. Devenish, F. Gibson, S.M. Howitt and P. Nagley (1992) in Molecular Mechanisms in Bioenergetics (Ernster, L., ed.), pp. 283–315, Elsevier, Amsterdam.
81   H.S.Penefsky and R.L. Cross, Adv. Enzymology 64 (1991) 173–214.
82   R.H. Fillingame, Biochim. Biophys. Acta 1101 (1992) 240–243.
83   M. Beadle, These Ruins are Inhabited, 1963 Robert Hale, London.
84   E. Holmes, The Metabolism of Living Tissues, 1938 Cambridge University Press, London.
85   F. Gibson (1991) (1992) The Exploration of Living Membranes. In: Public Lectures 1991–1992, pp. 1–21.

E.C. Slater, R. Jaenicke and G. Semenza (Eds.)
Selected Topics in the History of Biochemistry: Personal Recollections, IV
(Comprehensive Biochemistry Vol. 38) © 1995 Elsevier Science B.V.

Chapter 7

Charge Separation: A Personal Involvement in a Fundamental Biological Process

R.N. ROBERTSON

P.O. Box 9, Binalong, NSW 2584 (Australia)

It is a long way from ideas about membranes in 1936 to the sophisticated pictures of today, with detailed knowledge of transmembrane proteins which behave as specific channels or pumps, interacting with charge separation systems across the membranes. This article gives a history of one person's association with the development of some of the concepts which have led to the greatly increased but still incomplete understanding of these and other properties of biological systems. It is not intended to be an account of all my research or of my career, which I have written briefly elsewhere [1]. Part of this story involved a modest interaction with the famous Peter Mitchell, who received the Nobel Prize in Chemistry in 1978 for his outstanding original contributions to this field, especially his theory of chemiosmosis. The development of the ideas, sometimes right and sometimes wrong, illustrates the way that science advances, sometimes confidently striding along, sometimes stumbling and sometimes crawling slowly.

The original charges – H⁺ and HCO₃⁻

About 57 years ago I became fascinated with the problem of how living cells can use the energy from respiration to bring about the energy-requiring processes of moving substances, particularly ions. When I went to Cambridge as a graduate student (research student, we called them then) in 1936, my supervisor, G.E. Briggs, suggested that I might work on a problem of particular interest to him: was the movement of ions into plant cells, against a concentration gradient, a consequence of the production of carbon dioxide which resulted in the steady diffusion of H^+ and HCO_3^- to the exterior? S.C. Brooks [2] had suggested that the hydrogen ion might exchange for a positive ion entering the cell and the bicarbonate ion might exchange for a negative ion, through a membrane which was a mosaic of cation and anion permeable areas. Briggs [3] had pointed out that this simple hypothesis would not work because it is impossible, with positive and negative ions exchanging simultaneously, to build up a concentration gradient of both ions. He suggested that such an exchange could work only if there were alternations of positive and negative permeability in time. With this background, it is not surprising that the introduction to my PhD Dissertation states that 'the mechanism of the process is still obscure and, in spite of the formidable amount of literature which has accumulated, it is only by obtaining more data that we shall be able to formulate a picture of the mechanism' [4].

Much of the work done with Briggs was involved in a better understanding of the plant 'guinea-pig' (carrot root), with which we did our experiments, and is not directly relevant to my subsequent ideas on the fundamental importance of charge separation. Our techniques were to measure simultaneously the uptake of the ions of a salt (usually potassium and chloride) and the carbon dioxide output from respiration. The uptake of the salt was measured by following the change in conductivity of the solution surrounding the aerated plant tissue. The respiration was measured by passing CO_2-free air over the tissue and then bubbling it through NaOH solution which absorbed the CO_2 given off by the tissue. The amount of CO_2 was calculated from the change in the conductivity of the solution as the highly mobile hydroxyl ions were replaced by the less mobile bicarbonates. At that time we were responsible for the design and construction of much of the equipment we used and I enjoyed the experience of designing and putting together the best absorption vessels for the CO_2. Fortunately I had become reasonably proficient at glassblowing and was able to make the appropriate vessels, though the ones eventually used were obtained from professionals.

The tissue used in the Cambridge plant physiology laboratories consisted of discs of parenchyma cut from the central region of storage roots of carrot, and a number of my experiments characterised the physical properties of this tissue. It turned out to be a fortunate choice because some time after it is cut and kept in aerated distilled water, its respiration drops to a low level but is stimulated by the addition of salt so that the amount absorbed is proportional to the increase in respiration, a relationship which I was able to use later.

We soon abandoned any idea that the uptake of the ions was, except for a small fraction, due to the exchange with H^+ and HCO_3^-. The uptake was related to the respiration rate and could be increased by the addition of sugar which accelerated the respiration. By contrast, the addition of methylene blue, a redox substance, which greatly enhanced the output of CO_2, depressed the salt accumulation. It was clear that the salt accumulation was associated with some processes about which we had no real evidence except that stimulating an abnormal production of CO_2 with methylene blue destroyed the ion accumulation mechanism. I was aware of the work of Lundegårdh, including that which he had done with Burström [5], which showed that the respiration of barley roots was increased on the addition of salts. This increase in respiration they termed the 'anion respiration' and showed that the amount of anion absorbed was proportional to the rate of this respiration. Furthermore, when this respiration was inhibited by cyanide, recognised at that time as an inhibitor of the cytochrome system, the salt accumulation was stopped. Though I referred to this while I was writing my thesis, its significance had not yet dawned on me and, in any case, speculation about how the salt accumulation process might be linked to the somewhat vague hypotheses about the mechanism of respiration was not encouraged, despite the fact that we research students had been taken to the Molteno Institute to be fascinated with Keilin's demonstration of the spectroscopic changes which accompanied the oxidation and reduction of the cytochromes.

Charge separation – respiration and H$^+$ and e$^-$

Having completed my PhD thesis in December 1938 and satisfied the examiners, E.J. Maskell and T.A. Bennet-Clark, I returned to Australia at the beginning of 1939. There, as an assistant lecturer in the Botany Department of the University of Sydney, I quickly set up apparatus similar to that which I had used in Cambridge and began to follow up some of the interactions between the salt respiration rate and the accumulation of ions in carrot tissue. About this time, Lundegårdh [6] put forward his ideas in a letter to *Nature* titled 'An electrochemical theory of salt absorption and respiration'. He was immediately subjected to criticism from Hoagland and Steward [7] and replied (Lundegårdh [8]). Hoagland and Steward [9] replied again. By this time I had further results on my own which supported the idea of salt or anion respiration being related to the rate of salt accumulation and I also joined in the discussions in *Nature*. I presented data from my experiments done in Sydney which were consistent with those of Lundegårdh on barley roots but suggested that it was necessary to introduce a time factor in interpreting longer experiments be-

cause the ratio of anion absorption to salt respiration decreased with time [10]. My agreement with the general principles of Lundegårdh's interpretation had begun.

During the war years, my time for this kind of research was
limited, for I was involved, not only in an enjoyable load of
teaching science, agriculture and forestry students, but also in
work on storage of apples and wheat, investigations which
arose out of Australia's lack of shipping to overseas markets. I
did have a chance to clarify some of the characteristics of the
salt accumulation and accompanying respiration, such as the
way they drifted with time, the temperature coefficients of the
two, interpreted as being dependent on chemical rather than
physical processes and the reversibility of the salt respiration.
In 1944, I had the opportunity to take a busman's holiday in
Melbourne where, in collaboration with my friend from Cambridge days, J.S. Turner, and assisted by Jean Mathieson
(later Mayo), we did a series of experiments on his Warburg
respirometers, further characterising the dependence of the
accumulation on the salt respiration and especially establishing the inhibition of both processes by cyanide [11].

My thinking about the mechanism of salt accumulation had
been greatly influenced by Lundegårdh's 1940 paper [12] entitled 'Investigations as to the absorption and accumulation of
inorganic ions'. Though this paper had attracted further criticism, particularly from Steward and Preston [13], results of
my experiments published in 1940 [14] added support to the
Lundegårdh hypothesis of a relationship in some tissues between a special fraction of the respiration and the salt accumulation. That Lundegårdh paper had done much more and,
looking back now, it can be seen as a remarkable accomplish-
ment of experiment and interpretation for that time. As far as
I was concerned, the principal message was that the mechanism of ion accumulation was related to part of the respiration
dependent on what was probably a cytochrome system and I
was speculating on how such a respiration might move ions.
Lundegårdh had suggested that the redox system, working on

an oxygen gradient between the outside and inside of the cell, might set up a gradient along which the anion of a salt might move inside. In 1945, I had been thinking of the possibility that the oxidised cytochrome with its ferric iron might have a positive charge and pick up a negative ion. Then, if the complex could move in a barrier which was impermeable to a free ion, the ion would be trapped behind the barrier after it was freed from the cytochrome reduced to its neutral ferrous state by the next electron. In other words, if a cytochrome could take an electron in one direction, it might take an anion in the opposite direction. When Lundegårdh's paper [15] reached me later in that year, I found that he had developed the same idea.

It is not often that one remembers exactly when a good idea comes into one's consciousness but I recall perfectly my next step in the reasoning. It was the Easter holiday in 1945 and I was sitting in our house on a perfect afternoon overlooking one of the Sydney surf beaches, when I realised that if the mechanism depended on cytochrome as an anion carrier, there would be a stoichiometric relationship between the amount of the oxygen absorbed and the amount of salt accumulated. I immediately consulted such results of experiments as I had with me and found, to my great surprise, that the number of gram equivalents of salt accumulated was of the same order of magnitude as the number of gram mols of oxygen absorbed. Though Lundegårdh had reached the same idea about the cytochrome as a carrier and had a quantitative relation between the oxygen uptake and the ion absorption, he did not, even in his 1945 paper, think of checking the stoichiometry. There were two things wrong with my thoughts on that afternoon: the first was that I had used the old equation for the overall process of respiration, i.e. $C_6H_{12}O_6 + 6O_2 = 6CO_2 + 6H_2O$; the second was that most of my experiments had been done with lower salt concentrations which were rate limiting to both the salt respiration and accumulation. Nevertheless, the remarkable thing was that the two rates were of the same order of magnitude.

We needed better data and I was able to follow up these ideas because of a change in my occupation which increased my research opportunities. That year was my last as a full time member of staff at Sydney University for I accepted an invitation to become a Senior Research Officer in the Division of Food Preservation and Transport of the Commonwealth Scientific and Industrial Research Organization (CSIRO). My duty was to take charge of the Section of Plant Physiology and Fruit Storage. Under the generous provisions of this arrangement, I was expected to continue with fundamental research in plant physiology and to be responsible for the applied research on fruit storage which, fortunately, was carried out by a very capable staff. I was also permitted to continue to run the third year course in plant physiology in Sydney University. This joint arrangement between the CSIRO and Sydney University subsequently led to the establishment in 1952 of the Plant Physiology Unit under the leadership of Dr F.V. Mercer and myself. This Unit, which had considerable influence on plant physiological research in Australia, continued after my departure and was later located at Macquarie University [1].

In the laboratories of the CSIRO, assisted by one of my Research Officers, Marjorie Wilkins, I set about examining the stoichiometric relations between the salt respiration and accumulation in our carrot tissue. By this time too, we had acquired Warburg respirometers and were measuring oxygen uptake directly; salt uptake was calculated from conductivity measurements of the solutions surrounding replicate sets of tissue in flasks attached to the shaker of the Warburg apparatus to ensure aeration. By then we were using the correct equation for respiration:

$$C_6H_{12}O_6 + 6\ \underline{O}_2 + 6\ H_2O = 6\ CO_2 + 12\ H_2\underline{O}$$

which recognises that all the oxygen absorbed (underlined), appears in the water formed, none in the CO_2. Thus the separation of protons and electrons for each 6 molecules of oxygen taken up would be 24 H^+ and 24 OH^- and, if an anion entered the cell each time an electron left the cytochrome to combine with oxygen, the maximum rate of salt uptake would be:

$$\frac{\text{g equiv salt absorbed}}{\text{g mol oxygen absorbed}} = 4$$

Our experiments, completed in 1947 and published in 1948 [16, 17], gave quantitative results which were consistent with the hypothesis that the driving force for the energy-requiring process of salt accumulation could be liberation of one electron resulting in the uptake of the anion and liberation of one proton resulting in the uptake of one cation. Of course there were many more factors to be taken into account to explain the complicated phenomenon and these will be discussed later. At this stage it is historically important to note that we had stumbled on material that opened up the possible basic importance of charge separation as a consequence of oxidation reactions. Poole [18] pointed out that '...... salt respiration, while of interest in itself, has not yet thrown much light on the processes of ion transport'. It might be claimed that its principal contribution was to stimulate the idea of charge separation.

This was especially relevant at that time to the work which was being done on the secretion of hydrochloric acid by the gastric mucosa. Various people were looking at the quantitative relations between the stimulated oxygen uptake of the gastric mucosa and the amount of hydrochloric acid secreted simultaneously [19–21]. Conway and Brady were first to publish the suggestion that protons released at the cytochrome system might be the source of the hydrogen ions of hydrochloric acid. Crane and Davies, who had the same idea, were first to publish satisfactory measurements. Using an ingenious technique with isolated frogs' stomachs respiring in Warburg flasks, they showed that the amount of g equiv HCl secreted into the lumen of the stomach was about four times the g mol of oxygen absorbed in the stimulated respiration, accompanied by an equivalent amount of HCO_3^- passing to the exterior of the stomach. This result was very exciting because it was consistent with our ideas of the protons being derived by their separation from the electrons, which formed hydroxyls with oxygen. The hydroxyls were converted to bicarbonate by carbonic anhydrase, known to be active in the gastric mucosa.

In 1948 when all this excitement was brewing, I had the opportunity to travel from Australia to spend about five months in England with Cambridge as headquarters and three months' travelling in North America. In addition to the many broadening contacts that this trip provided, it was important to my developing ideas about the central position of charge separation. First, I met R.E. Davies who was then in Sheffield but arranged to meet me in London where we had a long and, for me, convincing talk about the likelihood of the H^+ ions of the HCl secretion being due to the charge separation at the cytochromes of the gastric mucosa. Second, I had a memorable visit to see Professor Lundegård who did his research in a private laboratory at his home in Penningby which was a short distance from his Institute of Plant Physiology at Uppsala. I returned to Australia even more convinced that the basis of ion movement would be found in its connection with charge separation.

Since the specific test for the involvement of the cytochrome oxidase system in those days was to inhibit it with carbon monoxide in the dark and to reverse the inhibition with light, I set about the necessary experiments in collaboration with D.C. Weeks, then of the Botany School, Melbourne University. We were able to show our conclusion based on the cyanide inhibition, that both salt absorption and salt respiration were dependent on cytochrome oxidase, was probably correct since they were both inhibited by CO in the dark but not in the light [22].

Energy-rich phosphate and mitochondria

But new ideas about the energy-distributing mechanisms in living cells had been introduced by the recognition of the outstanding importance of adenosine triphosphate (ATP) as the energy currency of cells, and the dependence of many energy-requiring processes on the 'energy-rich phosphates' or, in the shorthand of those times, '~ P' pronounced 'Squiggle P'. In 1950, Davies and Ogston [23] had suggested that some of Davies' results, where the ratio of HCl to O_2 uptake exceeded 4, might be due to ions being derived from the phosphate carriers. At that time, the method of looking for dependence of a process upon energy-rich phosphate was to use an uncoupler, so called because it would uncouple the oxidation process in respiration from the production of ATP. These observations raised the possibility that, like so many energy-requiring processes, the ion accumulation process or 'active transport' as we were calling it then, might depend on energy from ATP. With Wilkins and Weeks, I investigated the effect of the then commonly-used uncoupler, 2,4-dinitrophenol (DNP). We found that the accumulation of ions by carrot cells was indeed inhibited, despite the fact that the respiration was markedly stimulated and, judged by the light-reversible inhibition by CO, was proceeding through the cytochrome oxidase system [24]. Clearly the accumulation mechanism was not simply dependent on what happened at the cytochrome oxidase respiration site and our ideas had to be revised.

From 1948, another important development depended on the realisation that the sites of oxidative respiration were the mitochondria where the Krebs acid cycle, the cytochrome system and the oxidative phosphorylations were shown to occur. The initial work was done with animal mitochondria but by 1950, DuBuy, Woods and Lackey [25] had shown that the cytochrome components of the respiratory system are carried on the mitochondria of *Nicotiana* and *Lonicera*. Since I subscribed to the hypothesis that active transport of ions was linked to the cytochrome system, our experiments predictably moved to isolated mitochondria and their properties. Though some of our motivation was related to our work on active transport, an equally important aspect of our interests, the nature of respiration itself, was essential to our understanding of the processes occurring in the development and ripening of fruit. Our work on mitochondria was to lead back to another form of charge separation and its relation to the reaction of ATP.

Starting work on isolated mitochondria was fraught with difficulty as far as we were concerned. The essential equipment, a refrigerated centrifuge, was not available to us in Sydney at that time. In our CSIRO laboratories, we had good cold rooms for the experiments on the storage of fruit. We began by trying to run our bench centrifuges in the cold rooms but, alas, the cooling was not rapid enough to overcome the heat generated by the centrifuge motor. We had to wait to purchase a refrigerated model and then set about a systematic examination of the morphology, respiration and ionic characteristics of the mitochondria of red beetroot. The morphology of mitochondria at this time was very poorly understood, for the section-cutting techniques for electron microscopy did not really become available until Palade [26]. Before that we had been trying to interpret their morphology with the technique of 'shadowing', work done in collaboration with our CSIRO colleagues in the Division of Industrial Chemistry in Melbourne. There we had to use a borrowed centrifuge at the University of Melbourne and take our mitochondrial preparations some ten miles to the electron microscope at CSIRO. While we were doing this work, there was still controversy as to whether there are membranes in mitochondria. We could let our extracted mitochondria swell in dilute salt solution and then collapse on drying out, strongly suggesting that they each had a surrounding membrane. Using the electron microscope technique of 'shadowing' with uranium, we were able to show that a shadow on the edge of a collapsed mitochondrion could corre-

spond to a membrane or membranes about 270 Å in thickness
[27].

While we were doing this work, the techniques for embedding and cutting sections for the electron microscope became available and we were then able to distinguish not only the external membrane but also the internal membranes which came to be known as cristae. Other workers also showed the outer membranes surrounding the particles to be double. When we published the completed work, we concluded that the outer membranes consisted of two dense layers about 50 Å in thickness, separated by a less dense space of about 70 Å [28]. These were the pioneering days of the sections for the electron microscope and we did not commit ourselves to continuing to get better pictures of the mitochondria. We were more interested in their biochemical functions.

In my review [29] of 1951, I had suggested that mitochondria might act not only as the source of respiratory energy but also as carriers of ions within the cell, based on the somewhat naive idea that if the ions had to be transported across lipid membranes, the high lipid content of the mitochondria might make them suitable carriers, especially as they are known to move with protoplasmic streaming. Our experiments then attempted to determine whether ions are actively transported into the mitochondria. Using beetroot and carrot mitochondria, Wilkins, Hope and I investigated the absorption of different ions and their relations to oxygen uptake. At that time, we were characterising the physical properties of these organelles, realising that the complex structure containing indiffusible anions giving a predominance of negative charge, could result in mobile cations (Na^+, K^+ and Ca^{++}) being held in concentrations considerably greater than those of the external solutions. The mobile anions supplied to the external solutions could also be in higher concentration inside than outside, suggesting that there might be an accumulatory mechanism operating, particularly as the prevalence of indiffusible anions inside would be expected to depress the concentration of mobile anions. By measuring the time of adjustment to changes of external concentration of Cl⁻ and assuming a thickness of 200 Å for the membranes surrounding the mitochondrion, we calculated the apparent diffusion constant of Cl⁻ to be about 10^{-13} cm² sec. which agreed with that found for heart muscle sarcosomes and would be that expected of a lipoprotein membrane [30].

When it came to measuring the oxygen uptake of our mitochondria, we found that they behaved similarly to those of plant mitochondria being investigated by others at that time. But we were interested in the possibility that they were capable of active transport and chose to find whether there was evidence that the chloride ion, which, unlike the organic acid anions, is not metabolised, was actively transported; if so whether its rate of transport was quantitatively related to the oxygen uptake. When we used potassium chloride in the absence of organic acid substrates, we did find a relation between oxygen uptake rate and the concentration of Cl⁻ in the mitochondria which was not inconsistent with the hypothesis that the active transport could be dependent on the cytochrome being a possible anion carrier in the direction opposite to that in which it was carrying an electron [31]. However, we were left with the problem of how the accumulation mechanism in the cell as a whole was related to what was happening in the mitochondria.

At this time, thanks to the presence in our laboratory of S.I. Honda, a Fulbright post-doctoral, we set out to learn much more about the ionic relations of mitochondria and, as so often happens in research, found our system was more complicated than we had thought and the capacity to accumulate chloride varied with the time of year that the beetroot was harvested. With the variability of our experimental material, we were less confident in drawing conclusions about the mitochondrial mechanism to accumulate chloride and concluded that it might be a very sensitive process, strongly dependent on the intact nature of the mitochondria after their extraction from their cells [32].

We decided to investigate the electron transport chain of our mitochondria using both diphosphopyridine nucleotide (DPNH) and cytochrome c, each of which can supply electrons to the membrane-bound cytochrome oxidase system. We found that potassium, sodium, magnesium, calcium, chloride and orthophosphate consistently increased the rate of DPNH oxidation in beetroot mitochondria which had been isolated in a sucrose medium. The rate of DPNH oxidation was increased with increasing concentrations of chloride but decreased at higher concentrations. Higher salt concentrations were optimal for the stimulation of cytochrome c oxidation. We concluded that the effect of salt on DPNH oxidation could explain salt respiration in plant tissue [33].

Better understanding of the electron transport chain in our mitochondria became possible with the collaboration of R.K. Morton and J.T. Wiskich. I had taught Bob Morton as a second year student of agriculture at Sydney University in 1940, after which he spent the war years in the Australian Navy and later returned to complete his degree. After experience in Cambridge, he came back to Australia, first to the University of Melbourne and later to the Waite Institute, University of Adelaide. Joe Wiskich was then a graduate student in Sydney and was supervised jointly by Morton and myself. We set out to determine the sequence of reactions in the respiratory chain of beetroot mitochondria. We found evidence that the respiratory chain normal in intact cells was probably maintained in the extracted particles with cytochromes b, c_1 and c identified spectroscopically. The presence of an absorption band near $600\,\mu$ was attributed to cytochromes (a + a_3). We compared our mitochondrial properties with others which had been published at that time, finding mostly similarities but some differences [34]. In the light of present day ideas about the electron chain, a footnote in this paper is worth quoting: 'Lester and Crane [35] have reported substantial amounts of naphthoqui-

nones related to ubiquinones (R.A. Morton [36]) in plant mitochondria; whether ubiquinone is a component of the respiratory chain is unknown.'

Respiration and fruit ripening

By now we were approaching the end of the fifties which were
exciting times because speculation about the far-reaching effects of ATP was rife and we were interested in its possible role
in fruit physiology. As early as 1948, when I was in Cambridge
and attended some of the lectures given by E.J. Maskell, I was
impressed with the hypothesis that the relative amounts of
ATP and ADP in cells at any one time would have a controlling
effect on rates of reactions which were dependent on phosphorylation. In the analyses that my colleague J.F. Turner had
done on developing apple fruits, we had found that rates of
respiration, which had been increasing in proportion to the
protein nitrogen, increased more rapidly than protein nitrogen when starch was being broken down and unknown acids
were increasing rapidly compared with malic and citric. Maskell's suggestion that carbohydrate content, protein synthesis
and respiration might be connected through a phosphate
transfer mechanism, offered us a possible explanation. Starch
content was thought to be a measure of excess energy-rich
phosphates. With high content of energy-rich compounds
(ATP), the level of ADP would be kept low and the loss of phosphate from the phosphorylated compounds of the glycolytic
and acid cycles would be restricted. If however, owing to increased demand for phosphorylations in protein synthesis and
maintenance, less phosphorylations of sugar resulted in decrease in glucose-1-phosphate, the equilibrium would drift towards glucose and a decrease in starch. Simultaneously, the
more rapid turnover of phosphate carriers in protein synthesis
and maintenance would result in more phosphate acceptors to

pick up energy-rich phosphates from respiration and an increase in its rate could be expected [37].

When Judith Pearson (then Fraser) joined the group in
1950, we looked for ways to test the hypothesis that the climacteric rise in respiration characteristic of many developing
or ripening fruits, might be the result of the changes in ATP/
ADP ratio such as we had postulated for apples. Once again we
experimented with the uncoupler 2,4-dinitrophenol (DNP) to
see whether it would result in an increase in the respiration
rate by releasing the control due to the shortage of ADP as the
phosphate acceptor in the respiration process. Using discs of
the parenchyma tissue from the apple fruits and measuring
the respiration in Warburg apparatus, we showed that the
respiration of the cut tissue six weeks before the climacteric
rise was about four times that of the same weight of tissue in
the intact fruit. When the whole fruit went through its climacteric, the respiration per unit weight of the cut tissue did not
increase but was then only twice and, later, only one and a half
times that of the whole fruit. In the first sample of cut tissue
taken seven weeks before the climacteric, the addition of DNP
doubled the respiration rate. As the fruit went through the
climacteric rise, the effect of DNP decreased progressively
until, after the climacteric in the whole fruit, the oxygen uptake of the cut tissue was only slightly greater than in the
controls without the DNP (Pearson and Robertson [38, 39]). It
was also shown that the oxidative activity of particles (mitochondria) extracted from the tissue, and supplied with succinate, citrate or malate as substrate, increased during the climacteric rise. We felt justified in concluding at that time that
the climacteric rise might be associated with the decreased
control due to a change in the ATP/ADP ratio in the tissue.
Collaboration with M.D. Hatch and Adele Millerd enabled us
to demonstrate the presence of the Krebs acid cycle as part of
the oxidative system [40].

However, this attractive simple explanation of the climacteric was not to be generally applicable. With H.K. Pratt of the

320   R.N. ROBERTSON

University of California, Davis, who was with us as a
Fulbright Scholar, K.S. Rowan and I investigated the relationship of high-energy phosphate content, protein synthesis and
the climacteric rise in the respiration of ripening avocado and
tomato fruits. We found that the climacteric rise in avocado
ripened after picking was accompanied by a rise in total high-
energy phosphate. In tomatoes ripening on the plant, the
change in high-energy phosphate was not significant. The
ratio of protein nitrogen to total nitrogen increased in both
fruits. We had to conclude that, while the rates of respiration
and synthetic processes might be controlled by the phosphate
transfer system, the evidence suggested that uncoupling of
respiration and phosphorylation did not explain the respiratory climacteric in these fruits [41].

Our work on the hypotheses of the climacteric rise illustrates one of the common experiences of biological research:
starting with the simplest hypothesis to be tested (known as
applying Occam's razor), much experimentation often leads to
a more complicated theory! At that time we had not yet arrived
at clear ideas connecting the ATP formation in respiration
with the charge separation in mitochondria.

Charge separation and oxidative phosphorylation

During the 1950s, when I was preoccupied with charge separation in relation to active transport and with ATP/ADP ratios in
fruit, it was to be expected that we would have a lively interest
in the vigorous controversy of that time as to the mechanism of
the formation of ATP from ADP and inorganic phosphate in
oxidative phosphorylation. To understand this nowadays, it is
necessary to explain that the formation of high-energy phosphates from two of the glycolytic reactions were then thoroughly well understood; one comes from the oxidation of glyceraldehyde phosphate in the presence of phosphate resulting
in formation of 1,3-diphosphoglycerate and the other from the

conversion of 3-phosphoglycerate to phosphoenol pyruvate. The high-energy phosphates from these two compounds are passed to ADP to form ATP. In the oxidative phosphorylations, it was expected that some such intermediates would be formed and much work went into looking for them. Furthermore it was observed that in mitochondria with 2,4-DNP and azide as uncouplers, both secretion of H^+ ions and phosphorylation were inhibited. I was puzzled by this dual effect of uncouplers.

At this time, another important change in my career took place. I was invited to become a Visiting Professor in the Department of Horticultural Crops at the University of California at Los Angeles. Professor Jacob Biale who was in charge of post-harvest physiology in that Department was going to Israel for a year and I was asked to look after his responsibilities and give some lectures to the graduate students. CSIRO was agreeable to my taking leave without pay and the result was a most enjoyable and scientifically rewarding experience. I had time to spend in the library and studied all the then recent work on oxidative phosphorylation and its relation to secretion and active transport. The result was a review paper [42] submitted to *Biological Reviews* in August 1959 and published in 1960.

This gave me the opportunity to discuss the confused ideas about the relations of H^+ secretion, ion transport and phosphorylation. The principal contribution was to point out that the separation of positive and negative charge was probably *the* fundamental process in electron transport and that it might *precede both* the movement of ions and the phosphorylation of ADP and ATP. Further, I suggested that the two consequences of charge separation, i.e. secretion of H^+ and OH^- ions and phosphorylation might be alternatives thus:

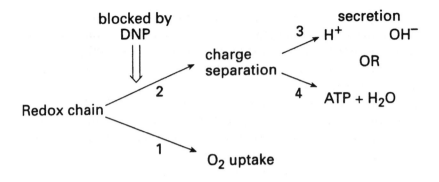

My conclusion, that the separation or secretion of hydrogen ions from the surfaces of isolated mitochondria was the direct result of the oxidation chain in a relatively impermeable membrane, was consistent with much evidence which had accumulated at that time from various workers (Chance, Lehninger, Slater and others). However, many had assumed that the oxidation/reduction had given rise to the ATP which then brought about the secretion of H^+ and that DNP was blocking the ATP formation. Another possible explanation was that secretion and ATP formation were both independent results of oxidation/reduction and that DNP blocked both. There was also the possibility that oxidation reduction brought about the separation of charge and that secretion and ATP formation both resulted simultaneously. It was an important part of my hypothesis that the secretion and the ATP formation were *alternative* results of the charge separation and that neither could occur if DNP or any other uncoupler had prevented that charge separation. My review went on to speculate about the sorts of reactions which might be involved in the phosphorylation of ADP to ATP but since we were then so ignorant of the nature of ATPases, about which we now know so much, the speculation was not very profitable. Looking back at that review now it is interesting to realise that I did not think of the possibility that breakdown of ATP might result in the separation of charge and hence a movement of ions such as H^+ secretion. As we now know the membrane-bound ATPases are elegant systems which can catalyse formation or breakdown of ATP, a very important property which Peter Mitchell realised about that time.

Immediately after my nine months at UCLA and a short period in Cambridge to work with G.E. Briggs on the book with A.B. Hope titled *Electrolytes and Plant Cells* [43], I returned to Australia to another career change. At that time the CSIRO was run by an Executive of four members and I was asked to be one of the four. This full-time administrative position involved much office work, meetings and travel to the Organization's laboratories all over Australia. My time for research was reduced virtually to zero. Leaving my laboratory at the Plant Physiology Unit, I first had an office in Sydney and then moved to Melbourne in 1961 where I was when Peter Mitchell announced his important hypothesis which was to change the way we thought about charge separation and oxidative phosphorylation.

Mitchell's chemiosmotic hypothesis

I had been aware of Mitchell's work on group transfer systems, particularly relating to bacterial membranes and the important property of membranes to be able to react differently on their two sides but it was not until he outlined his hypothesis in an article in *Nature* in 1961, titled *Coupling of phosphorylation to electron and hydrogen transfer by a chemi-osmotic type of mechanism* [44] that I saw this new idea must be basically right and added to the way I had been thinking of the relationships. In that outline he referred to my review of 1960, which I felt would offer support to his hypothesis but he did not refer to my suggestion, which I think was then original, that after charge separation, proton secretion or phosphorylation of ADP are alternative consequences. I wrote to him congratulating him on his hypothesis and thereafter I used to receive his writings, including those which were produced in his laboratory and not published.

In retrospect, I regret that I did not have enough insight to make the ATP hydrolysis liberate H^+ on one side of the membrane and OH^- on the other, but I did not; Mitchell did and now, of course, the ATPase is a well-known transmembrane protein complex. After my 1960 review, apart from not having time for research, I was preoccupied with the problem of ion transport linked to charge separation but I had seen that the Mitchell hypothesis was the probable explanation of ATP formation and that looking for possible intermediates in oxidative phosphorylation was on the wrong track. I was completely surprised that the hypothesis which, in its essentials seemed so likely to me, was not readily accepted by workers in the mitochondrial oxidative phosphorylation field.

At the beginning of 1962, having finished my obligation to the Executive of CSIRO, I moved to be Professor of Botany at the University of Adelaide. There I gave talks and lectures in which the Mitchell hypothesis was included as the probable mechanism. The Department of Botany in Adelaide was strengthened on the plant physiological side by the appointment of Michael Pitman who continued the work on ion uptake which he had been doing with G.E. Briggs in Cambridge and, a little later, J.T. Wiskich returned from the USA to join the staff and strengthen our research on mitochondria. I found the atmosphere of the Department, not only in plant physiology but also in its other disciplines, delightfully stimulating.

In 1964, I had the opportunity to visit the northern hemisphere and began with a short visit to Britton Chance's Johnson Research Foundation where, in that very lively research atmosphere, I enjoyed doing some experiments with Walter Bonner. As the International Biochemical Congress was about to be held in New York, many biochemists were converging from around the world and Chance had the splendid idea of gathering a number of them for an informal discussion (The Compost Heap Symposium) at his farm, where we sat comfortably on bales of hay in a large barn and talked about mitochondria. Topics included their origins, correlation of electron microscopy and biochemistry, structural and chemical compartmentation, swelling and shrinking. Among those present of particular interest to me were Azzone, Boyer, Chappell, Estabrook, Greville, Klingenberg, Lehninger, Palade, Parsons, Pressman, Slater, Stoeckenius and van Dam. I remember most of the discussion as being involved with mitochondrial structures, including the newly discovered knobs on the inside of the cristae membranes, then of unknown function, later known to be the ATPases, and with the electron and ion transports as affected by inhibitors and uncouplers. I do not remember any attention being given to Mitchell's hypothesis.

I went on immediately to the Congress in New York. The most exciting thing for me at that meeting was the paper on chloroplasts in which André Jagendorf announced their discovery that ATP could be made in the dark with chloroplast lamellae and a pH gradient. I like to quote a subsequent paper of Jagendorf and Uribe [45]: 'At the time when our confidence in a high energy state of an electron carrier was waning, Dr Hind spent some profitable time in the library and there discovered the chemi-osmotic hypothesis presented in the 1961 paper by Mitchell'.........'we decided to consider seriously, at least for the moment, the possibility that our energetic intermediate was really a pH gradient. One immediate prediction from the theory was that the pH of the external medium ought to increase as [the intermediate] forms in light, if the protons are moved into the internal space of the grana disk sacs and become inaccessible to the glass electrode. This rise in pH caused by illuminating a suspension of unbuffered or weakly buffered chloroplasts was indeed found to occur. With this discovery our opinion as to the probability of the chemi-osmotic hypothesis was revised upwards several notches and the nature of our experiments has never been the same since.'

I had gone to one of the other concurrent sessions so I first heard of what Jagendorf was reporting when he sought me out at lunch and said that he owed me an apology because he wanted to refer to some of my ideas and at the moment of doing so could not remember the name, so had to say 'the man from Australia', with someone in the audience supplying the 'Robertson'. Needless to say I did not need an apology and was delighted to learn of this clear demonstration of the validity of the Mitchell hypothesis which I could then support even more strongly.

Following the Congress in New York, I went on to the International Botanical Congress in Edinburgh, Scotland, where I gave a paper on the relation between the accumulation of ions and respiration. My own work in Adelaide at that time, assisted by Mary Kerr Grant, had been involved in ion uptake by carrot tissue and the effects of inhibitors. Some of this was published later in collaboration with M.R. Atkinson and G. Eckermann [46]. My paper in Edinburgh on our results supported the hypothesis that accumulation of ions in carrot tissue was probably coupled directly to the charge separation resulting from electron transport and did not require the intervention of ATP. This conclusion was based on our observation that salt reduced the ATP level, perhaps by preventing its formation, and that oligomycin, known to inhibit phosphorylation of ADP which enters mitochondria, did not inhibit ion accumulation. At this time I was also collaborating with Wiskich and Diane Millard; we were particularly interested in the entry of phosphate into our beetroot mitochondria. In phosphorylating conditions (supply of substrate, co-factors, Ca or Mg, ADP and Pi), mitochondria will take up ADP and phosphate to make ATP. If, however, everything is supplied except ADP, phosphate is taken up and accumulates inside. Magnesium in the external solution is also taken up and magnesium phosphate is precipitated inside the particle. Other ions such as Na^+ or K^+, accompanied by anions such as succinate, interfere with the uptake of the phosphate and magnesium, as do uncouplers. We were happy to interpret these results as due to the charge separation from respiration extruding H^+ ions to the exterior and OH^- ions to the interior [47, 48]. The principal message in that talk might be summarised as emphasising the central function of charge separation in the mitochondrial membrane which, depending on the medium around the mitochondrion, might lead to either ion accumulation or ATP formation. I had to express our ignorance of how those possibilities in the mitochondria were linked in the behaviour of the cell as a whole.

I was enjoying life in Adelaide which suited my wife and my son very well, when an unexpected development took me away from the kind of academic life I had planned. Early in 1965, I was asked by the relevant Commonwealth Minister to set up the Australian Research Grants Committee to advise on allocation of government money to research workers in universities. Grants were to go to individuals or teams, judged on their merits and on the quality of their proposals. This was a part-time job which I accepted with the consent of my University. Much travelling was involved which, because I kept up my teaching and adminstrative duties in the Botany Department, was largely at the expense of my research.

Protons, electrons, phosphorylation and active transport

Despite this heavy load for the next four years, I was fortunate in being able to take about six months study leave in the first half of 1967 and, thanks to the hospitality of the Plant Physiology Department of Imperial College, London University, I was able to spend time in thinking about the developments in the chemiosmotic theory. This was greatly helped by my first meeting with Peter Mitchell who by then had moved from Edinburgh to set up the Glynn Research Institute at Bodmin in Cornwall. My visit, accompanied by my wife, was most enjoyable and we saw the results of the development of Glynn House as a private research establishment. As I remember the story of the decision to purchase the property, it was that Peter had been spending a holiday in Bodmin when he was shown Glynn House and, though he liked the thought of doing up the early nineteenth century building, decided it would not be practicable especially at the price being asked. Back in Edinburgh he received word from the estate agent that the price was now £ 6000 and would he be interested? He replied that he would not be interested if it was offered for £ 3000. A telegram followed which said that because of some fault, the agent had reduced the offer to £ 2800 which had been accepted and advice was required. So Mitchell had acquired Glynn House. We admired the spiral staircase which Peter had designed and had made up by the local foundry.

Our discussions during that day were very pleasing. By that time he and Jennifer Moyle had, with impeccable experimental techniques, determined the stoichiometry of the proton translocation through the respiratory chains and adenosine triphosphatase systems of rat liver mitochondria. It was about a year after he had contributed his comprehensive review in *Biological Reviews* [49] and I found that I was in agreement with practically all that he had concluded, at the same time marvelling at the sweep of his ideas and their consequences. Important experimental work carried out by others and often only tentatively related to charge separation in the electron transport chain of mitochondria or chloroplasts were brought plausibly into his reasoning. After I returned to Australia, Peter sent me the draft of the paper '*Chemiosmotic Coupling and Energy Transduction*' which was 'scheduled to appear in *Theoretical and Experimental Biophysics 2*, in a few months time'. This draft was inscribed with:

'*To Bob Robertson*, with admiration from your willing student, Peter Mitchell, January 1968.'

That paper was published about a year later (Mitchell [50]).

At Imperial College, I worked on a short article setting out
my views in the light of the chemiosmotic hypothesis, published in *Endeavour* [51] and titled *The Separation of Protons
and Electrons as a Fundamental Biological Process.*' I was spe-
cially pleased when I received a letter from Hans Krebs saying
that he had liked my historical account. In this article I recognised that the properties of charge separation might not be
confined to mitochondria and chloroplast membranes but
might also occur in other membranes of the cell.

I was invited to give five postgraduate lectures which were
attended by visitors from other London Colleges and from
Cambridge. The subjects of those lectures became the basis of
the small book which I completed after returning to Adelaide,
published in 1968 under the title of *Protons, Electrons, Phosphorylation and Active Transport* [52]. The book was well reviewed and I think that it was appreciated because it helped
many, who had been confused by the controversy over the hypothetical intermediates in oxidative phosphorylation, to appreciate the essentials of the Mitchell hypothesis and its elegant simplicity. That book went into a second impression
(1970) and was translated into German [53]. Years later, V.P.
Skulachev, Moscow University, told me that he had tried to
have it translated into Russian, along with Mitchell's work
published in 1966, but senior biochemists in the USSR said
they were not worth translating because most of the world's
leading biochemists did not accept the Mitchell hypothesis.
That was an indication of how slowly the chemiosmotic hypothesis was adopted by many senior biochemists of that time,
I have often thought that it was not until 1977 that many
people came to accept Mitchell's hypothesis, particularly after
the appearance of the multiauthored review, written by six
invited investigators, Boyer, Chance, Ernster, Mitchell,
Racker and Slater and published by *Annual Reviews of Biochemistry* [54]. In his concluding remarks, Slater says 'The
feasibility of the essential features of the Mitchell chemios-

motic hypothesis is now firmly supported experimentally.' The next year Mitchell was awarded the Nobel Prize.

Mitchell's reaction to my book was particularly gratifying. In a letter of 16 October 1968 he said:

'Thank you very much for sending me a copy of your monograph. I have greatly enjoyed reading and re-reading your artistic survey of the role of charge separation in biology. I must say that I envy your ability to use such a wide canvas and yet to obtain such a nicely balanced composition without distorting the perspectives. I am sure your little book will be a landmark, like your beautiful review of 1960, and that many students of biochemistry and physiology will take inspiration from it.'

Needless to say the book is dated now because the Mitchell hypothesis and the relevant experimental evidence have progressed considerably since.

I was embarrassed to receive a letter from R.J.P. Williams pointing out that I had not cited two relevant papers of his which had been published in 1961 and 1963 in *The Journal of Theoretical Biology* [55, 56], in which he had proposed that the protons in the lipid phase of the membranes were the factor which resulted in the formation of an anhydride of a phenol or of a phosphate. This interpretation had features in common with Mitchell's hypothesis and I should have known about it to quote but I had missed seeing the papers. It is interesting that E.C. Slater had told me that there was some work by 'Williams' which was relevant and I had looked for it but unfortunately looked in the work of G.R. Williams instead of R.J.P. Williams whom I did not know. When my book went into a second impression, I was able to insert a brief reference to the R.J.P. Williams papers.

In 1969, I changed my affiliation once again when I accepted an invitation to become Master of University House at the Australian National University in Canberra. University House was then similar to an Oxford or Cambridge College and the Master had a senior academic position in the University, with generous time and some assistance to pursue his own research; I had laboratory space in the Research School of

Biological Sciences. Alas, my plans for pursuing a quiet research life were much curtailed by my being persuaded to become President of the the Australian Academy of Science (the previous President having died in office). That honorary part-time job was no sinecure and I had four very busy years from 1970 to 1974, which played havoc with my time for research. In 1973 I was invited by the University to leave the Mastership of the House and to become Director of the Research School of Biological Sciences, a position which I occupied until my retirement in 1978.

One advantage of my position as President of the Academy of Science was that I had to make a quick trip to London early in 1971 for meetings at the Royal Society. I did not have time to go to Bodmin but Peter very kindly came to London so that we could have lunch together and have a long talk about the state of his ideas at that time.

Speculation on the energy-transducing bilayer

I was fortunate to be able to take about six months study leave in 1974 and to spend nearly all that time in Cambridge where I took the opportunity to catch up on some of the literature, especially that relating to increased knowledge of what happens in the bilayers of lipid molecules in which the membrane-bound functional protein molecules occur. At this time, I was impressed by the evidence that the tumbling of a molecule of lipid from one side of a bilayer to the other – usually referred to as 'flip-flop' – was a very rare event. Despite this, Pagano and Thompson [57] had reported that an electrically silent chloride flux across a tetradecane and egg phosphatidylcholine (PC) bilayer could be very much faster than the Cl^- current calculated from the electrical parameters of the system. If $^{36}Cl^-$ is put into a KCl solution on one side of the PC bilayer, the Cl^- and $^{36}Cl^-$ exchange across the bilayer about 1000 times faster than KCl can diffuse. To explain this and similar results obtained by others, Bangham [58] had suggested that the flux of ^{36}Cl across bilayers might be due to the permeation of H^{36}Cl. Toyoshima and Thompson [59, 60] thought that this penetration could make only a minor contribution and suggested that the ^{36}Cl flux might be due to a flip-flop of PC acting as Cl$^-$ carrier. I had been thinking of a hypothesis for the interaction between the H$^+$, the Cl$^-$ and the PC and believed it likely that the zwitterion of the PC, with its positive and negative charges, would have its polar groups in the water; if a proton combined with the negative group and a Cl$^-$ combined with the positively charged choline, the resulting neutral molecule would decrease in hydrophilicity and 'bob down' into the lipophilic region of the bilayer. There, I postulated, the H$^+$ and the Cl$^-$ would be likely to form HCl which would not dissociate in the lipophilic environment and would diffuse quickly to the other side of the bilayer where it would dissolve in the water as H$^+$ and Cl$^-$ again. This hypothesis was subsequently refined in collaboration with T.E. Thompson and published [61].

While I was thinking about the difficulties of flip-flop in membranes, I became concerned that the essential proton carriers of the chemiosmotic hypothesis, ubiquinone in mitochondria and bacteria and plastoquinone in chloroplasts, might not be capable of flipping from one side of their respective membranes to the other. I therefore began to speculate on the idea that these molecules might be like my picture of PC in that, having bobbed up in a membrane when negatively charged, they might become neutral when they picked up a proton and, with their resulting hydrophilicity reduced and lipophilicity increased, bob down again into the lipophilic environment. There they would, if an anion such as Cl$^-$ were available, lose their protons to form HCl, which would then diffuse to the other side of the membrane or, I suggested, straight to the lipophilic part, the F$_0$ of the ATP-synthetase. From my point of view, this possibility had another advantage: it would fit with R.J.P. Wiliams' suggestion that protons might move to the

ATPase in the lipophilic region in the bilayer. Furthermore, I was attracted to the idea that a non-hydrated proton (a so-called 'dry' proton) attached to chloride would be a more efficient dehydrating reagent than a hydroxonium ion entering from the aqueous phase, to bring about the condensation of ADP and inorganic P to form ATP.

About this time I had the opportunity to visit Peter Mitchell again. I arrived with the intention of talking over the current state of my thinking with him but when I mentioned my interest in Williams' suggestion of a proton in the lipophilic region, his reaction was immediate and very definite! If I remember correctly, he said 'I won't have a bar of that idea of Bob Williams!' Such was my respect for Peter's opinion in this field that I thought I had better think again and I did not even mention my own speculation.

I went back to Cambridge and, with the idea developed somewhat further, returned to Australia via the USA where I tried out my explanation on a number of people with interests in the field. Among those I talked to were people in Britton Chance's laboratory, University of Pennsylvania, Sterling Hendricks at USDA, Beltsville, Al Lehninger, Baltimore, André Jagendorf, Cornell, Paul Stumpf, Davis and, most importantly, Paul Boyer at UCLA. I don't think that I convinced anyone but the evidence for or against at that time was meagre. Back home, I put my ideas to a number of colleagues and found considerable interest from Fred Chow and Keith Boardman of Division of Plant Industry, CSIRO. They had been working on photosynthesis in green plants and considered that the idea of protons moving in a small lipophilic molecule to the thylakoid ATPase where the ATP is synthesised, was not unreasonable. The idea was developed for two outlets: first, I had been invited to give the Burnet Lecture of the Australian Academy of Science which I titled *Molecules, Membranes and Imagination* [62] and second, Boardman and I decided to write it up as a communication to *Nature*. In the Lecture, I explained in popular terms the nature of molecules in an energy-transducing membrane and tried to bring the molecules to life for the general audience by showing an animated film made from molecular models. The word 'Imagination' was in the title because I likened my pictures to those which Kekulé had in mind when he was thinking out the theory of molecular structure. I attempted to outline the Mitchell hypothesis with my modification.

Different interpretation of Mitchell's work

Boardman and I prepared two short papers for submission to *Nature*: the first, of which I was the sole author, dealt with the energy transduction mechanism in the mitochondrial membrane and the second, jointly with Boardman, was concerned with the thylakoids of chloroplasts. When the papers were drafted, we sent a copy each to Mitchell and to Williams on 7 May, 1975, both of whom responded with some criticisms. As might be expected, Williams saw that our suggestions were somewhat in accord with his ideas and, after making some helpful points of detail, said of the first paper 'I hope it is published'. Of the second paper, he said: 'Overall there can be no logical objection to your views that I know. Experiment has to decide'.

Not surprisingly, Mitchell's reaction was quite different and he wrote (15 May, 1975) with a number of criticisms to which I replied later. He regarded our ideas of the hydrophobic lipid phase with a high steady-state concentration of a small proton-carrying molecule, usually HCl under natural conditions, as overlooking or rejecting 'the notions of group translocation and vectorial metabolism, on which the chemiosmotic theory was based, and you even seem to relegate the enzymes and catalytic carriers to a secondary position in controlling the overall biochemistry of the reactions in the membrane'. To this, my reply was: 'Not so; the vectorial component is a directional effect due to the orientation of the enzyme systems involved across the membrane. Ubiquinone, for instance, picks up protons only from the matrix side, liberating them into the membrane whence they move to the outside or to the ATPase; that is vectorial'.

He commented on the way my thesis seemed to go along with the view of Williams that H^+ really causes phosphorylation by acting simply as a dehydrating agent and questioned my suggestion that a proton attached to Cl in the hydrophobic environment would be any more energetically favourable than the proton which has passed out of the membrane into water, where it would become a hydroxonium ion before taking part in the phosphorylation reaction in the ATPase. He questioned my discussion of protons having to move along inside the membrane to the ATPase because I seemed to be, in his words 'describing something different from the process by which the redox chain generates proticity by being plugged through the membrane so that the protons move through the aqueous proton-conducting media on either side of the membrane to the ATPase.' To this my reply was: 'I cannot see why you believe that protons *must* move to the outside of the membrane' ...... 'I would think that thereby they lose efficiency because they become hydrated and must lose that water of hydration (an energy-requiring process) to become good dehydrating agents again.' Mitchell also said: 'Your idea of HCl in the membrane is not analogous to the transmission of power by proticity as you seem to suggest'. I was surprised at this because I had thought there was no inconsistency. We then had some differences of interpretation of the different sites of the electron transport chain and their relation to ubiquinone but these reflected the uncertainties of the time and are not worth repeating here.

His disagreement with my view that, in his words, 'the high electrically silent Cl^- permeability of lipid films is due to dissolved HCl in the lipid does not seem to be tenable in view of the recent work of Toyoshima and Thompson (1975)', did not concern me. As mentioned earlier, I did not agree with their interpretation and later collaborated with Thompson in rein-terpretation of their data [61].

Mitchell's letter said: 'I would hate, for reasons for our long-standing friendship, to seem unsympathetic and over-critical, but as you solicited my comments, I feel bound to reply straightforwardly'. In concluding that letter he said 'Mean-while I am sorry that my reaction to your manuscripts was so critical, and sincerely hope you will not take this to heart or feel that I have been unwarrantably harsh'. I replied: 'I found your criticisms very stimulating. You were certainly not 'un-warrantably harsh'; I expect my friends to be critical when I propose ideas and see no reason to think criticism should af-fect friendship'.

The papers, amended in the light of the comments of Wil-liams and Mitchell, were sent to *Nature* on 21 July, 1975 and copies of the final drafts were sent to both. Though I thought we were making a modification of the Mitchell chemiosmotic hypothesis, it was now clear that Peter did not have any belief in my approach. He (letter of 28 July, 1975) regarded my hy-pothesis that HCl is the energy-rich chemical intermediate, common to the redox chain system and the reversible ATPase system, as essentially a chemical coupling hypothesis. After mentioning some new work on the cytochromes, he said 'but I see from your letter and manuscript – which I have found very interesting reading, and for which many thanks – that, as you say, it is clear that you and I look at these problems very differ-ently at present. So I only mention, and will not labour this point about the progress of the chemiosmotic theory to the cytochrome b complexes because I am afraid that it will serve to emphasise the divergence of our viewpoints – which, I must admit, I accept with a mixture of anguish and sadness.'

We did not continue the correspondence at that time, partly because there might be some comment from *Nature*. In the event, that journal used a referee who made much the same comments as I think Peter would have made. We had the anti-climax of our papers being rejected because they were 'of insufficiently wide interest to compete successfully for our limited publication space'. We were advised to try a specialist journal. Boardman and I decided to redraft our account in one joint paper suitable for *FEBS Letters*. This was published late in 1975 [63]. On 16 March, 1976, Peter wrote saying that he had only just realised that our paper had appeared because copies of *FEBS Letters* had been delayed in delivery. At the same time he enclosed the manuscript of the review to be published in *J. Theoret. Biol.* [64]. He said 'I continue to feel a mixture of sadness and anguish that your view, now publicly expressed in *FEBS Letters*, is essentially contrary to the one developed in the enclosed.' After taking time to digest (at least partially!) the new developments, I replied that I was not yet ready to abandon my simple alternative even though he had turned it down earlier because it was not sufficiently chemiosmotic. Boardman and I still regarded the hypothesis as a variant of his, which would have been impossible without his developments.

In his reply, Peter (27 April, 1976) generously said that he was sorry that our paper was rejected by *Nature* and 'their reasons seem to have been typically idiotic. Your point of view obviously deserves to be known and discussed'. His next sentence is important because it sums up his conviction about this theory: 'I was only agonised about it because it seemed, and still seems, to me to turn right back from the vectorial expression of chemical reactions in oxidative and photosynthetic phosphorylation, which, in my opinion, is just beginning to provide a revitalising conceptual background for the rationalisation of bioenergetics'. He generously went on to say: 'It has been a special pleasure to me to see the marvellous work of my old teacher and friend, David Keilin, on the cytochrome system, emerging as the foundation of the notion of the connection between spatial directiveness and chemical reaction in biology – linking up with the ideas of Fritz Lipmann and other biochemists and integrating the work and speculations of Lundegårdh and of yourself'. He then went on to regret that

I had espoused 'an essentially non-directional central concept that is purely chemical and runs contrary to the fundamentally vectorial view of the cytochrome system and of the other interrelated protonmotive and protonmotivated systems'.

From then on we did not continue the discussion. It seemed necessary to wait to see whether we received any support from other workers in the field. The only one that I have on record was that of Barr who had suggested molecular HCl as a transport form of Cl⁻ [65]. I spent such time as I had for research in trying to think of possible experiments which would provide evidence for or against our interpretation, without any success. One important contribution would have been information about the behaviour of HCl in a lipid environment and I attempted to get some collaboration from an NMR chemist but that ran into difficulties of technique and interpretation. I increased my interest in the properties of the lipids of the bilayer and particularly in relation to their resistance to various types of diffusion. In May 1977, I wrote to congratulate Peter on his election as a Foreign Associate of the US National Academy of Sciences and mentioned that I was 'battling with the problems of how to determine what happens inside the bilayer'. In his letter of thanks, he wrote: 'About your question of what happens in the bilayer – my feeling is that the short answer for the protonmotive chemiosmotic system is: insulation'. He then went on to stress the importance of the catalytic proteins of enzymes and carriers, with the 'stuff in between acting as a diffusion barrier'.

In October 1978, Peter was awarded the Nobel Prize for Chemistry. In writing to congratulate him, it gave me great pleasure to be able to say: 'I have never made predictions about Nobel Prize winners before, so it is some satisfaction to me to recall that I told my audience at Imperial College in 1967 that I was sure that the proper recognition of your work – which was lagging at that time because of the ingrained ideas of some of the traditional biochemists – would lead to the Nobel Prize'.

Charge separation and ion transport

With some of my time, I tried to further the understanding of our original theory of the connection between charge separation and ion movement in plant cells. Experiments, in collaboration with Peter Anderson and Barbara Wright [66], carried out on carrot root cells, measured cell membrane potentials as influenced by respiratory inhibitors and uncouplers. We found that carrot cells seemed to have two electrogenic ion transport systems. A large electrogenic polarising transport was coupled to the cytochrome energy supply of the cell since it was inhibited by cyanide and carbon monoxide in the dark, and by the uncoupler carbonyl cyanide m-chlorophenylhydrazone. Because we then believed that the cytochrome system was located in the mitochondria, we postulated a mitochondrial extrusion of H^+ located close to the plasmalemma interior surface to which the mitochondria appeared to be contiguous. At that time the presence of cytochrome oxidase in the plasmalemma was unknown, an important advance which will be referred to later (see Crane, Morré and Löw [67]). The second electrogenic exchange was coupled to a Cl^- influx; we were unable to characterise the metabolic link in this system.

I was not able to think of any way to obtain experimental evidence for the idea that HCl might be put into the lipophilic region of the bilayer so I thought that it would be interesting to investigate whether bacteriorhodopsin, which brings about a charge separation in the membrane of *Halobacterium halobium*, could transfer a hydrogen accompanied by a chlorine into the hydrocarbon region of an emulsion. In starting this work, my aim was not to regard the system as a model of what happens in mitochondrial and chloroplast membranes, but rather to see whether HCl could be transferred into the non-aqueous environment. Preliminary experiments, carried out with Susan Young in Canberra, were promising and the work was continued in Sydney.

After retirement, continuing interest in membranes

The year 1978 was my last at the Australian National University and we moved from Canberra to Sydney where I was happy to accept an invitation from Michael Pitman to be an honorary visitor in the plant physiology laboratories of the University of Sydney. There, with Adele Post as a graduate assistant, I was able to work on bacteriorhodopsin which was introduced into emulsions of octane in water with soybean phospholipids as the emulsifying agent. Freeze-fracture electron microscopy indicated that the bacteriorhodopsin was at the interface of the octane droplets, possibly in a phospholipid monolayer. When this emulsion was illuminated, protons were removed from the aqueous medium with a half time of 30 sec. When the light was switched off, the protons returned to the water. As fitted the hypothesis, the proton uptake was obtained with sodium chloride in the aqueous solution. We were interested to find that sodium nitrate was even more effective. The uptake of protons increased with increasing concentration of the solution, up to about 2M. Potassium salts were less effective than sodium [68].

As so often happens with research, a serendipitous opportunity presented itself to let us study an unexpected property of bacteriorhodopsin. At this time I was on an advisory committee concerned with providing money for 'wool-harvesting' research, methods of removing wool from sheep, as an alternative to the conventional method of shearing. Ideally, a substance was being sought which would, if injected into a sheep, interrupt temporarily the growth of the wool fibre so that a weakness would develop and the fleece could be pushed off with the hands. Before a solution to this problem had been found, many different chemical compounds had been administered to experimental sheep. One day I was listening to Dr Liepa of the CSIRO describing the substances he had tested. One, which he described as very good could not be used because it made the sheep blind. When I asked why, he said that he thought that it formed a Schiff's base with the retinal which is an essential molecule that normally attaches to rhodopsin in eyes to make sight possible. We were fascinated to find that this o-toluamide, which has no shorter name than N-[5-(4-aminophenoxy)pentyl]-2-methylbenzamide, also inhibited the uptake of protons by our bacteriorhodopsin emulsion system, probably by forming a stable Schiff's base with retinal. This inhibitor caused a decrease in the light absorbance in the region of the bacteriorhodopsin peak, consistent with its being effective at the retinal site. We were not able to take this work any further [68, 69].

After retirement, I worked on another book in an attempt to give readers a picture of how lively the remarkable structures of membranes are at molecular level. As mentioned earlier, I was always influenced by Kekulé's ability to visualise molecules in motion and had produced a film titled *A Vision of Membranes* made from space-filling molecular models with animation to show their presumed motions and changes in reactions. The book, *The Lively Membranes* [70], attempted to convey some of the same ideas with descriptive text and line diagrams. It was reasonably well reviewed and some of the mistakes I had made or differences in interpretation, were pointed out. For one I feel I should apologise: the well-known Nernst equation applied to an example in Chapter 7 was wrongly transcribed and is in fact a nonsense. The book is now out of print.

In some ways I was writing this book at the wrong time. The transmembrane proteins in which the most important reactions of membrane processes occur, were only just beginning to be understood. At that time, bacteriorhodopsin, thanks to the brilliant work of Henderson and his collaborators, was better understood than any other [71, 72]. Though I appreciated the importance of the hydrophobic α-helices, which not only held the protein molecule orientated in the right way in the membrane but also provided, in some unknown manner, the channel through which the protons were separated from the electrons and allowed to move across the membrane, I did not have enough insight to recognise that this kind of structure was going to be the pattern for energy-transducing molecules. Indeed, I showed complete ignorance of the F_0 part of the ATP-ase which is its membrane-bound hydrophobic structure. Within a few years, the powerful techniques of site-directed mutagenesis were being used to determine the particular amino acids involved. It is not too much to say our knowledge of membrane behaviour was waiting for our understanding of the trans-membrane proteins and, equally importantly, our ability to manipulate their synthesis. These developments, especially relating to ATP synthase, are very well summarised by Gibson [73][1] and Cox, Devenish, Gibson, Howitt and Nagley [74].

Molecular movements within the lipid bilayer

I continued to be concerned about whether the ubiquinone had sufficient mobility in the bilayer to eliminate the need for a small proton carrier such as the hypothetical HCl. At this time I was fortunate to be able to take part in the research programme of Bruce Cornell and his group (at the CSIRO Division of Food Research in Sydney), using nuclear magnetic resonance (NMR) techniques. We investigated the location of ubiquinone (Q_{10}) and deuterated analogues of Q_{10} in model lipid-bilayer membranes of dimyristoyl-phosphatidylcholine (DMPC), in plasma membranes of E. coli and in beetroot mitochondrial membranes. The Q_{10} analogues were also incorporated into membranes of an E. coli mutant (AN750) which had no native ubiquinone; the synthetic Q_{10} increased the membrane transport of electrons. We also studied the properties of a short chain analogue. Concurrently we measured the ability

---

1 See Chapter 6 in this volume.

of native and deuterated ubiquinone analogues to restore electron-transport activity to Q_{10}-depleted membranes of both *E. coli* and beetroot mitochondria. Our results were interpreted as showing that most Q_{10} was aggregated into a mobile environment which was physically separate from the orientational constraints of the bilayer lipid chains. By contrast, a short straight chain analogue of Q_{10} in which the ten isoprene groups of the chain had been replaced by a perdeuterated tridecyl chain, showed NMR spectra typical of an ordered lipid which was intercalated into a bilayer. The NADH oxidase acivity and oxygen uptake were independent of which analogue was incorporated into the membrane. Thus, despite the major difference in their physical association with membranes, the electron transport function of long and short chain ubiquinones was similar. We concluded that the intercalation of ubiquinone into the membrane lipids is not of primary importance in its biological transport function (Cornell, Kenary, Post, Robertson, Weir and Westerman [75].

In subsequent work (Cornell, Kenary, Knott, Post, Robertson, Separovic, Weir and Westerman [76]), two techniques were used: NMR to give information on the dynamic disorder of Q_{10} in the membrane and neutron diffraction for information on the distribution and location of Q_{10} in relation to the surrounding lipid. Enzyme assays (NADH oxidase activity) were carried out on *E. coli* membranes from which Q_{10} had been extracted and with membranes reconstituted with deuterated analogues of Q_{10}, which restored activity to > 30% of the original. Both the NMR and neutron diffraction data were interpreted as showing that the Q_{10} was not orientated with the membrane lipid but was tumbling isotropically on a time scale shorter than a μs. This lack of dependence of the motion of the Q_{10} analogues on the lipid phase was interpreted as showing that the Q_{10} is physically separate from the membrane lipid. We concluded that the Q_{10} aggregates as a separate phase and is not systematically organised relative to the lipid structure, though it is possible that a small quantity

(< 10%), organised in a regular manner relative to the lipid, would not be detected. Thus, despite the major difference in their physical association with the membrane lipids, the electron transport functions of long and short chain ubiquinones were similar. These results could suggest that the proton and electron transporting function performed by Q_{10} is carried out in conjunction with another membrane molecule and does not arise from the physical interaction of Q_{10} with the membrane lipid.

Thus this work, with its unexplained location of most of the ubiquinone, did not resolve my question of whether an additional small proton carrier was necessary for the transfer of protons. Following a heart attack in 1986, I found it necessary to decrease my involvement in active research and have not kept up with all the considerable work in this field. However, one of my principal reasons for suggesting that a small molecule was involved in addition to the ubiquinone or plastoquinone, was that I did not believe that such long chain molecules were mobile enough in the bilayer, especially in their transverse diffusion. The work of Hackenbrock and his associates may have answered my misgivings because they have shown that ubiquinone in model lipid membranes is highly mobile laterally and transversely [77].

So it may be that my friend Peter Mitchell was right in believing that the quinones were adequate and I was wrong in believing in the need for a small proton carrier such as my suggested HCl. A complete understanding of the system does require detailed knowledge based on the sorts of techniques which measure the locations and movements of the reactant molecules in the living bilayers. Such knowledge takes the explanations a step deeper than considering the lipophilic region as mere insulation.

A life influenced by charge separation

I have been fortunate to enjoy the stimulating development of
thought due to realisation that charge separation across membranes less than 100 Å in thickness is a fundamental process
in living things. Much of my pleasure was as a spectator ad-
miring two stars of the game – Lundegårdh and Mitchell. I
make no attempt to summarise the great body of work and
development of ideas for which Mitchell was responsible since
it is done so well in the symposium *Perspectives in Vectorial
Metabolism and Osmochemistry* [78], both in Mitchell's own
article and in those of the other authors. In his chapter, Mitchell quite objectively recalls his mistake in rejecting the
evidence for the protonmotive function of the cytochrome c oxidase which, he said, set back his own interest in this important problem, and perhaps diverted the attention of some others for about seven years. He then wrote: 'However, I have
subsequently tried to make up for lost time'. The quality of
both his character and his scientific contributions was summarised in masterly fashion by Garland in *Nature* [79].

In my own activities, my interest in research was inter-
spersed with modest attempts to help science be better used in
community welfare, sometimes with success and sometimes
with failure. Overall, it has been a fortunate life and a happy
interaction with many friendly colleagues and a helpful and
unselfish wife who aided the preparation of this chapter.

REFERENCES

1 R.N. Robertson, Ann. Rev. Plant Physiol. Plant Mol. Biol., 43 (1992)
 1–24.
2 S.C. Brooks, Protoplasma, 8 (1929) 389–412.
3 G.E. Briggs, Proc. Roy. Soc. B, 107 (1930) 248–269.
4 R.N. Robertson, PhD Thesis, Univ. Cambridge (1938).
5 H. Lundegårdh and H. Burström, Biochem. Z., 261 (1933) 235–251.

6 H. Lundegårdh, Nature, 143 (1939) 203.
7 D.R. Hoagland and F.C. Steward, Nature, 143 (1939) 1031.
8 H. Lundegårdh, Nature, 145 (1940) 937.
9 D.R. Hoagland and F.C. Steward, Nature, 145 (1940) 116.
10 R.N. Robertson, Nature, 145 (1940) 937.
11 R.N. Robertson and J.S. Turner, Aust. J. Expt. Biol. Med. Sci., 23 (1945) 63–73.
12 H. Lundegårdh, Ann. Agric. Coll. Sweden, 8 (1940) 234–404.
13 F.C. Steward and C. Preston, Plant Physiol., 16 (1941) 85–116.
14 R.N. Robertson, Aust. J. Biol. Med. Sci., 19 (1941) 265–78.
15 H. Lundegårdh, Arkiv. Bot., 32A (1945) 1–139.
16 R.N. Robertson and M.J. Wilkins, Nature, 161 (1948) 101.
17 R.N. Robertson and M.J. Wilkins, Aust. J. Sci. Res. B, 1 (1948) 17–37.
18 R.J. Poole, Encycl. Plant Physiol., 2A (1976) 229–244.
19 E.J. Conway and T.C. Brady, Nature, 162 (1948) 456–457.
20 E.E. Crane and R.E. Davies, Biochem. J., 43 (1948) xlii.
21 E.E. Crane and R.E. Davies, Biochem. J., 43 (1948) xlii.
22 D.C. Weeks and R.N. Robertson, Aust. J. Sci. B., 3 (1950) 487–500.
23 R.E. Davies and A.G. Ogston, Biochem. J., 46 (1950) 324–333.
24 R.N. Robertson, M.J. Wilkins and D.C. Weeks, Aust. J. Sci. B., 4 (1951) 248–264.
25 H.G. Du Buy, M.W. Woods and M.D. Lackey, Science, 111 (1950) 572–574.
26 G.E. Palade, J. Histo. Cyto. Chem., 1 (1953) 188–211.
27 J.L. Farrant, R.N. Robertson and M.J. Wilkins, Nature, 171 (1953) 401–402.
28 J.L. Farrant, C. Potter, R.N. Robertson and M.J. Wilkins, Aust. J. Biol., 4 (1956) 117–124.
29 R.N. Robertson, Ann. Rev. Plant Physiol., 2 (1951) 1–24.
30 R.N. Robertson, M.J. Wilkins and A.B. Hope, Nature, 175 (1955) 640–641.
31 R.N. Robertson, M.J. Wilkins, A.B. Hope and L. Nesztel, Aust. J. Biol. Sci., 8 (1955) 164–185.
32 S.I. Honda and R.N. Robertson, Aust. J. Biol. Sci., 9 (1956) 305–320.
33 S.I. Honda, R.N. Robertson and J.M. Gregory, Aust. J. Biol. Sci., 11 (1958) 1–15.
34 J.T. Wiskich, R.K. Morton and R.N. Robertson, Aust. J. Biol. Sci., 13 (1960) 109–122.
35 R.L. Lester and F.L. Crane, J. Biol. Chem., 234 (1959) 2169–2175.
36 R.A. Morton, Nature, 182 (1958) 1764–1767.
37 R.N. Robertson and J.F. Turner, Aust. J. Sci. Res., 4 (1951) 92–107.
38 J.A. Pearson and R.N. Robertson, Aust. J. Sci., 15 (1952) 99–100.
39 J.A. Pearson and R.N. Robertson, Aust. J. Biol. Sci., 7 (1954) 1–17.
40 M.D. Hatch, J.A. Pearson, A. Millerd and R.N. Robertson, Aust. J. Biol. Sci., 12 (1959) 167–174.

41 K.S. Rowan, H.K. Pratt and R.N. Robertson, Aust. J. Biol. Sci., 11 (1958) 329–335.
42 R.N. Robertson, Biol. Rev., 35 (1960) 231–264.
43 G.E. Briggs, A.B. Hope and R.N. Robertson, *Electrolytes and Plant Cells*, (1961) Blackwell.
44 P. Mitchell, Nature, 191 1961, 144–148.
45 A.T. Jagendorf and E. Uribe, Proc. Natn. Acad. Sci. U.S.A. 55 (1965) 170–177.
46 M.R. Atkinson, G. Eckerman, M. Grant and R.N. Robertson, Proc. Natn. Acad. Sci. U.S.A., 55 (1966) 560–564.
47 D.L. Millard, J.T. Wiskich and R.N. Robertson, Proc. Natn. Acad. Sci. U.S.A., 52 (1964) 996–1004.
48 D.L. Millard, J.T. Wiskich and R.N. Robertson, Plant Physiol., 40 (1965) 1129–1135.
49 P. Mitchell, Biol. Rev., 41 (1966) 445–602.
50 P. Mitchell, Theor. Exp. Biophys., 2 (1969) 159–216.
51 R.N. Robertson, Endeavour, 26 (1967) 134–135.
52 R.N. Robertson, *Protons, Electrons, Phosphorylation and Active Transport* 1968, Cambridge University Press.
53 R.N. Robertson, *Protonen. Electronen. Phosphorylierung. Aktiver Transport* 1971, Wilhelm Goldmann, München.
54 P.D. Boyer, B. Chance, L. Ernster, P. Mitchell, E. Racker and E.C. Slater, Ann. Rev. Biochem., 46 (1977) 955–1026.
55 R.J.P. Williams, J. Theoret. Biol., 1 (1961) 1–17.
56 R.J.P. Williams, J. Theoret. Biol., 3 (1963) 209–229.
57 R. Pagano and T.E. Thompson, J. Mol. Biol., 38 (1968) 41–57.
58 A.D. Bangham, Ann. Rev. Biochem., 41 (1972) 753–776.
59 T. Toyoshima and T.E. Thompson, Biochemistry, 14 (1975) 1518–1525.
60 T. Toyoshima and T.E. Thompson, Biochemistry, 14 (1975) 1525–1531.
61 R.N. Robertson and T.E. Thompson, FEBS Letters, 76 (1977) 16–19.
62 R.N. Robertson, Records Aust. Acad. Sci., 3 (1976) 88–91.
63 N.K. Boardman and R.N. Robertson, FEBS Letters, 60 (1975) 1–6.
64 P. Mitchell, J. Theoret. Biol., 52 (1976) 327–360.
65 D.G. Spear, J.K. Barr and C.E. Barr, J. Gen. Physiol., 54 (1969) 397–414.
66 W.P. Anderson, R.N. Robertson and B.J. Wright, Aust. J. Plant Physiol., 4 (1977) 241–52.
67 F.L. Crane, D.J. Morré and H.E. Löw, *Oxidoreduction at the Plasmamembrane: Relation to Growth and Transport vol. 11 Plants.* 1991, CRC Press.
68 A. Post, S.E. Young and R.N. Robertson, Photobiochem. and Photobiophys., 8 (1984) 153–162.
69 R.N. Robertson, Proc. Amer. Philosophical. Soc., 130 (1986) 374–381.
70 R.N. Robertson, *The Lively Membranes* (1983) Cambridge University Press.
71 R. Henderson, Ann. Rev. Biophys. and Bioeng., 6 (1977) 87–109.

72 Engleman, D.M., Henderson, R., McLachlan, A.D. and Wallace, B.A.
    Proc. U.S. Nat. Acad. Sci., 7 (1980) 2023–2027.
73 F. Gibson, *The Exploration of Living Membranes*, Aust. Acad. Sci. Pub-
    lic Lectures 1991, 1–21.
74 G.B. Cox, R.J. Devenish, F. Gibson, S.M. Howitt and P. Nagley in *Mo-
    lecular Mechanisms in Bioenergetics* ed. L. Ernster, 1992 283–315.
75 B.A. Cornell, M.A. Kenary, A. Post, R.N. Robertson, L.E. Weir and P.W.
    Westerman, Biochem. 26 (1987) 7702–7707.
76 B.A. Cornell, M.A. Kenary, R. Knott, A. Post, R.N. Robertson, F. Sepa-
    rovic, L.E. Weir and P.W. Westerman, in *Biochemistry Bioenergetics
    and Clinical Applications of Quinone* 1990 pp 27–32, Taylor and
    Francis.
77 B. Chazotte, E-S. Wu and C.R. Hackenbrock, Bᵢ ᵢim. Biophys. Acta,
    1058 (1991) 400–409.
78 P. Mitchell and C.A. Pasternak, *Perspectives   Vectorial Metabolism
    and Osmochemistry* 1992, Plenum.
79 P. Garland, Nature, 356 (1992) 747.

E.C. Slater, R. Jaenicke and G. Semenza (Eds.)
Selected Topics in the History of Biochemistry: Personal Recollections, IV
(Comprehensive Biochemistry Vol. 38) © 1995 Elsevier Science B.V.

Chapter 8

A Biochemical Autobiography

CLAUDE RIMINGTON[†]

The Norwegian Radium Hospital, Oslo, (Norway)

Introduction

Biochemistry is the study at molecular and atomic levels, by the use of chemical procedures and concepts, of the organization and function of biological systems. In particular, it deals with the study of (*i*) the nature of the chemical constituents of living matter and of chemical substances produced by living things, (*ii*) the functions and transformations of these chemical entities in biological systems and (*iii*) the chemical and energetic changes associated with these entities in the course of the activity of living matter.

The attempt to describe life processes in physical and chemical terms presupposes that the laws of physics and chemistry are obeyed by living systems, just as they are by inanimate matter, and that it is feasible to study processes not only in intact organisms but even in the components of the cells which make up organisms. The origins of biochemistry can be ascribed to the validation of both these assumptions. That the chemistry of living matter is susceptible to study in the laboratory, and that it does not require the intervention of a mysterious 'vital force', was shown largely by the work of 19th-century

---

† Deceased August 8, 1993.

chemists such as Liebig, Wöhler and Kolbe, who synthesized
from simple starting materials substances previously known
to have been elaborated only by living organisms. The additional demonstration that (non-living) extracts of living organisms could catalyse, in the test-tube, the reactions normally
occurring in the cell rests on the observations of Buchner*.

Biochemistry was established as a separate discipline at the
turn of the present century. In Great Britain, Halliburton
built up the first research school in biochemistry (1910–1925)
and in the words of Gowland Hopkins 'was the first ... to secure
for biochemists general recognition and respect'. Hopkins' professorship in Cambridge dates from 1922. Others were already
teaching Physiological Chemistry at various universities in
the U.K.

The major centres for biochemical research between 1840
and 1870 were the German medical schools, as described by
Kohler. Hoppe-Seyler founded his *Zeitschrift für Physiologi-
sche Chemie* in 1877, but there still lingered among German
organic chemists a contemptuous reference towards biochemistry as 'Schweinchemie' as opposed to 'Reinchemie'. Surprisingly enough, such an attitude was still to be occasionally
found later in England as I well remember personally when, as
a young Cambridge graduate in 1924, I called on the then
Registrar of the Institute of Chemistry in London to ask his
advice concerning my future career. He tried to dissuade me
from studying biochemistry which he described as 'analysing
urine and faeces and suchlike things'. Fortunately I rejected
his advice and, instead, enrolled under Hopkins for his new
Part II Tripos in Biochemistry.

But to return to the early days, the most significant development in the U.K. was the establishment of the first Chair in
Biochemistry at Liverpool in 1902 to which Benjamin Moore

---

* Prior discovery of cell-free fermentation has been claimed for M. Manasseina (see J.
Lagnado, *The Biochemist, Oct. /Nov. 1992, pp. 21–22.*

was elected, an extraordinarily versatile person whose work included studies in diabetes, photosynthesis, renal calculi, heavy metal toxicity and marine algae. He also had a fairly definite conception of membrane equilibria.

Much has already been written about Gowland Hopkins' (Fig. 1) struggle to found a real school of biochemistry in Cambridge and his many disappointments and frustrations until in 1922 he became the Sir William Dunn Professor of Biochemistry with an Institute opened in 1924 and acceptance of Biochemistry as a Part II (Honours) subject. He was the third professor of biochemistry in England, Arthur Harden (Fig. 2) at the Lister Institute (1912) being the second after Benjamin Moore (1902).

I have written these words in my ninetieth year and have therefore witnessed vast and impressive developments in the science of biochemistry, at times revolutionary in character, and still continuing.

To fulfill the invitation to contribute a biochemical autobiography to the 'History of Biochemistry', I propose to record my life's scientific journey together with allusions to the changing thought and concepts of biochemistry and also to some of the notable characters I have met on the way.

Personal early years

As a child I grew up in close contact with nature since we lived in country surroundings in Hertfordshire. I had very few companions, but to me the woods and fields offered abundant recompense. I roamed far and wide in all seasons of the year, delighting in the wild flowers, birds and animals; there was not a stream or pond which I did not know intimately as well as the newts, frogs and smaller creatures to be found in it. I collected flowers, grasses and butterflies, but more as a pastime than for scholarship. I must have been attracted more artistically than scientifically.

At the age of 9 I was sent to the Haberdashers' Aske's Boys School, then at Cricklewood, and on the long train journeys back and forth from home I began to read anything I could get hold of in our school books, much poetry, a selection of French authors and a German classic – Goethe's Faust! For preventing my interest in literature at this period from becoming absolutely chaotic, I will be ever grateful to my teacher in English Literature who put before me R.L. Stevenson, William Morris, Joseph Conrad, Thomas Hardy and others.

Chemistry was my other great attraction. It fascinated me and we had an excellent teacher who very thoughtfully gave me the run of the laboratory on Saturdays. I had already borrowed Perkin and Kipping's 'Organic Chemistry' from the school library for leisure reading and had taught myself quite a lot before we began any class work in organic chemistry. With this knowledge and a copy of Weston's 'Carbon Compounds' which I had acquired, I set to work to identify several chemicals whose bottles had lost their labels, and to carry out small text-book experiments – on the whole with a fair amount of success. All this involved the long, time-consuming train journeys every Saturday but I still roamed the woods and fields and set up a small laboratory at home in which I attempted to isolate various materials which had been reported to occur in plants. Years before, I can remember, I had pottered about in our garden and made decoctions from some of the plants I had found growing there.

Among our school prizes was one called 'The Humphrey Originality Prize' which was usually awarded for a collection of something or other. I wrote up some of my experiments and submitted, with specimens, an effort entitled 'Original Chemical Research'. As my prize, I chose a copy of Green's 'Natural and synthetic colouring matters'.

The 1914–1918 war

The war of 1914–1918 occurred at a critical stage in my schooling. These were the years when I should have learned most. In
point of fact, the teaching deteriorated lamentably. Several of
the staff enlisted and we in the upper forms were left much on
our own, without guidance. I can only regret that we wasted so
much of our time! Conditions of life were also uninspiring, not
to say depressing. We still lived out in the country where one
of our neighbours, an elderly maiden lady, made her contribution to the war effort by serving once a week on the night shift
of a canteen, a hut erected between King's Cross and St. Pan-
cras stations and run, I believe, by the Church Army. This was
no show place with neat little tables and white tablecloths but
a most primitive shelter where exhausted, mud-caked men,
straight from the trenches, could get a fried egg and sausage
and a mug of hot tea before being taken to another station for
a further leave-train.

My sister and I, sixteen and fourteen years old, begged and
implored until we were given permission to accompany this
lady. One night each week, we worked from about 9 p.m. until
6 a.m. with scarcely any respite, mostly frying mountains of
sausages and eggs on the huge stove or 'washing up' in caul-
drons of hot but greasy water. Often there were air-raids over-
head, the Ack-Ack being just audible above the general clatter,
but we had no time to think or worry about danger. Dead tired,
we slept in the train on the way home, washed, got some
breakfast and set off again for the day's school. It was perhaps
crazy but it opened my eyes to some of the grim realities of a
war – men so dead with fatigue that they slept on the floor in
their muddy trench coats, on chairs, even on tables if there
was not room on the floor. I also saw, one night, from my bed-
room window the Cuffley Zeppelin falling in flames!

The immediate post-war years brought fresh problems. The
family moved from our home in the country and we lived for a
time in a rather depressing boarding-house in Bloomsbury be-

fore finally settling in Golders Green. It was easier to get about without the long train journeys but I missed the freedom and healthy activity of the countryside.

Cambridge

In my last year at school in 1920, I took the London Intermediate Examination with First Class honours and intended to sit for an open scholarship to Cambridge but instead suffered a serious breakdown in health, necessitating a year's recuperation in South Devon. Here I once more lived an open-air life, roaming widely, particularly over the lovely and then unspoiled stretch of coastline between Bolt Tail and Bolt Head. I made friends with the local fishermen and drank cider with them in the evenings in their cottages. It is the year of my youth which I remember best and look back upon with nostalgia – a care-free, happy existence. I composed a lot of poetry in a rather Swinburnian style and soon felt well enough to tackle a more serious job – the plugging of Latin for the Cambridge 'Little-Go'. This I passed successfully, and also the entrance examination to Emmanuel College, and went up in 1921 with chemistry, physics and botany as my subjects for Part I of the Science Tripos – and great expectations. These were not realized. Physics I had never liked and the chemistry teaching struck me as dull and uninspired. Sir William Pope was then Professor and delivered his lectures almost word for word from 'Perkin and Kipping'. In fact, I often sat with this book in front of me to check his delivery. The only breath of the new spirit stirring in organic chemistry came from a more junior lecturer named Palmer. Botany was more exciting under Seward and Blackman and the field excursions with Gilbert Carter were an intellectual romp. I also joined an Easter-holiday expedition from the Botany School to study sea-weeds in the laboratory at Port Erin, Isle of Man. The red algae, with their beautiful anatomical construction, fascinate me to this day.

I got only a Second Class in the 'Mays' examinations, which I felt was not good enough. As luck would have it, one of my friends had switched over from science to medicine and I decided to include physiology as a fourth Tripos subject. This was a decisive step; I had found the opening for which I had been blindly searching! It also brought me, later, under the Tutorship of T.S. Hele (Fig. 3) who steered me towards biochemistry. My teachers included Barcroft, Adrian and Hopkins.

The next year's examination brought me a First Class with highest marks in physiology and an Emmanuel College Scholarship. Part I in the four subjects followed in 1924 with a First Class and high marks in both chemistry and physiology, and I went on to sit for a London Honours B.Sc. that autumn, offering chemistry, physiology and physics as my three subjects. I was confident of having done well in chemistry and particularly in physiology where I had been fortunate enough of have as an essay subject 'The Cell Theory'. It so happened that I had prepared a paper on this theme for the Emmanuel Science Society and was therefore well equipped. I thought my physics had been a dismal failure, so much so, in fact, that I confidently expected to plough the whole examination. My Physiology and Chemistry must have saved me, however, because I appeared in the First Class Honours List. During my comparatively easy third year for Part I, I had listened to many of the Part II physiology lectures, some of those in chemistry, and had made lasting friendships with biochemists such as Robin Hill and Malcolm Dixon, and later on Thaddeus Mann and Max Perutz. Several eminent visitors from abroad attended meetings or gave lectures; among them Svedberg, Tiselius and Wieland. This was all good experience for a young scientist.

Now it was a question of the future. Hele had encouraged me to plan my time so that I could take the new Part II Tripos in Biochemistry which Gowland Hopkins was launching. Being not quite sure what biochemistry would lead to, I went to see Pilcher, Registrar of the Institute of Chemistry (now the Royal Institute of Chemistry) in London. As mentioned above, he did all he could to dissuade me, pointing out that biochemists worked with such stuff as urine and faeces and that a sound training in analytical chemistry would be much wiser. I thanked him for his advice which I promptly and emphatically disregarded, entering for the Part II in Biochemistry instead. There were three of us taking the examination and several others, mostly visitors, including John T. Edsall from Harvard, who attended the lectures and practicals; we became good friends.

There was a good deal of improvisation in this first year before things got properly sorted out but the course ran to a triumphant conclusion under the tireless energy of the Reader in Biochemistry, J.B.S. Haldane. He was also one of the examiners, castigated me afterwards for my horrible hand-writing, but nevertheless gave me a First. This was in 1925.

First biochemical research

During the preceding summer, after the Part I examination was over, I had begun my first piece of serious research. I did not know what to do and Haldane suggested that I analysed a stock of urines which he had from one of his self-experiments on mineral balance. Providentially this was overheard by H.D. Kay (Fig. 4), then working on non-protein phosphorus compounds in milk. He told us he had made a pilot experiment to see what happened to the phosphorus in caseinogen when the latter was acted upon by trypsin. This was in the days when phosphorus was 'the Cinderella of the elements', as he called it, brought to notice only by Robison's work with Harden on the sugar esters and their possible role in yeast fermentation. I liked the idea of investigating this further and Haldane magnanimously agreed. It is to H.D. Kay that I owe my real introduction into biochemical research.

Working alone in the 'Part II lab.', I quickly found that there were two stages in the liberation of phosphorus from caseino-

Fig. 1.
F. Gowland Hopkins

Fig. 2.
Arthur Harden

Fig. 3. T.S. Hele
(right) with
R. Coombs

Fig. 4.
H.D. Kay

Fig. 5.
S.P.L. Sørensen and
wife

Fig. 6. K. Linderstrøm-Lang (left),
Carlsberg Laboratory Copenhagen

gen by the crude trypsin preparations then available – first, removal of a peptone-like moiety which still retained the greater part of the organically bound phosphorus, and then a further decomposition of this with the liberation of inorganic phosphate. By suitably adjusting the conditions and following the process analytically, one could prepare solutions rich in the organic fraction. To isolate this was my next problem. By the successive use of precipitating agents, lead, uranium, copper, etc., I found that I could obtain the phosphorus-containing peptone as its barium salt and with a fairly constant analytical reproducibility (P/N ratio). My first paper on the determination of phosphate in the presence of ammonium sulphate was published in *Biochemical Journal* in 1924 [1].

It was at about this stage when, one hot summer afternoon, Haldane wandered into the Part II lab. and listened to my account of what I was doing. He departed rather abruptly but reappeared a few minutes later with Hopkins to whom I had to show my curves and figures and to tell the story over again. There was not much chance of going any further until the Part II was out of the way and I could take up the research once more in earnest.

This manipulation of protein and peptone mixtures was not arrived at without a lot of gruelling reading of the literature. I had to seek examples of what others had done, Hopkins himself, of course, in the isolation of tryptophan but also the extensive work of Emil Abderhalden on proteins. This was in German. I slogged and struggled through his innumerable papers with the help of a 'German Dictionary for Chemists' and at last acquired reasonable proficiency. It was a quip among us students that Abderhalden had in one year published 365 papers!

Being a science student, I was unable to take up rowing or other sports requiring training and practice in the afternoons, but I made up for this by exploring far and wide on my bicycle, usually doing a bit of botanizing on the way, and by becoming an adept punter on the river. I soon knew where to find gravel on the river-bottom nearly all the way up to Granchester. During holidays, I often revisited my old haunts in Devonshire, carrying a home-made tent on my bicycle and cycling prodigious distances each day – the first day's goal was 100 miles to Mere on the western edge of Salisbury Plain! Dartmoor and the Cotswolds also came in for exploration and it never occurred to me to take the train between London and Cambridge.

Norway and Denmark: Linderstrøm-Lang and S.P.L. Sørensen at the Carlsberg Laboratory

It was in 1925 that, together with another Emmanuel College friend, I made my first visit to Norway. We crossed from Newcastle to Bergen and set off with a tent in our rucksacks to work up the West coast by a combination of walking, fjord steamers and buses where necessary. The latter were in those days open 4-seater Ford cars! We followed no scheduled route and eventually took a train from Åndalsnes to Oslo and then, after a few days, a boat to Copenhagen. This unpremediated extension of our tour had important results for, arriving in Copenhagen, I got the idea of calling upon Professor S.P.L. Sørensen (Fig. 5) at the Carlsberg Laboratory. He received me most kindly and, after having heard about my work in Cambridge, invited me to spend some months later on in Denmark. This I did during a long vacation and acquired a new insight into the discipline of research under him and K. Linderstrøm-Lang who became a life-long friend. The atmosphere of the Carlsberg was stimulating and happy, sometimes almost hilarious because of Linderstrøm-Lang's (Fig. 6) puckish sense of humor! It was also he who introduced me to Danish poetry and prose for I had by then acquired a reading knowledge, at any rate, of Dano-Norwegian. On one memorable occasion, I heard Mme. Curie give a lecture in the University.

Phosphorus in caseinogen

Having concluded from my experiments that the combination
of phosphorus in caseinogen was in the form of phosphoric
esters of hydroxyl compounds, among which serine was obviously a likely candidate, I determined to see if other proteins
could be chemically phosphorylated. It seemed that this might
be achieved by a kind of Schotten-Baumann reaction and this
was the programme for my work in Copenhagen. Sørensen
provided some purified serum albumin which was vigorously
stirred mechanically in solution while $POCl_3$ dissolved in CCl_4
was added drop-wise for some hours together with sufficient
sodium hydroxide to maintain a pH of about 7.5. The product
was then isolated, washed free from phosphate and analysed
to determine its P/N ratio and the action of bone and kidney
phosphatases upon it. Synthetic phosphoproteins were readily
produced in this way [2].

Back in Cambridge, I wrestled with the problem of the constitution of my 'phosphopeptone' from caseinogen. In those
days, a full amino acid analysis of a protein could only be obtained by Emil Fischer's ester distillation procedure and less
complete information by Foreman's calcium salt separation of
the dicarboxylic acids and other equally formidable techniques for the bases etc.; there was no 'Moore and Stein' or
paper chromatography to do the work for one.

After hydrolysis of my phosphopeptone, I saturated the
solution with HCl gas but no glutamic acid hydrochloride crystallized out which led me to believe (erroneously it later transpired) that the dicarboxylic acid was present as 5-hydroxy-
glutamic acid, together with serine and some threonine. I believe it must have been the presence of so much phosphoric
acid – and possibly some still unhydrolysed phosphoserine –
which prevented crystallization of the glutamic acid hydrochloride. Treatment with bone and kidney phosphatases established the phosphoric ester linkages in the peptone and I
arrived at a provisional structure which I took to Harden for

publication in the *Biochemical Journal* [3, 4]. He realised the tentative nature of the conclusions, as we discussed them at the Lister Institute, but generously allowed reproduction of the proposed structure. It was not entirely correct but was the best I could do with the cumbersome methods then available. Years later, Perlman in America and others took up the phosphoprotein problem, completely overlooking my pioneering efforts, a fact which still makes me rather sad – although it was not to be the last example of its kind.

Isolation of a carbohydrate moiety from serum proteins

Towards the end of my time in Cambridge, I came across a paper by Fränkel and Jellinek in the *Biochemische Zeitschrift* reporting that they had found a carbohydrate moiety to be an integral part of a protein. This interested me and I determined to find out if other proteins such as serum albumin and globulin could also contain such chemically bound carbohydrate complexes. They proved indeed to do so and from them I isolated a glucosamino-dimannose complex after alkaline hydrolysis of the purified proteins. The work appeared in the *Biochemical Journal* in 1929 [5] and 1931 [6] and was really a pioneering discovery. But once again, subsequent workers who entered this field ignored my contribution.

Personalities I met in Cambridge included Gilbert Adair, Prof. Barcroft and Rudolf Peters who had shown that there was a stoichiometric relation between the oxygen capacity of haemoglobins and their iron content.

As relaxation from my research work, I was reading very widely and indiscriminately in more than one language – H.C. Andersen in Danish, some German novels, Bernard Shaw, a set of Joseph Conrad which I had got as a College prize, Kipling, Dostoievsky and a lot of poetry. It was an awful mixture but I had a good memory and I also began to take delight in good music. Classical music was not an entirely new experience, for our next-door neighbour when we lived near Elstree had been Thomas (later Sir Thomas) Beecham and his gramophone records could be heard nearly all day long. But at Emmanuel we then had E.W. Naylor who gave informal lectures on music, especially on Wagner whose operas he illustrated with astonishing effect on the piano. He came of a musical Yorkshire family who had known Wagner and I remember him telling us that the score of 'Tristan' was rehearsed in his parents home before it had been officially published. Sailing was another pastime. I had done quite a bit of sailing on the Norfolk Broads when I saw one day on the Union notice board an advertisement suggesting 'A sailing holiday in the Baltic with opportunity to learn some navigation'. It gave an address in Kiel but no other details. A South African research worker and I were thrilled! We made arrangements to meet the yacht's owner on a certain day in Kiel and set off on a Danish cargo ship from Hay's Wharf in London – having practically no idea of where we would go or what we would have to pay.

Sailing in the Baltic with Herr Dibbern in 'Te Rapunga'

We eventually met Herr Dibbern (Fig. 7) who proved to be as charmingly unpractical as his letters had suggested. He had ran away to sea as a boy, served on four-masters and learnt a wonderful repertoire of sea-shanties until he got tired of it and worked on a sheep farm in New Zealand. In 1914, being a German, he was promptly interned there and only dumped back in Germany when the war was over! Being penniless and without a job, he decided to get married! During the period of inflation, he did any odd work, putting his wages at once into the building of a fine little five-ton cutter to which he gave the Maori name 'Te Rapunga' – longing (Fig. 8).

In this boat we set off along the north German coast, visiting Lübeck and Warnemünde and then turning northwards where we cruised for some glorious weeks among the Danish islands.

Fig. 7. Herr Dibbern

Fig. 8. 'Te Rapunga'

Fig. 9. D. Keilin

Fig. 12. Douw Steyn

Fig. 13. John Quin

*Fig. 10.
Sir Arnold Theiler*

*Fig. 11. Onderstepoort Veteri-
nary Research Laboratory, S.A.*

We stopped whenever we felt inclined, bought provisions in tiny village stores, swam in the clear water and sang shanties as we sailed by night up the sounds. Dibbern was one of those gloriously romantic characters without which the world would be a poorer place. We all got on well together and he even tried to persuade my friend and me to sail 'Te Rapunga' with him to Iceland in the following year for the 1000 years Allting celebration; my parents refused point-blank to let us go!

Back to Cambridge. Personalities there

Sunday afternoons at Emmanuel were invariably spent in T.S. Hele's house where he, together with three or four of us research students, played model trains on the sitting room floor with his daughter Priscilla, then about five or six; Mrs. Hele (A.V. Hill's sister) produced quantities of tea and buns and the romp went on until the Chapel bell sounded and we had hastily to make ourselves a bit more presentable. Oh, those days!

In the biochemical laboratory, Hopkins had gathered quite a few personalities about him. There were Robin Hill and Mrs. Onslow, Quastel, Holden, Hicks, Joe Needham, the 'enfant terrible' J.B.S. Haldane and among the visitors Lemberg and Szent-Györgyi. The latter showed me one day the crystals of an hexuronic acid he had just succeeded in isolating from adrenal glands – vitamin C! Keilin (Fig. 9) at the Molteno Institute was unfolding the story of cytochrome. It was a good atmosphere, disturbed by minor quarrels at times, but also having its lighter side. Majorie Stephenson, the bacteriologist, objected to Quastel's incursions into this field and protested to Hopkins, 'either he must go – or I will go, there is no other solution!' – to which Hopkins replied, 'Oh yes there is; why don't you marry him?'

My first job. Research on wool in Leeds: 'Lanaurin' and contact with Hans Fischer

The time had come for me to think again about the future. I felt I could not stay in Cambridge indefinitely.

A few other things having fallen through, I eventually accepted an offer from the Woollen Industries Research Association in Leeds to set up a Biochemical Department at a salary of £ 425 a year. The Institute was a large converted house, 'Torridon', where I should have one large room and a smaller one adjoining it. On a preliminary visit, I was instructed to prepare a detailed plan for the conversion of my space into a laboratory – benches, desks, lists for equipment, etc. – the whole to be ready in about three hours' time as the contractor wanted to start the work as soon as possible! Well, it didn't turn out so badly after all.

The break with Cambridge in 1928 was less painful for me because the opportunities there were less appreciated at the time. Looking back on those seven years, I realise that what I lacked was understanding guidance. The greater part of my chemistry was self-taught, biochemistry too; I wasted a good deal of time through not channelling my energies more systematically, through not realising and not grasping the opportunities which surrounded me. Perhaps it is inevitable; one lives to regret. But my whole life has been characteristically unplanned.

In Leeds, I plunged into the problem of the form in which sulphur was combined in the keratin molecule. There was going on, just then, a fierce dispute between the English workers on wool and Hedley Marston in Australia, the latter insisting that all wool had the same fixed sulphur content, whereas to us it was clear beyond dispute that the percentage content of sulphur varied from one wool type to another. My contribution was to relate total sulphur to the cystine determinable after acid hydrolysis. There was a very close correspondence, the small difference being due, it turned out later, to methionine which had not then been discovered. Various other studies of a more technical nature also came out of my laboratory.

W.T. Astbury had just then produced the first exciting X-ray diffraction photographs of wool fibres, which indicated some regularly repeating configuration. This, I felt and pointed out, could only be the diketopiperazine structure – which was not far from the actual truth.

Working temporarily at Torridon was A.M. Stewart from Perth, Western Australia, who had brought with him some fleeces having a yellow discolouration. This was not removed by washing and was not due to bacterial action; it was an inherited characteristic and the nature of pigment was unknown. Stewart and I tackled this problem, using pilot-plant large-scale equipment at the University. The pigment was extractable from the defatted wool by dilute ammonia, was acid in nature and could be purified by repeated precipitation. It appeared to resemble urobilin in many of its characteristics and I sent some to Hans Fischer requesting his opinion. This was my first contact with Fischer and with his published work which it had been necessary to study. Stewart and I named our pigment: 'Lanaurin'. Lemberg and Legge note that from our empirical formula and other observations it is identical with bilifuscin and therefore a dipyrrolic pigment.

Poisonous plants in South Africa. Sir Arnold Theiler, Onderstepoort, D. Steyn and J. Quin

Several South Africans visited Torridon, mostly veterinarians, with whom I became friendly. When it so happened that an advertisement appeared in *Nature* offering three Special Research Fellowships, sponsored by the Empire Marketing Board and tenable at the Onderstepoort Veterinary Research Laboratory in Pretoria, these friends urged me to apply for one – that for the chemical investigation of plants poisonous to livestock. In vain I pointed out that I had absolutely no experience in this field; they were insistent! There were also other considerations. I had got married some two years earlier and my wife and I had bought and furnished a small house in Leeds with the meagre funds at our disposal. Leaving for South Africa, if I was offered the post, would mean selling everything – and times were bad, the beginning of the great depression of the 1930's. Nevertheless, the idea appealed to me. We had been in many ways unlucky in Leeds, which with its factories was a dismal, dirty town, very different from my wife's native Norway. So I took a sporting chance, sent in an application and was quite non-plussed when I learned that I had been selected!

Our belongings were almost given away, but fortunately the Empire Marketing Board had agreed that before sailing I should visit several countries in Europe where work on plant constituents was proceeding and also to meet in Switzerland Sir Arnold Theiler, the Founder and only recently-retired Director of Onderstepoort. My first call was upon George Barger in Edinburgh who gave me excellent advice as to whom I should see, what reference books I should need, some valuable practical tips, and much else besides. I was to meet him again later in Heidelberg. What success I had in South Africa owed much to his great kindness on this first occasion. Barger was a notable linguist and on hearing that my wife was Norwegian, he at once wished to know all about the differences between Riksmål and Landsmål. Needless to say, my knowledge of Norwegian and its dialects was hopelessly insufficient at that time.

On the Continent, I visited several laboratories but found that nearly all the chemists were interested only in the structure of alkaloids and in sources of raw material for their researches. Exceptions were Freudenberg in Heidelberg and Ruzicka in Vienna and pharmacists such as Kofler and Casparis in Switzerland. In some Institutions, despite my Letters of Introduction bearing the insignia of the Union of South Africa, I was clearly regarded as a nuisance and disposed of briefly by some junior.

The climax of the tour was when I met Sir Arnold Theiler (Fig. 10) in his flat in Lucerne. This dedicated scientist and great-hearted man placed me in a corner of the little sitting room, drew up his chair to bar any possible escape and discoursed for over an hour on South African poisonous plants, their effects on stock and the efforts which he and others had so far made to elucidate the nature of the active principles. I came away a convert, a convert to his scientific devotion and with an admiration which I tried years later to convey when in 1965 I had the honour to give in Pretoria the First Arnold Theiler Memorial Lecture.

Travel to South Africa in those days was, of course, by ship; it took about seventeen days. From Cape Town to Pretoria was a two-day train journey passing through the Hex River Mountains and then endless miles of the Karroo with its koppies and stunted bushes until finally one reached the Transvaal Highveld.

The Onderstepoort Laboratory (Fig. 11) lay some ten miles north of Pretoria, but those living in the city took turns in using their cars. The Director was P.J. Du Toit and my immediate associate D.G. Steyn (Fig. 12), the toxicologist, who has remained my very close and dear friend during the years until his death. My work brought me later into collaboration with the head of the Physiology section, J.I. Quin, as I shall describe below; we shared many experiences together. Quin's life was cut short at an early age by a heart attack shortly after he had assumed the Directorship in succession to Du Toit. For both science and for Onderstepoort, in particular, this was a tragic loss.

I found equipment to be of the most primitive nature, judged by the requirements of a biochemical laboratory. As an example, the only pipettes in the store were of 100 ml capacity or more and there were very few organic chemicals. In those days, everything had to come by ship from England and at

least six weeks passed between ordering and receipt. Nevertheless, I had brought my own books with me and the library
had ordered for me essential works such as Wehmer's 'Pflan-
zenstoffe', Van Rijn's 'Glucoside', Czapek's 'Biochemie der
Pflanzen' and some others including Klein's 'Handbuch der
Pflanzenanalyse'. This latter was invaluable. Its several volumes covered, with typical German thoroughness, much general organic and physical chemistry as well as more specialized photochemistry.

We made a start with a pretty aster-like Composite, *Dimor-
photheca spectabilis*, which was strongly cyanogenetic and
which Steyn and I collected near Pretoria. The cyanogenetic
glucoside which turned out to be linamarin [7] was easily isolated because of the enormous amount present in the plant. It
was an encouraging start.

We had arrived in the South African summer which in Pretoria was not only terribly hot but also brought with it a dysen-
tery-like infection which was probably conveyed by milk. This
I caught, but refusing to go sick after only a few weeks at the
laboratory, I struggled on until hardly able to stand and was
compelled, at last, to seek medical advice. I got over it, but it
left me with a weakness for years; for a long time I never travelled anywhere without a good supply of purified kaolin which
is absolutely the best medicament.

After some months, I was given as an assistant a young man
named Roets, seconded for the purpose from the Chemistry
section which was working almost exclusively on the mineral
metabolism of cattle. Roets was rather suspicious of me at first
since his friends had been taunting him with having to go and
work with a 'Rooi-nek' – literally red-neck, the Afrikaans slang
for an Englishman. But mutual trust and respect were soon
established between us and he proved to be the most loyal of
collaborators. There was still some anti-British feeling among
the Afrikaans-speaking Nationalists; from Roets I learned
much and derived some understanding of the bitterness left by
the Boer War and the tragic loss of life, mainly through typhoid,

in the concentration camps to which the displaced Veld-farm-
ers' families were sent. The whole campaign was guerilla war-
fare, the embers of which it takes a long time to extinguish!

Our work together on the poisonous plants prospered. One
fact quickly emerged, that very few indeed of the active principles were alkaloids; we had to do with bitter principles, new
types of cyanogenic glycosides and a bizarre assortment of un-
classifiable but pharmacologically active materials as strange
as the South African flora itself.

I had to plough a lonely furrow, there were very few people
to whom I could turn for advice or with whom to discuss our
findings, and they were widely separated in different parts of
the Union.

One active constituent defied all attempts at isolation, that
of *Dichaepetalum cymosum*, now known to be fluoroacetate!
This plant is also a freak in its manner of growth. It is one of
a family of tropical and sub-tropical climbers but, with migration into the Transvaal, it has adopted an underground habit,
only small tufts of leaves, like rabbits' ears, being visible and
poisoning animals which eat them. Seeking a means of eradi-
cation, the Botanical Department decided to follow one plant
from the surface downwards. I remember seeing the excava-
tion, a great hole 60 feet deep – and even at that depth, the
tissue was morphologically still a *stem* and not a root!

We used all our ingenuity on this plant, even trying separation by diffusion through collodion membranes, but all we
could establish was that the poison resisted oxidation, reduction and hydrolysis, was thermostable, non-precipitable, and
always followed the acetate residues introduced by such reagents as lead acetate. How the fluorine was missed, I can not,
to my chagrin, imagine to this day!

Geel dik-kop of sheep and photosensitization

About this time, I became drawn into work with Quin (Fig.

13), the physiologist, on the sheep disease 'Geel dik-kop' which
in Afrikaans means 'Yellow thick head' and is descriptive of
the main symptoms. Theiler's earlier work had shown almost
conclusively that it followed ingestion by sheep of the procum-
bent, spreading plant *Tribulus terrestris* but only at certain
times when rapid growth after early spring rains had been
followed by drought and wilting. Nothing was known about
the active principle and, in fact, attempts to reproduce poisoning by administration of the plant or of extracts were almost
invariably unsuccessful, being without noticable effect or rapidly causing death from severe methaemoglobinaemia. The
latter effect we traced to the presence in some batches of the
plant of large concentrations of nitrate which became reduced
to nitrite by the ruminal micro-organisms [8].

In the field, affected sheep were first noticed to cease feeding; the face and ears became distended by yellow fluid, the
eye-lids and lips became hard and stiff, the eye-balls even
bursting on account of fixation by the lids. In this miserable
condition, the animals usually died and post-mortem examination revealed an intensely yellow, bile-stained carcass, ruminal stasis and a large intestine filled with compacted
masses of hard faeces. The gall-bladder was enlarged and distended with bile.

Quin and I made several excursions into the Karroo region
to study outbreaks reported either by field veterinary officers
or by farmers themselves. Reaching these districts involved
about two days' journey by train and then car – or even horse-
back in the more remote parts – and often we arrived only to
find that fresh rains had fallen and the disease had ceased.
Nevertheless, we conducted post-mortems, brought back samples and gleaned all information we could from the farmers.
Some insisted that a little yellow grub often found in the *Tribulus* stems was the cause of the whole affair!

The work was trying and exhausting. Often, after spending
many hours in the blazing sun cutting up dead, jaundiced
sheep, we sat down with the family to their inevitable dish –

curried boiled mutton! But there were compensations. Early
morning in the Karroo can almost be likened to eternity –
infinity of space, air and sky and a purity as of the earth's
beginning. In the evening, after sundown, I have watched the
sheep coming down to the dam to drink, milk-white rivulets
winding through the bushy scrub. It was all so lovely!

Quin rightly believed that the symptoms of swelling etc.
were due to phososensitisation and to identify the photoactive
agent was our first task. *Tribulus* contained nothing like hy-
pericin, so I turned to affected sheeps' blood and identified
phylloerythrin in it [9,10]. Now, phylloerythrin is a porphyrin
derived from chlorophyll and it had been shown by March-
lewski to be a regular constituent of sheeps' bile and faeces.
Animals with geel dik-kop had enormous amounts of phylloerythrin in their gall-bladder bile and it was also present in their
urine. Our hypothesis was that some active agent caused a
biliary stasis as a result of which the phylloerythrin-contain-
ing bile diffused back into the blood stream thereby rendering
the animal photosensitive. To put this hypothesis to the test,
Quin performed surgical ligature of the common bile duct on
stock sheep at Onderstepoort and exposed them to the sun –
they became jaundiced and photosensitive. Next, we selected
two sheep, fed one on a diet of fresh green lucerne while the
other was given only old straw devoid of chlorophyll. After
some days, operative ligature was performed on each. The first
animal became yellow and extremely photosensitive; the second only jaundiced. The diets were then switched round,
which caused a corresponding reversal of the symptoms. As
final proof, I prepared phylloerythrin chemically and this was
injected intravenously with resulting photosensitisation but,
of course, no jaundice.

Phylloerythrin and icterogenin

Where and how was chlorophyll degraded into phylloerythrin

in the ruminant? Quantitative determination throughout the alimentary tract of the sheep showed that the prime seat of formation was the rumen while absorption from the small intestine could also be demonstrated. The ruminal flora consists mainly of protozoa. We concentrated these and showed that they contained phylloerythrin and were able to bring about the decomposition [11]. Bacteria from the rumen played little, if any, part. This explained why it is only in ruminants that cholestasis leads to photosensitisation. Shortly after seeing our publications, the New Zealand workers were able to show that it was also phylloerythrin which was responsible for their 'Facial Eczema in Sheep'.

Quin carried out further excellent research on photo-active dyestuffs but our next important joint problem was to identify the icterogenic factor. As already said, *Tribulus* was unsuitable for this purpose owing to the ephemeral nature of its toxicity. Around Pretoria, on the other hand, there grew *Lippia* and *Lantana* species which had been reported to have caused jaundice although these plants are rarely eaten by stock on account of the highly aromatic oils which they contain. We selected *Lippia rehmanni*, a shrub with stout woody stems, rather leathery, lemon-smelling leaves and a fleshy root. Extracts fed to sheep caused jaundice – and photosensitisation if the diet was rich in chlorophyll. The fleshy root was most efficacious, then the young leaves.

Isolation of the icterogenic principle was an extremely difficult matter owing to the multitude of terpene compounds which were also present. It was achieved, however, by a device using the sparing solubility of its sodium salt in concentrated sodium chloride [12]. The acid was named 'Icterogenin' and a tentative structure proposed which was completed later by Barton and De Mayo in London who also identified another closely-related minor constituent which they named 'Rehmannic acid'. When given to sheep, icterogenin reproduced the typical signs and symptoms of geel dik-kop.

The mechanism by which icterogenin produces intrahepatic

cholestasis has been studied intensively and in detail during more recent years in my laboratory in London by myself and my colleagues Heikel, Brown, T.F. Slater and Sawyer. In South Africa, Quin and I established its inhibitory action upon the plain muscle of the intestine, thereby explaining the intense constipation from which poisoned sheep suffered. Our final experiment with *Lippia rehmanni* was to induce sprouting of fresh leaves by pruning and to show that the concentration of icterogenin was highest in these; it is gradually transferred to the root system where it accumulates in the fleshy rind. Strangely enough, exactly the same syndrome of jaundice and photosensitisation can, at times, follow the ingestion by sheep of young lucerne or even pasture grass; the chemical peculiarity seems, therefore, not to be confined to *Lippia* species.

Ruminal digestion

The process of digestion by the ruminant is accomplished mainly by fermentation brought about in the rumen by its protozoal fauna. Quin carried out some pioneering investigations of ruminant digestive physiology which have not received the credit they deserve. Permanent fistulae were operatively established in sheep whereby samples of ruminal contents could be withdrawn and the effect of introducing nutrients be studied; contractions of the musculature could also be recorded. In order to follow the process of digestion biochemically, a scheme of analysis was required and this I provided for him. Not only did this enable determinations to be made of volatile and non-volatile acids and bases, titratable acidity etc., but also the ruminal gases could be analysed, including hydrogen and methane through catalytic combustion. The whole procedure was successfully employed by a young research student named Arnold working with Quin; his thesis was submitted to Potchefstroom Agricultural College. Shortly afterwards, Quin presented the results to the International Veterinary Congress held in Cambridge and thereby stimulated the work of the Cambridge school and the Rowett Research Institute in this subject. The ruminant digestive process produces largely acetic acid and similar lower fatty acids which are absorbed and transported in the blood stream.

Porphyria and porphyrins

The year 1935 proved to be of momentous significance in my life, directing my research interests, as it did, for the next forty years.

Sir Arnold Theiler was back at Onderstepoort as a guest worker and one day he called a few of the staff to the Anatomy Department to see some bones which had been sent in by one of the Field Veterinary Officers from Swaziland. These bones were chocolate-brown in colour. They came from a heifer which had been slaughtered as a 'poor-doer' by the farmer who reported that he had had several such animals since using a certain bull to upgrade his stock.

The reddish-brown bone looked exactly like illustrations of the skeletal discolouration seen in the extremely rare human disease congenital erythropoietic porphyria. One famous case, of the 50 or 60 reported in the world literature, a man named Petry, had been studied by Hans Fischer in Munich. Petry produced and excreted enormous quantities of series I porphyrins of a physiologically abnormal type and these were also deposited in calcified structures such as his bones and teeth, giving them a dark reddish-brown colour. I asked for some of the bovine bone and soon showed in the laboratory that the discolouration was due to the same porphyrins. Up to that time, no living animals had ever been discovered suffering from congenital erythropoietic porphyria although there were one or two reports of similar discolouration in abbatoir material. Since human congenital erythropoietic porphyria

Fig. 14.   Fig. 15. John Falk   Fig. 16.
Hans Fischer   Elizabeth Dresel

Fig. 17. First International Conference on Porphyrins at The Ciba Foundation, London (1951)

(Günther's disease) appeared to be a recessively inherited 'inborn error of metabolism' a field investigation seemed well worth while and so Onderstepoort's haematologist, P.J.J. Fourie, and I set off on the long car journey to Swaziland, taking with us various items of equipment. We found in the herd no less than 13 living animals with the disease, recognition being easy by the pink teeth and red urine. The bull appeared normal but was evidently a carrier of the recessive gene since breeding records showed that cases had only arisen after crossing him with his own female progeny. We selected one affected animal for slaughter, withdrew all blood, urine and bile, wrapped the various organs and bones in cloths soaked in formalin as a preservative and set off for Pretoria. The distance was about 300 miles over bad roads and in summer heat, so that we were nearly choked by the fumes of formalin when we reached the laboratory.

There ensued some two to three months of intensive work determining and characterizing the porphyrins present, but results [13] showed almost complete identity with the findings of Fischer on the man Petry. Onderstepoort Laboratory purchased the bull and several females of the Swaziland herd and, by controlled breeding experiments, the inheritance of congenital erythropoietic porphyria as a Mendelian recessive character was established with a high degree of probability.

The discovery of these first living animal cases was followed some time afterwards by the finding of others in quite unrelated herds. It was remarkable to observe how quickly information from the big breeding societies dried up when our examination of their records pointed, in one instance, to a valuable imported bull as being a highly suspect carrier! The recessive character of the inheritance was confirmed through discovery, some years later, by Amoroso, Loosemore, myself and Tooth of the disease in bovines in England. It is also known and has been studied in Denmark by T.K. With.

During a vacation in Europe, taken as this work was nearing completion, I took the opportunity of visiting Hans Fischer

(Fig. 14) and telling him of my results, and also of meeting the pathologist Max Borst in Munich who, together with Könings-dörfer, had carried out a masterly post-mortem histopatho-logical study of Petry's organs, using a technique of ultraviolet fluorescence microscopy, as described in their celebrated mon-ograph. Borst was a fine gentleman of the old school; Köningsdörfer had moved to a small town near the Czechoslo-vak border and arrangements were made for me to meet him there. On dismounting from the train, I was confronted by a Nazi salute from a young man in brown uniform! He took me to his home but all my attempts to talk about porphyria were futile; he was much more interested in politics and Hitler's 'mission' to liberate Czechoslovakia! Continuing to Prague, I was left in no doubt of the tenseness of the situation.

After attempting unsuccessfully to revisit some of the Ger-man biochemists and plant physiologists I had previously met – I was merely told that their whereabouts were unknown – I paid a final call upon Otto Schumm in Hamburg, then an old man. He had contributed much to porphyrin chemistry and received me courteously. It was pathetic, however, to realize how completely blind he was to Hitler's intentions. 'Why', he asked me, 'are you British so jealous of Germany and her right to existence?' I do not know the date of his death, but I hope he did not have to suffer through disillusionment and despair.

The National Institute for Medical Research, London

Before returning to South Africa from England, I went to see Sir Henry Dale, Director of the Medical Research Council's National Institute for Medical Research. My situation at On-derstepoort had become strained. When the Empire Market-ing Board, whose Fellowship I held, was dissolved, the De-partment of Agriculture and Veterinary Services of the Union Government had offered to keep us three Research Fellows by appointing us as Civil Servants – my two colleagues, who were both qualified veterinarians, at their existing salaries and me at a considerably reduced salary, notwithstanding the success of the poison-plant work. I felt this to be unfair and when Dale offered me a position on his staff as biochemist in succession to Dudley, who had died, I gladly accepted. To fulfill my contract, I had to work another year in South Africa and eventually left in 1937.

So ended a rich and varied period of my life. We had seen much of the country, the Krüger National Park, the beautiful Knysna coastal region and much else, while my work had also brought me out into the desert regions bordering on South West Africa. I had come into contact with notable scientific personalities in General Smuts, the archaeologists Robert Broom and Raymond Dart, and had made many friends. Our daughter had been born in Pretoria. To return to an atmosphere of keen intellectual activity after having been for all intents and purposes a lone worker in a foreign land was not easy; adjustment took time and courage but I was helped by the goodwill of my colleagues Harold King, Otto Rosenheim and Patrick Laidlaw, to mention only three, and of course Sir Henry Dale himself. I well remember Rosenheim showing me the first paper chromatographic separation of leucine and 'isoleucine'. Synge and Martin were revolutionising the amino acid analysis of proteins by their new technique – previously, this could only be accomplished by fractional distillation of the amino acid esters according to Willstätter. King and Rosenheim together formulated the correct structure of cholesterol and steroids in general. It was an exciting time.

My work on porphyrins and porphyrin metabolism continued. I made a closer study of turacin, the red copper-containing porphyrin complex in the wing-feathers of the *Turacos*. I had been unable to prepare in South Africa good specimens from them of uroporphyrin I which, according to Fischer, was the porphyrin involved. The Natural History Museum gave me plumage from several species of these birds and I examined them individually only to find that in each case the resulting

uroporphyrin belonged to the isomeric series III, partial decarboxylation affording indisputable coproporphyrin III. I published these results in the Royal Society's 'Proceedings B' in
1939 [14]. It took no little courage to challenge Fischer's erroneous identification and I was gratified when my work was
fully corroborated soon afterwards by With and others. Incidentally, this finding made uroporphyrin III available for the
first time for metabolic experiments; the so-called Waldenström porphyrin is a mixture of isomers.

During the years that followed, I steadily widened my interests in the metabolism of porphyrins and haemoproteins and
also in the porphyria diseases. The sulphonamides were
shown to be porphyrinogenic and porphyrin biosynthesis to
occur in incubated yeast press-juice.

The 1939–1945 war

With the outbreak of war in 1939, all my plans were upset. We
had been for our summer holiday to the island of Askeröy in
Norway where we had built a small house two years previously. Leaving my wife and daughter, aged six, to enjoy a few
more weeks there, I returned to England at the end of August
only to discover the seriousness of the international situation.
The whole staff of the National Institute for Medical Research
in Hampstead was busy erecting sand-bag walls and con-
structing gas-proof air locks. On September 3rd we were at
war and expecting an aerial blitz on London at any time. I
therefore cabled my wife advising her to stay with our daughter in Norway. We had rented and furnished a house in Hen-
don on our return from South Africa and I now packed up and
sent all our belongings to a furniture storage depot in Hol-
loway while I myself with a South African colleague, Van den
Ende, arranged to share the flat belonging to the Institute's
caretaker whose family had been evacuated to the country. A
fire-fighting squad, led by Alan Parkes, was formed at the In-

stitute and during air raids Van den Ende and I had responsibility for patrolling the top floor and roof of the building. We had an excellent view of the pyrotechnics.

As the 'phoney war' dragged on, it was possible once more to attempt some serious research but by March 1940 it was obvious that Norway was becoming threatened. The telegrams and letters I sent urging her return failed to reach my wife in time and on April 9th the German invasion of Norway severed all communication between us and made her return impossible.

Volunteers with a knowledge of Norwegian were being urgently recruited to go as interpreters with the British expeditionary forces and I asked Sir Henry Dale if I could offer my services. He was most sympathetic in his understanding and, whilst urging me not to be too impetuous, he allowed me, after a few days, to go to the War Office. There my knowledge of Norwegian was tested, found adequate, and I was taken to the Posting Officer. One condition for appointment as an interpreter was apparently essential, namely that I had had some previous experience in the Forces. This condition I was unable to fulfill; I had been too young during the 1914 war and the school Officers Training Corps was not counted as sufficient. The Officer told me, as we parted, that had he been able to claim for me a single day's service, I should have been off that very night to Narvik! I had to accept what fate had dealt out to me and tried to settle down to my research work again.

The memory of these years is all rather confused. After some time, the caretaker's family wished to return home so I went to live with my brother-in-law at Brookmans Park in Hertfordshire, my sister and their children having been evacuated to Shropshire. Each day and in all weathers, I bicycled the 36 odd miles to Hampstead and back and so happened to be in very good training when the physiologists at the Institute wanted a volunteer for some investigations involving working on an ergometer whilst breathing different gas mixtures through a mask. Experience had shown that naval divers working at

great depths occasionally suffered unaccountable psychological disturbances of a severe kind. The work performed in the
laboratory tests had to be hard and, so far, no one at the Institute had been found capable of continuing for longer than 1 to
2 minutes under the most extreme conditions. They asked me
to try and I became their guinea pig, achieving about 6 minutes. The various oxygen and carbon dioxide mixtures did, indeed, induce strange psychological states, sometimes a feeling
of exhilaration, sometimes intense anxiety and 'black outs'.

Another contribution I was able to make to the general war
effort was an investigation of a synthetic plasma-expander,
'Periston', which had been captured from the Germans in
North Africa. It was found to be a nitrogen-containing poly-
mer, not unlike a protein, but it could cause hepatic injury.

I was still at Brookmans Park when the Home Guard was
called into being. Starting in the ranks with a double-bar-
relled shot gun, kindly lent to me by Hamilton Fairley, I eventually became Intelligence Officer and finally Captain, Second
in Command of my Company in whose area the BBC European
Transmitter was located – a prominent target for German
bombs. I took this Home Guard work very seriously, learning
a lot about military tactics and intelligence work and devoting
long hours to it each evening and every week-end. It helped to
ease the enforced separation from my wife and daughter with
whom I could only communicate officially by Red Cross letters,
a maximum of 25 inconsequential words which took six weeks
or more on their journey via Portugal. I had learned, after six
months of silence that my family were still alive and living on
the island, but knew no more details until we established an
'underground' route through Denmark and Sweden with the
help of a lady in the Danish Consulate in London.

As for scientific work, I had become involved in some investigation of TNT poisoning in shell-filling factories and had also
commenced with I.W. Rowlands work on the purification and
biochemical properties of the gonadotrophin of pregnant
mares' serum [15]. This material turned out to be a glycopro-

tein or glycopeptide but, although we carried its purification very far, its separation from other closely similar compounds in the serum proved to be extremely difficult with the techniques then available. Rowlands and I wished to continue but Sir Henry Dale ruled that the cost involved would be too great. I regret that this project was never carried to completion. We had evidence of a sialic acid-like linkage upon which hormonal activity depended, and our preparations turned out to have been 90–95% pure [16].

With Van den Ende, I was also investigating the bound carbohydrate of the serum proteins, sero-mucoid, serum albumin and the so-called globo-glycoid. But these studies were in addition to those on porphyrins and porphyria. I may be wrong, but I always had the impression that Dale looked upon the porphyrin work as biochemical dilettantism.

The psychological stresses imposed by the war were not conducive to serious reading but during this time I did write some poems, some of which were published; most remain to this day in manuscript form.

When Sir Henry Dale retired as Director of the National Institute, Sir Charles Harington was appointed to succeed him from the close of 1942. A. Neuberger had followed Harington from the University College Hospital Medical School (UCHMS) and I was surprised, after a short time, to learn that he proposed to take porphyrin biochemistry as his own field of investigation. Fortunately at this juncture I was offered the Chair in Chemical Pathology vacated by Harington so that complications were avoided!

Chair of Chemical Pathology at UCHMS

I took up my new appointment in May 1945. The first problem was to get the rooms comprising the Department fit to live and work in and to equip them. During the London blitz the Medical School had been evacuated to the country. The windows of the laboratories had been shattered by bombs and everything had been exposed for 3 years to wind and rain. Whatever was of value seemed to have been 'borrowed' by others still working in the Hospital and to make matters worse, the empty rooms had been found to be a convenient dumping place for unwanted discarded materials and furniture. It was indeed a grim sight which confronted me!

There was in London at that time a friend of mine, Major P. Symons, a chemist whose Unit was standing by until ordered out to Hong Kong to tackle the public health problems arising after the Japanese capitulation. He generously gave me of his time and help. Together we scrubbed floors, cupboards, benches and tables, and cleared out from the drawers the masses of broken equipment and rubbish which filled them. We worked week-ends and holidays in addition to regular hours and after a few weeks the place began to look usable once again. Restocking was almost as great a problem since all laboratory equipment was extremely scarce, even simple glassware such as test-tubes and beakers was practically unobtainable. However, within a year the Department was turning out some research and my first guest worker was Otto Rosenheim who had then retired from the National Institute for Medical Research.

Following the liberation of Norway, my wife and daughter returned to England in June 1945 and we moved into a house near Barnet which I had bought. The furniture depository where I had stored our belongings at the beginning of the war had been bombed and set on fire during an air raid. Our furniture happened to be in the basement into which water from the fire-hoses and the rain accumulated to the depth of 3 to 4 feet. By the time the contents were got out, there was nothing much left that could be used. Chairs and tables were warped and falling to pieces, carpets, mattresses, curtains, etc. were mouldy, sodden masses and many of my books were in a similar condition.

For the first months in our house we lived under the most

primitive conditions while what repairs that were possible were carried out since new furnishings were practically unobtainable. Some articles were patently stolen by the removal men on the way from the store to the house; I saw a usable carpet and our sewing machine among the goods but they never turned up at the house! Of course, there was no such thing as an inventory, one had to be thankful that anything at all was saved from the wreckage.

I gave my inaugural lecture at the Medical School on October 4th, 1945, choosing as my subject 'Geel dik-kop – the record of an investigation' since I felt that this work in South Africa offered a good example of the contribution of chemistry, biochemistry, physiology and pathology to the unravelling of a complex veterinary-medical problem. It had not been the custom for a Professor in the Medical School to deliver an inaugural lecture; I was told that as this was an academic occasion my wife would not be invited but what was much more serious was that by some oversight my old teacher J.B.S. Haldane, then Professor of Biometrics in University College, was overlooked by the School Office. This he took as a personal affront and suspected that I was responsible for it, something which has caused me deep distress.

About this time, biochemistry was enormously fortified by Schoenheimer's introduction of isotopes. Shemin and Rittenberg's first paper, using ^{15}N labelling indicating that glycine was a specific precursor of the haem of haemoglobin appeared early in 1945 and was followed soon afterwards by the more detailed evidence of the participation of the tricarboxylic acid cycle in haem biosynthesis. These were the clues I had been watching for in my porphyrin work. In 1949 we had successfully applied paper chromatography to the separation of the porphyrins and had developed quantitative methods for porphyrin investigations. J.E. Falk had also joined me that year from Lemberg's laboratory in Sydney while W.A. Rawlinson from Melbourne was also in the Department. The time seemed ripe for a concerted attack upon the problem of the detailed

Fig. 18. Signatures - Porphyrin Conference participants

Fig. 20. A Porphyrin and Porphyria Conference in South Africa

biosynthesis of the porphyrins and haem. For this to become effective, one needed to gather together a group of people practising different disciplines and whose future could be financially guaranteed for some years. After discussing the matter with Dame Janet Vaughan, I approached Farrer-Brown of the Nuffield Foundation which generously granted me sufficient funds, spread over five years, to set up a 'Nuffield Unit for Research in Pyrrole Pigment Metabolism'. I cannot express strongly enough my indebtedness to the Foundation which made possible all that was subsequently achieved.

Nuffield Unit for Research in Pyrrole Metabolism; First Ciba Conference on Porphyrin

Falk (Fig. 15) became my leader of the Unit with Elizabeth Dresel (Fig. 16) from February 1951 until November 1955 when he returned to Australia as Head of the Biochemistry Section of the Division of Plant Industry, CSIRO, Canberra. Together we organised in 1951 the First Ciba Conference on Porphyrins, an international gathering which brought together all those then active in porphyrin research (Figs. 17 and 18). My appreciation of Falk's leadership is quoted in the Memorial Notice by Lemberg (Fig. 22) and Frankel (in *Records of the Australian Academy of Science, Vol. 2, No. 3, Canberra, 1972*).

Porphobilinogen

The key to the whole problem of porphyrin biosynthesis turned out to be porphobilinogen, the substance excreted in the urine by patients suffering from acute intermittent porphyria. Waldenström suspected it was a dipyrrylmethene but all attempts to isolate it had failed. In 1953 a patient excreting large quantities of porphobilinogen was admitted to University College Hospital. Westall, working in the Medical Unit, found that it could be precipitated as a mercury complex and that it formed a hydrochloride sparingly soluble in hydrochloric acid of specified concentration. He proceeded with attempts at purification by the use of ion-exchange resins and I carried out the quantitative estimations on his fractions. There was considerable decomposition into deeply coloured products. Small amounts of the unstable material were obtained but the elementary microanalysis of these was disappointingly irregular. In order to minimise the time during which decomposition could take place, I took each sample immediately by car to Oliver, the microanalyst at Imperial College, who had been warned of my coming and was waiting to proceed with the analysis without delay. On one occasion as he weighed out the sample he remarked to me that it appeared to be gaining in weight. This gave us the unsuspected clue that indeed anhydrous porphobilinogen took up moisture spontaneously to form a monohydrate. The previous discrepancies in analytical results were thereby explained. I then devised a manipulative procedure by which I could crystallize and recrystallize a small quantity of the hydrochloride which was more stable and more suitable for analysis. This gave good results.

I record these particulars in order to put on record the contribution made to Westall's isolation of porphobilinogen. Dent asked me if Westall, to support an application for improved status, might be allowed to publish the note in 'Nature' as sole author and to this I agreed. A very convenient large-scale method, based on precipitation procedures, for the isolation of porphobilinogen from pathological urines was later included in the paper on its structure by Cookson and myself.

As a monopyrrole with acetic and propionic acid side chains, it fulfilled the postulated requirements of Shemin's 'pyrrolic precursor'. Falk, Dresel and I were able to show that it was indeed a precursor of uro-, copro- and protoporphyrins when added as substrate to a chicken red-cell haemolysate. The chemical sequence from glycine and succinate to haem was thus established. Shortly afterwards, Falk and Dresel were able to demonstrate that 5-aminolaevulic acid was also a substrate for the biosynthesis of porphobilinogen and porphyrins.

A large volume of work in the porphyrin and porphyria fields and on photosensitisation issued from the Department as the years went on. We introduced the first paper chromatographic separation of porphyrins which was of enormous value in our investigations and incidentally revealed the occurrence of porphyrins containing 1 to 8 carboxylic functions. One significant finding was made in 1960, the recognition of erythropoietic protoporphyria, a syndrome with urticaria due to protoporphyrinaemia. Conferences were frequently held in S. Africa (Fig. 21).

Whilst the biochemistry of the porphyrins and porphyrias continued to be the main preoccupation of the department, with F. De Matteis (Fig. 19) other lines of research were not excluded. Work on antithyroid drugs led to the introduction of carbimazole (neomercazole) for the treatment of thyrotoxicosis.

Biliary secretion and *Sporidesmin*

The synthesis of ergothioneine was accomplished and its biosynthesis elucidated. In addition, a large volume of work was devoted to the investigation of biliary secretion and the action of cholestatic drugs such as icterogenin and sporidesmin. The latter, a fungal product, had been found to be responsible for the New Zealand 'Facial Eczema of Sheep', photosensitisation of affected animals being occasioned by phylloerythrin as in geel dik-kop.

George III and the 'Royal Malady' hypothesis

In 1968 appeared the joint paper with Macalpine and Hunter setting forth evidence that the hereditary illness affecting sev-

Fig. 19. F. De Matteis   *Fig. 22. Max Rudolf Lemberg*

Fig. 21. A group with L. Eales in Cape Town. Left to right: S. Sano, T.K. With,
C. Rimington, L. Eales, N.R. Primstone, S.M. Joubert

eral members of the Royal houses of Stuart, Hanover and Prussia was probably a form of porphyria similar to variegate porphyria. The background history of this collaboration is as follows. Some time before, the Editor of the British Medical Journal had sent to me for comment the typescript of Macalpine and Hunter's first paper on the illness of George III. I did not find this altogether satisfactory and it was returned to the authors for further work and revision along lines I had indicated. The amended manuscript presented a more convincing case, suitable for publication as a most interesting hypothesis.

Soon after their paper appeared, I had an opportunity of meeting the authors personally and of course we discussed it at length. There were two ways in which such a unique hypothesis could be further tested. First, by examining records of the subject's antecedents for evidence of such a transmissable disease – as was done by Dean in South Africa who traced the variegate porphyria gene from the present day back to 1688. Secondly, one could look for clinical and biochemical evidence in descendants. The latter task was by far the more formidable of the two since the persons involved would be royalty or close relatives of royalty and the specimens required of them would be faeces. Nevertheless, both lines of investigation were embarked upon. The British Museum and other libraries and owners of private papers afforded an immense volume of evidence recorded at the time by royal physicians or by the subjects themselves. Thus James I described his urine as having at times 'the colour of Alicante wine', a fact confirmed by his physician Mayerne who also noted his puzzlement because the discolouration could not be attributed to blood from a diseased kidney, there being no renal pain. The King also remarked that he had inherited his illness from his mother, Mary Queen of Scots.

Macalpine was able by the use of great tact to obtain the cooperation of two living persons related to the Hanoverians whose clinical histories suggested that they might have inher-

ited the condition. They did provide specimens which I analysed. The results were suggestive but disappointing in that
they were not grossly abnormal. However, a European colleague of mine who was well acquainted with porphyria was
able to tell us of an elderly patient of his whom he had attended during an attack and whose urine he found to be
strongly positive for porphobilinogen. He was fully convinced
that she was porphyrinuric; her pedigree from the Hanoverian
stock revealed every possibility of inheritance of the family
disease. At the physician's request, neither his name nor that
of his distinguished patient was revealed. Further than this it
was impossible to go. The number of George III's direct descendants is quite small; so many died at an early age with mysterious symptoms and the further out one comes among living
descendants, the more uncertain does the chance of genetic
inheritance become. A number of the aristocrats whom Dr.
Macalpine approached indignantly refused to give us any
help; some were unrestrainedly abusive!

Reviewing all the information at our disposal, the historical
evidence seemed to be very strong – sufficiently strong to support and extend the hypothesis originally put forward concerning George III although with the correction that the clinical
features corresponded more closely with variegate porphyria
than with acute porphyria. The three of us therefore published
together the second paper on 'The Royal Malady' [18]. This had
a mixed reception. I was not greatly surprised when vitriolic
criticism came from certain quarters; I had rather expected it.
Other colleagues were much more favourably disposed. Two
things I would like to emphasize, however, firstly that what
we put forward is an hypothesis – and one as difficult to establish as the hypothesis of the geological evolution of the earth's
crust – the other that no critic has offered any reasonable alternative interpretation whatsoever of the recorded facts we
have presented.

Peptide-conjugated porphyrins

A gratifying climax to my work over 30 years on the porphyrins and porphyrias was the discovery with Lockwood and Belcher just before I retired in 1968 of peptide-conjugated porphyrins present in the excreta, normally in small quantities but in much larger amounts in subjects with porphyria, especially variegate porphyria [19]. These peptidoporphyrins are hydrophilic and had therefore been missed completely following acetic acid-ether extractions or had been mistakenly regarded as uroporphyrin in the residues. They are present in the bile of variegate porphyrics and are also now known through the work of Eales and Grosser in South Africa to be specifically labelled after administration of labelled 5-aminolaevulic acid, indicating that they play some definite role in intermediary porphyrin metabolism.

The investigation started with a medical student admitted to hospital with a severe attack of porphyria. His urine contained much hydrophilic porphyrin which was not uro- or heptacarboxylic porphyrin and his faecal residue after ether extraction still had bright porphyrin-like fluorescence. New methods were devised for the quantitative recovery of these materials, for their fractionation and purification as, for example, by filtration through columns of Sephadex. Their conjugate nature was demonstrated by reaction with a radioactive reagent known to combine with proteins and peptides, followed by paper chromatography and radioautography. One purified conjugate showed homogeneity by physical criteria and on hydrolysis yielded 15 different amino-acids. The molecular weight was approximately 8000.

The protein moiety was split off from these conjugates by treatment with silver or mercury salts and the porphyrin appeared as haematoporphyrin, hence the union with the peptide must be through thio-ether linkages as in cytochrome c. Any direct connection with cytochrome c metabolism could be ruled out, however, since the amino acids surrounding the porphyrin structure were predominantly acidic and not basic as in cytochrome c. Porphyrin c was synthesized together with Sano in 1964 (Fig. 21).

The medical student's family was investigated as well as other variegate porphyric families and the occurrence and nature of these porphyrin conjugates shown to be common to them all.

In view of the obvious importance of these findings for an understanding not only of the disease process but also of normal porphyrin metabolism, I have been surprised that so little further research has appeared since their publication. Eales and Grosser in Cape Town have done work on the problem but their results, so far, have been largely confirmatory of our findings. It may be that the extraordinary difficulty of handling these materials can account for an apparent lack of interest in them; they require quite new and specialised skills in porphyrin chemistry.

Retirement in 1968

Since my retirement in 1968, I have visited several colleagues in different countries and have worked with them on porphyrin problems – with Eales in South Africa, Torben K. With in Denmark and Sveinsson in Norway.

Perhaps the most interesting study was in Israel on the Dubin-Johnson syndrome. This extremely rare condition is characterized biochemically by predominance of the coproporphyrin I isomer in the urine in place of the III isomer, by dark pigmentation of the liver and associated disturbances. It is an inherited disease and in Israel a surprisingly large number of cases had been found among Jewish families of Iranian origin. Dr. Ch. Sheba, Director of the Tel Hashomer Hospital, where I worked, had a masterly knowledge of medical history and of ancient Hebrew writings and tradition. He explained that in the time of King Darius a group of Jews were given permission to settle at Isphahan in his kingdom. There this community lived in virtual isolation from the rest of Jewry for generations. Their history is recorded in the Book of Esther. It would seem that the genetic abnormality causing the Dubin-Johnson syndrome must have been present in some of the original settlers and have been further disseminated by inter-marriage. Only in recent times have some of these Iranian Jews returned to Israel and have brought this clinical syndrome with them, thus accounting for the relative frequency locally of the disease.

The leisure of my retirement in 1968 I have spent in Norway enjoying the opportunity of watching the pageantry of the changing seasons enriched with flowers, birds and beasts, in reading more widely and indulging my unfailing pleasure in making things.

During this peaceful time I also renewed my literary interests, translating into English verse some of Norway's distinguished poets, Bjørnson, Wergeland, Øverland and others.

Another contribution has been the Biographical Memoir on Max Rudolf Lemberg (Fig. 22) written in collaboration with Charles Gray for the Royal Society in 1975/6. This work gave me much pleasure, in the first place for enabling me to pay my tribute to an esteemed scientist and friend and secondly because it involved much sympathetic study relative to Lemberg's philosophical and religious convictions.

The Norwegian Cancer Hospital

My leisure proved to be of short duration, however, for in the autumn of 1983 I was approached by a group working at the Norwegian Cancer Hospital's Research Institute on the application of porphyrins in the photodynamic therapy of cancer. I was invited to join them as a guest worker at the Institute and began working there in 1984.

This was an entirely new field for me and I had to do an

immense amount of reading to become familiar with it and to catch up on developments in biochemistry during the sixteen years since 1968. The work proved to be extremely interesting and the atmosphere of the Biophysics Department, which I joined, most stimulating and friendly under the leadership of Professor Johan Moan.

I soon realized that there was still much need for more purely chemical work in order to understand fully the factors concerned in this form of treatment. The usual porphyrin administered to the patient is produced empirically from haematoporphyrin and despite considerable investigation of this so-called 'haematoporphyrin derivative' its chemical constitution is still uncertain; it has been proposed that it is a mixture of isomers in which hematoporphyrin molecules are united in ether linkage or in ester linkage or both. The complex displays selective affinity for malignant cells. Irradiation by light of suitable wave length then initiates photochemical reactions which damage or kill tumor cells.

It appeared to me that this problem could be best approached by synthesizing a series of pure haematoporphyrin ethers from aliphatic or aromatic alcohols and testing their effect on malignant cells in vitro and in vivo. This has been done and shows that the more hydrophobic species, such as haematoporphyrin di-amyl ether, are considerably more efficient even than 'haematoporphyrin derivative' itself.

In the stimulating atmosphere of the Biophysics Department the programme has steadily expanded resulting in numerous publications; we have received generous support from the International Association for Cancer Research. Our contributions have received gracious recognition through the recent award to me of the Norwegian Order of Merit by King Olav V (Dec. 6th, 1989). I was elected F.R.S. in 1954 and a Member of the Norwegian Academy of Science and Letters in 1988, an honour which I prize together with the Honorary Fellowship of the Royal College of Physicians of Edinburgh, bestowed in 1966.

Fig. 23. Painting of me by Alcira Batlle, 1988

Fig. 24. Photograph of me, 1992

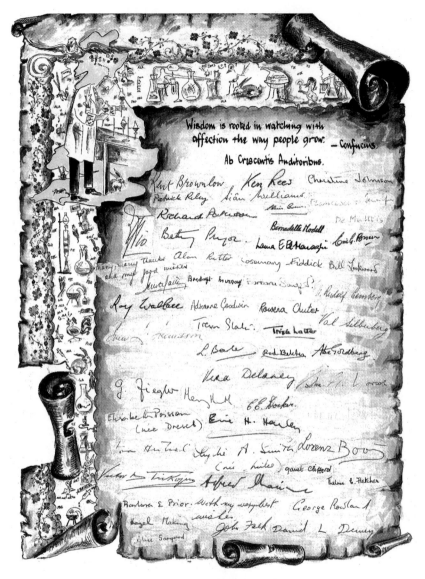

Fig. 25. Painting by Alcira Batlle for my retirement in 1968 with signatures
of some collaborators

Retrospect

Now I have come to the end of my story. I fear I have much indulged in personal recollections and have inadequately attempted to trace the development of biochemistry and its effect upon the men who practiced it – or *vice versa*. History is, after all, a record of men, not of events.

Hopkins said: 'The life of the cell is a dynamic equilibrium in a polyphasic system'. And yet, are we not 'such stuff as dreams are made on?' What is the mind in contrast to the brain?

We may continue to pursue our microdissection but still stand like Cortez' men in 'wild surmize' before the enigma 'What is life?' 'A flame unquenchable once it had birth, that filled the seas and clothed the naked earth – progenitor of Man'.

We can, each one, but say our line – and bow – before the curtain falls.

Acknowledgements

I am much indebted to the Biophysics Departmental secretaries, Mette Jebsen and Gerd Akerlind, for their expert and untiring typing of the manuscript, and to the Photographic Department for reducing all photographs to suitable uniform sizes. The photographs of Sir Arnold Theiler, Douw Steyn and John Quin were kindly supplied by the Director, Onderstepoort Veterinary Laboratory, S. Africa. The titles to the illustrations were kindly inserted by Aasa Dalen.

REFERENCES

1   C. Rimington Note on the effect of ammonium sulphate and other salts
    on the colorimetric estimation of phosphorus. Biochem. J. 18 (1924)
    1927.

2   C. Rimington. Phosphorylation of proteins. Biochem. J. 21 (1927) 272.
3   C. Rimington. The phosphorus of caseinogen. I. Isolation of a phospho-
    rus-containing peptone from tryptic digests of caseinogen. Biochem. J.
    21 (1927) 1179–1186.
4   C. Rimington. The phosphorus of caseinogen. II. Constitution of
    phosphopeptone. Biochem. J. 21 (1927) 1187.
5   C. Rimington. The isolation of a carbohydrate derivative from serum
    proteins. Biochem. J. 23 (1929) 430.
6   C. Rimington. The carbohydrate complex of serum proteins. II. Im-
    proved method for isolation and re-determination of structure. Isolation
    of glucosamino-dimannose from proteins of ox blood. Biochem. J. 25
    (1931) 1062.
7   C. Rimington. Isolation and chemical examination of the poisonous
    principles of *Dimorphotheca spectabilis* Schltr. and *Dimorphotheca
    Zeyheri* Sond. In: 18th Rep. of the Director of Vet. Services & Animal
    Industry, Union of South Africa, pp. 955–972.
8   C.   Rimington   and   J.I.   Quin.   Das   Vorkommen   eines
    Methämoglobinämie erzeugenden Princips bei einigen Arten der Pflan-
    zenfamilie *Tribulus*. Biochem. Z. 269 (1934) 4–13.
9   C. Rimington and J.I. Quin. The photosensitizing agent in 'geeldikkop',
    phylloerythrin. Nature 132 (1933) 178.
10 C. Rimington and J.I. Quin. Studies on the photosensitisation of ani-
    mals in South Africa. VII. The nature of the photosensitising agent in
    'geeldikkop'. Onderstepoort J. Vet. Sci. and Animal Industry 3 (1934)
    137–157.
11 J.I. Quin, C. Rimington and G.C.S. Roets. Studies on the photosensiti-
    zation of animals in South Africa. VIII. The biological formation of phyl-
    loerythrin in the digestive tracts of various domesticated animals. On-
    derstepoort J. Vet. Sci. and Animal Industry 4 (1935) 463–478.
12 C. Rimington, J.I. Quin and G.C.S. Roets. Studies on the photosensiti-
    zation of animals in South Africa. X. The icterogenic factor in Geel-
    dikkop. Isolation of active principles from *Lippia rehmanni* Pears. On-
    derstepoort J. Vet. Sci. and Animal Industry 9 (1937) 225–255.
13 C. Rimington. Some cases of congenital Porphyria in cattle. Chemical
    studies upon the living animals and post mortem material. Onderste-
    poort J. Vet. Sci. and Animal Industry 7 (1936) 567–609.
14 C. Rimington. A reinvestigation of turacin, the copper porphyrin pig-
    ment of certain birds belonging to the *Musophagidae*. Proc. Roy. Soc. B.
    127 (1939) 106–120.
15 C. Rimington and I.W. Rowlands. Serum gonadotrophin. II. Further
    purification of the active material. Biochem. J. 38 (1944) 54.
16 C. Rimington and I.W. Rowlands. Enzymatic inactivation of serum go-
    nadotrophin. Nature 165 (1950) 366–367.
17 G.H. Cookson and C. Rimington. Porphobilinogen. Biochem. J. 57
    (1953) 476–484.

18 I. Macalpine, R. Hunter and C. Rimington. Porphyria in the Royal
    Houses of Stuart, Hanover and Prussia. A follow-up study of George
    III's illness. Brit. Med. J. i (1968) 7–18.
19 C. Rimington, W.H. Lockwood and R.V. Belcher. The excretion of por-
    phyrin-peptide conjugates in Porphyria Variegata. Clin. Sci. 35 (1968)
    211–247.

E.C. Slater, R. Jaenicke and G. Semenza (Eds.)
Selected Topics in the History of Biochemistry: Personal Recollections, IV
(Comprehensive Biochemistry Vol. 38) © 1995 Elsevier Science B.V.

Chapter 9

The Importance of Asking Questions

PETER N. CAMPBELL

Department of Biochemistry and Molecular Biology, University College London, London (U.K.)

Introduction

My contribution to this series takes the form of two aspects of my life as a biochemist. First, I survey my work as a research biochemist and second, my efforts towards the organisation of biochemists both nationally and internationally. At the end I attempt to show how these two activities, which may appear to be disparate, are linked. In any event I feel very privileged to have been able to make some contribution during a most exciting time in the development of biochemistry.

In order to set the scene I briefly sketch my background prior to embarking on biochemical research. In no way is this meant to be a biography but there are certain matters which seem essential to mention.

My father went to a good private boarding school which he left when he was 16 to enter as an apprentice in a firm of builders in Dartford, just south of London. Soon after, he was called up for the 1914–1918 war and served in the cavalry. On demobilization in 1919 he quickly married and I was born in 1921. My mother also went to a traditional private school but was not educated professionally. My father rejoined the builders and remained in the same field all his life. The fact that neither my mother nor my father held a professional qualifica-

tion had a profound affect on their educational plans for their children (I have one sister) and they were determined to do their best for me. At the age of seven I was dispatched to a nearby boarding school where I stayed until I went to East-bourne College (a minor public school) in 1934. There were many effects from going to boarding school so young but per-haps the main one was that after the initial phases school became the centre of life for me. The holidays were rather boring for I had no local friends and most of the time seemed to be spent in attending the dentist, buying uniforms and at-tempts by my parents to correct the worst habits picked up at boarding school. At school I was slow in learning to read, but was interested in practical things and soon learnt to organise others.

At Eastbourne, which I much enjoyed, I got involved in many activities, especially rowing, and was busy at carpentry and metal work. The only subjects I was any good at were science, maths and English, but a turning point came when I entered for an essay competition run for the boys in the school holidays. Although a junior aged 15, I entered for the senior physics prize on cathode ray tubes for which I collected numer-ous books from the local library and copied out chunks of text I hardly comprehended. I won the prize much to everyone's surprise and for the first time I and others thought I was not so dim. I then set about writing many articles for the newspa-pers on morbid subjects such as 'why you need not be afraid of poison gas', for war was looming, but none of my pieces was ever accepted. My parents and teachers were still concerned as to whether I would ever earn a living for I had rejected my father's choice of a career as an architect. In view of the sub-jects in which I showed some ability I approached my mater-nal grandfather, W.B. Nelson, who was a pharmacist. He had moved from shop sales to become the works manager of Allen and Hanbury's factory at Bethnal Green and was the origina-tor of the trade mark 'Haliborange' a product of halibut oil and orange juice which is still sold. His advice in 1938 was not to plan a career in pharmacy for it had been ruined by companies like Boots who were more interested in selling handbags than pharmaceuticals. He predicted that the future lay with Biochemistry. At school, therefore, I was able to answer challenging questions as to my future, although the response was an embarrassing silence since nobody had ever heard of the subject.

My parents were always financially hard up so the obvious route for my higher education was to try and be accepted for a local university and University College London (UCL) was chosen. In those days there was no course for biochemistry at UCL so I entered to do chemistry with zoology as a subsidiary subject.

I was due to enter UCL in October 1939. War was declared on September 3 and UCL was turned into a fire station since it was on a direct line connecting the northern mainline railway stations. We were evacuated first to Bangor in North Wales and then to Aberystwyth in mid Wales. After obtaining my degree I was invited by the Professor, Christopher Ingold, to stay and do a Ph.D. in organic chemistry but I wanted either to study biochemistry or to do something more directly connected with the war. I went to see Professor Jack Drummond who as Professor of Biochemistry in the Physiology Department at UCL had been drafted to the Ministry of Food to advise on dietary matters in wartime Britain. He told me to forget biochemistry and to do whatever scientific work I was directed to by the government.

During a vacation while at university I had taken a job at Standard Telephones and Cables in their quality control lab. As a result of this they invited me on completion of my B.Sc. to join a new factory which was to make radar tubes for the Ministry of Aircraft Production. I became responsible for the preparation of the fluorescent material for the screens and the coating of the cathode ray tubes. (Interestingly, my essay came in useful.) I became a shop foreman in charge of more than a hundred girls and certainly learnt a lot about organisation. At Christmas 1945, following the end of the war, I had to make up my mind what to do. Standard Telephones and Cables were happy for me to be employed on the production of television tubes but I was still interested in biochemistry and I was put off by the idea of being employed by a multinational American company.

The Government offered me a grant to enable me to study for a Ph.D. in biochemistry at UCL. The grant came to £ 165 per annum which was even less than my salary of about £ 250 at that time. A major problem was that I hardly knew any biochemistry and I had to face an interview with the new Professor, F.G. Young, who was leaving St Thomas's Hospital Medical School at the end of 1945 to take up his post at UCL in succession to Jack Drummond. I had for long taken an interest in polarography which had been developed as an analytical tool by the Czech Heyrovsky. I guessed that Professor Young would know little about the subject so my aim was to turn the interview in that direction. I succeeded well in this but whether it influenced his decision to take me is doubtful. On the strength of my grant I married Mollie who as a school teacher was earning the vastly greater salary of £ 360 per annum. And so it was that I arrived at UCL in January 1946 to study for a Ph.D. in Biochemistry.

Research

Diving in at the deep end

I was in truth given a rather uninteresting project by my Professor. There had been much interest in the structure of glycogen and starch and Haworth at Birmingham was a king in the field. Although much remained to be learnt it was clear that the main glucoside linkage was between C-1 and C-4 which meant that C-3 had a free hydroxyl. Young had in mind that, if one fed a rat with 3-methylglucose, 3-methyl-glycogen might be synthesized in the liver. I was never quite sure of the importance of this work. Anyway I set to work synthesizing the sugar and learnt how to feed it by stomach tube to rats. I also set up the means of analyzing for methyl groups in the urine. I soon showed that the sugar was absorbed from the intestine and was largely unmetabolised with no sign of methylated glycogen [1, 2]. This was a relatively dull finding but a more important one was to emerge.

I learnt my biochemistry by conducting practical classes for medical students and also by attending lectures in biochemistry and physiology. I attended a seminar by Dr Hugh Davson who together with Danielli had in the mid 1930s proposed the model of the structure of biomembranes that is now widely accepted. Davson was interested in the absorption of substances across the blood eye barrier and I suggested to him that it might be interesting to study the uptake of 3-methylglucose by the intestine. Davson was a master in the preparation of intestinal loops in cats and so we set to work. I learnt a lot from Davson and together we showed that 3-methylglucose was (against a concentration gradient) actively absorbed, even more rapidly than glucose, and yet it was not metabolised. Verzar was the leader at that time of theories for active absorption in the intestine and had suggested that the sugars were first phosphorylated. This was clearly not so and we had in a neat way countered his theory. We prepared a paper for the *Biochemical Journal* which was accepted [3]. Unfortunately I forgot to tell my Professor, for I seldom saw him about my Ph.D. work, until I asked him how many reprints I should order. He was rightly quite annoyed.

I should add that our work with Davson was not as original as we had hoped, for it subsequently transpired that Csaky, working in Czechoslovakia during the war, had published a similar finding [4] but the German journal was not available to us until after the war.

In January 1947 I was appointed Assistant Lecturer in the department and my annual salary rose to £ 465, the biggest percentage rise I have ever had. After 3 years I obtained my Ph.D. and the matter of my future arose. My Professor told me that I had no future as an academic and that I was only fit for industry. In spite of this I was offered three posts and decided to take the one with Tommy Work. In view of his attitude I did not ask my Professor for a reference, which made him even angrier.

Move to the National Institute for Medical Research

Work was on the staff of the National Institute for Medical Research then located in an old building at Hampstead. The immediate objective was to isolate the toxic factor from agenized flour. Sir Edward Mellanby, then the Secretary of the Medical Research Council, had shown that dogs fed a diet rich in flour whitened by nitrogen trichloride (agene) developed fits and he later showed that this also occurred in ferrets. It appeared that a component of the gluten fraction of wheat flour was modified by the NCl_3 and it was our job to characterise the modification. At that time virtually all flour used for breadmaking, at least in the UK and USA, was treated by this agenization process.

The work on agenised flour entailed a lot of chromatography which I had to organise. An interesting side-line is that we used collidine for the paper chromatography but the Director at Hampstead, Sir Charles Harington, would not allow this smelly solvent to be used in his Institute, so it was my job to take the spotted chromatograms out on a bus to the site of the new institute at Mill Hill. In the event the characterisation of the agenized factor, as it came to be known, was achieved ahead of us by a group at The Flour Millers Research Institute at St Albans [5], working on the maize protein zein, but we were close behind and were the only ones working on wheat [6, 7]. The factor is methionine sulphoximine, subsequently shown to be extremely active when administered to animals.

Its action depends on its competition for glutamate in brain metabolism [5–7].

The main attraction of the move to the National Institute of Medical Research concerned the long-term plans of Work. He had explained to me from the start that his real interest lay in a study of the mechanism of protein synthesis, so having completed the work on agene we were free to turn our attention to this subject. Our first approach was to utilise our expertise in paper chromatography to try and detect the presence of peptides as possible intermediates in protein synthesis. We did this by running a chromatogram of extracts from liver and repeating the run after hydrolysing with acid. We failed to find any significant amount of peptide apart from glutathione but an interesting result did arise. In liver extracts we repeatedly detected a substance which by its mobility using two solvents appeared to be citrulline. However the substance seemed unusually unstable to heat and acid. While Work was away on holiday I identified it as glycerylphosphorylethanolamine [8]. I was able to confirm this through the kindness of Eric Baer of Ottawa who had just synthesized it and sent me a sample. Work and I did not pursue this finding further, for we did not wish to get involved in lipid biochemistry.

At that time George Popjak was working upstairs in the Institute on lipid biosynthesis for which he used rabbits. Work was friendly with Popjak and so it was decided to turn our attention to the biosynthesis of milk proteins by rabbits using radioactive amino acids. The object was to determine whether peptides from plasma proteins could be utilised after partial degradation for the synthesis of milk protein. If this was so then an enzyme was likely to be involved in their biosynthesis. We had to learn lots of techniques apart from how to milk rabbits. The biggest obstacle was the separation of amino acids from protein hydrolysates by cation exchange resin chromatography. Moore and Stein at the Rockefeller Institute were the experts in this technology. Stanford Moore used to visit us frequently – they say he always had the same room

overlooking the Thames at the Savoy Hotel when he came to
London. He was very generous with his time. The first experiments were successful and indeed showed that peptides were
unlikely to be intermediates in protein synthesis [9]. Work
and I published a review in *Nature* concerning the evidence for
a template which might be used for the assembly of the amino
acids in the newly synthesized polypeptide chain [10].

The next step was to isolate specific milk proteins and for
this we turned to goats [11, 12]. They proved to be more of a
handful and our friends at the Institute got lots of amusement
from our methods of driving the unwilling goat from the farm
at the bottom of the hill to our laboratory on the third floor
where we had a metabolism cage. Out of this work developed
my interest in immunology. We had been joined by Brigit (Ita)
Askonas, a Canadian who had studied for her Ph.D at Cambridge with Dr Malcolm Dixon. We had talked to John
Humphrey who was then the Head of the Department of Immunology in the Institute and was an oracle when it came to
matters of immunology. He suggested that we study whether
the lactating mammary gland was able to synthesize antibody. We, therefore, immunised our goat with pneumococcus
immunogen and then at the point of parturition injected her
with [^{35}S]methionine. We showed that the mammary gland
did not synthesize antibody but that this was passed from the
plasma to the milk. This was the first time this had been
shown [13].

Departure for the Courtauld Institute of Biochemistry

At the completion of 5 years at the Institute it was suggested
that I move on elsewhere and I was fortunate enough to be
offered a temporary post with Professor Frank Dickens at the
Courtauld Institute of Biochemistry. The Institute was the
Department of biochemistry and chemical pathology for the
Middlesex Hospital and Medical School. Dickens was profes-

sor of Experimental Biochemistry but the Director of the Institute was Professor Sir Charles Dodds who had in the mid 1930's discovered stilboestrol. Also working in the Institute were the Taits who discovered aldosterone, originally named 'electrocortin'.

My post was funded by the Cancer Research Campaign. The project was to study protein synthesis by incubated slices of rat liver tumour, and in particular to investigate the incorporation of ^{14}C from labelled glucose into the tumour protein and explain why this was much more rapid than the incorporation into slices of normal liver. I learnt to produce liver tumours in rats by feeding diazoaminoazobenzene in a diet prepared by hand. I dread to think what the modern safety officer would have thought of our procedures. The outcome was uninteresting in that the explanation was that in the case of normal liver there was a rapid breakdown of glycogen to glucose which diluted the radioactive glucose whereas this did not occur with the liver tumour [14].

I then engaged in further experiments to critically assess the idea that tumours were special in that they were able to utilise peptides from normal tissue for the synthesis of their protein. This idea was attractive in that many solid tumours had been shown to have a marked avidity for plasma proteins [15]. After my experience at the National Institute on protein synthesis, I decided the way forward was to study the synthesis of a specific protein and for this purpose I chose serum albumin. I was joined by Nancy Stone from Australia and we set about developing some of the original experiments of Anfinsen on the synthesis of serum albumin by liver slices. We used antibody to detect the albumin synthesized by rat liver. We also used regenerating liver and showed this to be more active. An important result was that hepatoma tissue in contrast to cholangioma (derived from bile duct canaliculi) retained some ability to synthesize serum albumin [16]. This has remained a criterion of the various minimal deviation hepatomas derived from parenchymal tumour cells.

It was then 1957 when Zamecnik and his group in Boston reported on the incorporation of radioactive amino acids into the isolated microsome fraction from rat liver. With our experience of serum albumin synthesis we set about trying to discover whether such a fraction retained the ability to synthesize serum albumin rather than merely the synthesis of uncharacterised protein. For this purpose we obtained very high specific activity [^{35}S]methionine from Fred Sanger at Cambridge and subsequently showed that the labelled peptides closely resembled those labelled in vivo. We concluded that although there was no net synthesis of protein by the incubated isolated microsome fraction the polypeptide chain of albumin was extended in such a system [17].

Although we considered this to be the first demonstration of the synthesis of a specific protein in a cell free system (about 1958) we were in competition with Dr Mel Simpson who originated the idea that mitochondria might be the site of synthesis of some proteins. He claimed to have demonstrated with Bates that incubated liver mitochondria produced a *net* synthesis of cytochrome *c*, in fact he claimed that you could weigh the increase. At the time this detracted from our achievements but subsequently it was shown that Bates had fabricated his results. Another diversion was the work of Webster who had claimed to show a net synthesis of protein by incubated pea ribosomes. Again this claim proved to be fraudulent. We subsequently went on with John Sargent to study the direction of assembly of the albumin polypeptide chain [18] after the elegant work of Dintzis published in 1961 had shown that globin chains were assembled from the amino terminus [19].

Diversions on autoimmunity and diabetes

While the work on protein synthesis was proceeding I was engaged in two other ventures, namely thyroid autoimmunity and the role of a serum peptide in antagonising insulin in diabetics. I will now describe the basis of these interests.

I have explained that John Humphrey at the National Institute of Medical Research had aroused the interest of Ita Askonas and myself in immunology. It was around 1955 that Medawar and his colleagues had made the discovery of 'immunological tolerance' which intrigued me and I used to talk about the work with another member of the Courtauld lab, Ivan Roitt, who was engaged in some rather boring work on carcinogenesis. In the summer of 1955 the Bland Sutton Institute of Pathology in our medical school held an open day to celebrate the birth of its founder Bland Sutton. I was attracted by an exhibit of a surgeon from New Zealand called Richardson. He had studied the prognosis for patients who had come to the Middlesex Hospital and had received surgical treatment for breast cancer. He was particularly interested in the 7% with medullary carcinoma with lymphoid infiltration. Of the 117 patients with this type of tumour he found a relatively favourable prognosis and he made the suggestion that the inflammatory infiltrate could be a visible sign of some slight tumour antigenicity [20]. I drew Ivan's attention to this exhibit and we talked about it extensively.

Our thoughts, which even at that time could not be considered original, were along the following lines: We wondered whether the reaction between antibody and antigen would have a deleterious effect on the growth of the tumour. If this were so and the tumour really did contain an antigen against which the body synthesized an antibody then we suggested the following procedure. The surgeon would remove the tumour from which the tumour-specific antigen would be isolated after removal of normal proteins by antisera. The purified antigen would then be administered to the patient in the hope that this would stimulate the production of antitumour antibody which might inhibit the reappearance of tumour. We wrote this up for Dickens and Dodds and suggested that what was wanted was a model experiment to determine whether the reaction of antibody and antigen within a tissue would inhibit growth.

The model we suggested concerned milk proteins for we postulated that they might be regarded by the host as 'foreign' since, according to the recently described theory of tolerance, a protein which is tolerated by the body must have been circulating in embryo. We argued that the milk proteins would not have circulated in embryo and hence would not be recognised as 'self' in the adult. This of course does not usually present any problems since the milk proteins are synthesized by a relatively 'isolated' organ and are immediately exported in the milk. Our plan, therefore, was to administer to lactating rabbits milk protein from rabbits and then attempt to check the effect on the functioning of the mammary gland. We thought that this would be a relatively easy way to assess the function of the gland.

Although our grants did not cover the proposal, Dickens supported us and in particular suggested that we use as our model thyroglobulin which is synthesized by the thyroid gland, normally retained there and possibly not recognised as 'self'. We liked the idea but felt we would be happier with the milk system with which I was more familiar. As we were setting up our experiments we met Deborah Doniach, a physician working on a diminutive grant with a thyroid surgeon, Ralph Vaughan Hudson. Deborah had a particular interest in lymphadenoid goitre, better known as Hashimoto's disease. It had recently been reported that patients with this disease had high levels of gamma-globulin in the serum. The highest values were found in untreated patients with large goitres. The raised gamma-globulin levels and the infiltration of the thyroid with lymphoid tissue, lymphocytes and numerous plasma cells suggested to Deborah and Vaughan Hudson that an immune response might be involved in the disease.

In view of the state of our thinking it was obvious that we should check whether the raised gamma-globulin represented anti-thyroglobulin. Ivan prepared saline extracts of normal and thyrotoxic human thyroid glands and added these carefully to the serum. A precipitin ring soon formed at the interface. I still remember this simple experiment which only required the serum, the extract and a test tube and pipette. The appropriate controls were done which showed that the antibodies were organ specific to thyroid and that they were only present in patients with Hashimoto's disease.

We naturally wanted to conclude that the destruction of the thyroid in Hashimoto's disease resulted from progressive interaction of thyroglobulin in the gland with the autoantibody present in the patients' circulation. We were encouraged in this view by the exciting work of Rose and Witebsky which had just been published in *Federation Proceedings*. A full paper was subsequently published [21]. They had removed a lobe from the thyroid of a rabbit and had reinjected the same rabbit with an extract of the lobe. The rabbits produced antibodies to thyroglobulin and the histological damage was proportional to the antibody titre. We published our results in *The Lancet* [22]. A longer paper was published in the *Biochemical Journal* [23]. Our clinical colleagues were scathing as was Rodney Porter who warned me that I had made a fool of myself. Medawar and his colleague David Newth were happy with our results and we were careful to have their views before publication. I think the story has many implications which are, briefly: The atmosphere in which research is conducted is important, research workers should have broad interests, biology should not be divided into rigid disciplines, advances come through collaboration between clinicians and basic scientists, and the young should not be inhibited by dogma.

John Vallance-Owen, a diabetologist at the Postgraduate Medical School, Hammersmith, was introduced to me by Stan Peart concerning an insulin antagonist, claimed to be present in the serum of type I diabetics. He used rat hemidiaphragms to detect the biological action of insulin on the synthesis of glycogen. It is a convenient preparation in that the other hemidiaphragm can be used as a control. He had found that in uncontrolled diabetics the serum inhibited the action of insulin. He wanted me to fractionate the serum proteins and identify the active factor. I obtained some indication that the inhibition was associated with serum albumin for I could isolate this by the trichloroacetic acid-ethanol procedure which we had used in our studies on albumin biosynthesis. The results, although not conclusive, were in accord with a peptide being bound to the albumin and so the factor became known as the 'synalbumin antagonist' [24, 25]. Later, it was shown by others that insulin was split in the liver to the two chains and that the B-chain was subsequently bound to albumin but it seemed clear that this was not the explanation for our results. Much more recently a hydrophobic peptide 'amylin' has been implicated in insulin action and it certainly remains a possibility that we were dealing with amylin [26].

Back to protein synthesis

I have mentioned previously the ill-fated work of Simpson on the biosynthesis of cytochrome c. After I left the National Institute for Medical Research, Work and others had shown that this mitochondrial protein was not synthesized in the mitochondria and so presumably was synthesized in the cytoplasm and subsequently transported into the mitochondria. Nestor Gonzalez-Cadavid joined me as a postdoctoral student from Argentina in 1964 and set about locating in the liver cells the site of the synthesis of the apoprotein, i.e., cytochrome c lacking its covalently bound haem group. First, extensive work was required on the purification of the protein from rat liver. The synthesis was followed after injection of rats with ^{14}C-labelled amino acids. Evidence was obtained that the cytochrome c was synthesized within the microsome fraction of rat liver. It seemed first to be bound rather tightly to a component of the microsomes and was then transferred to a looser microsomal binding before being transferred to the mitochondria

[27–29]. There has, of course, been much subsequent work on the details of the transfer of cytochrome c to its final mitochondrial location, particularly by Walter Neupert and his colleagues in Munich.

Another problem that occupied a lot of effort was that of the difference between membrane-free and membrane-bound polyribosomes. Today it is readily accepted that a cell such as the parenchymal liver cell contains two physiological types of ribosomes, free and bound, and that they each form the site of the synthesis of proteins which have different destinations, but it took many years for this to be clearly established. In this work I had many colleagues but especially Alexander von der Decken, Ross Lawford and Jo Stahl, who separated and studied the protein-synthesizing activity of the two fractions [30]. We were the first to show that serum albumin was synthesized uniquely by the membrane-bound polyribosomes [31].

In 1964 I returned again to a study of the synthesis of milk proteins with the arrival of Keith Brew as a Ph.D. student. A major attraction of milk proteins was that their synthesis was under hormonal control and so they were a good model for tumour proteins many of which were under similar control. We wanted to set up a cell-free system and picked on the guinea pig as a convenient animal with good mammary glands. We decided to start by studying whey proteins for at that time casein was difficult to characterise and we believed β-lactoglobulin to be the major whey protein in all milks.

Although the Cancer Research Campaign was supporting our work, we thought it would be good to get some more money from the US Department of Agriculture. We duly applied and were visited by one of their distinguished scientists, Dr Timasheff. He suggested that instead of the title of our application being 'The synthesis of whey proteins' it should be 'The synthesis of guinea pig β-lactoglobulin'. We were not happy about this but we were beggars. We got the grant; but soon after, Brew showed that there was no β-lactoglobulin in guinea-pig milk, only α-lactalbumin. We decided that the honest course of action was to inform the donors in the USA. They told us that in view of the change of title we would have to reapply and that there was no hope of a grant since the money had run out. So much for our honesty. We were relieved when the Agricultural Research Council in the UK agreed to support us.

α-Lactalbumin proved to be a splendid protein for the biosynthetic studies since it is small and not glycosylated [32]. By a lucky chance the guinea-pig protein also proved to have virtually the same structure as that in humans. (In fact β-lactoglobulin is confined to the milk of ruminants.) The work on biosynthesis went well but we and others were concerned that it had no known physiological role. I recall in 1967 David Phillips after he had lectured on lysozyme and I on α-lactalbumin telling me that my protein was boring. Brew did some work on the structure of α-lactalbumin and came to the conclusion that it resembled that of lysozyme.

The turning point came when in a paper published in 1967 we wrote 'To the extent that the properties mentioned reflect similar primary structures the α-lactalbumins may have evolved by gradual modification from lysozyme, which is found in the milk of many species. Guinea-pig milk α-lactalbumin had, however, no detectable lysozyme activity.' [33]. Rodney Porter, who was the referee for this paper, objected to this reference and asked us to remove it. We refused on the grounds that we should be allowed a little conjecture in the Discussion section of our paper. How right we were, for the relationship between the two proteins has given rise to an enormous literature (see e.g. a recent review of the subject by McKenzie and White [34]). Brew departed for a postdoctoral post with Robert Hill in the USA and determined the primary structure of α-lactalbumin. At that time Ebner showed that α-lactalbumin was a component of the enzyme system responsible for the synthesis of lactose. In fact the protein changed the substrate specificity of the galactosyltransferase [35].

Brew and his colleagues have continued their work on the structure and function of α-lactalbumin to the present time.

Appointment at the University of Leeds

In October 1967 I was appointed to the Chairmanship of the Department of Biochemistry at the University of Leeds. I was to spend a busy but very enjoyable 8 years in this post. In Leeds I continued studies on the synthesis of α-lactalbumin and together with David Tarin and Diana McIlreavy [36] we showed that the RNA isolated from the polysomes derived from lactating guinea-pig mammary gland, when injected into Xenopus oocytes, caused them on incubation to synthesize α-lactalbumin. This showed that it was possible to isolate an RNA containing mRNA for the milk protein and we had the means to detect it, since α-lactalbumin has only a single methionine; on treatment with CNBr it breaks down to give two characteristic peptides. There was then a difficult period before methods for the isolation of mRNA were devised. These largely depended on the finding that in eukaryotes the mRNA had a terminal chain of poly(A) which allowed it to be isolated by affinity chromatography. At this point Roger Craig joined us and he succeeded in the preparation of antibodies for both guinea-pig α-lactalbumin and the caseins. He and his group isolated the mRNA and translated it in a wheat-germ heterologous system and identified the products [37]. The mRNA was extremely active and the product was shown to be pre-α-lactalbumin with a 20 amino acid extension at the amino terminus in accord with the signal hypothesis for secreted proteins enunciated by Gunther Blobel.

Return to London

In 1976 I moved back to the Middlesex Hospital Medical School as the Director of the Courtauld Institute and Roger Craig joined me soon after. Extensive studies followed on the effect of the presence of the poly(A) chain on the turnover of the mRNA [38]. Burditt showed that in the guinea pig the synthesis of α-lactalbumin and the caseins was differentially controlled. The situation in this respect seems to differ among mammalian species [39].

In 1981 we worked on the isolation of mRNA from human mammary gland. Fortunately, as I have mentioned, the structure of the guinea-pig and human protein is very similar so that the same DNA probe could be used to detect the mRNA in the two species. Reports were appearing at that time that the presence of α-lactalbumin in the serum of patients with breast cancer could be used as a tumour marker. Craig and Hall were able to show that the tumour tissue did not contain mRNA for α-lactalbumin and that the protein detected although similar to the milk protein was not identical [40].

Len Hall, a former student from Leeds, came and worked on the determination of the structure of the genome for the milk proteins and made a comparison with the genome for lysozyme [41]. This work showed that the distribution of the exons and their length was very similar for the two proteins and supported their origin from a common ancestral gene. There were differences in size among the introns. It was possible to attribute these differences to the presence of repetitive sequences which must have arisen after duplication of the ancestral gene. This work has also initiated the identification of the presence of a sequence upstream of the 5' initiation site which is common to the genome for all the milk proteins so far studied. This in the terminology of Hennighausen has been named the milk box [42]. So far the exact sites for the control of the expression of the milk proteins have not been delineated.

Work on α-lactalbumin has proceeded with the ability to transfer the gene into mice and the identification of mRNA for α-lactalbumin but not its expression in the skin of guinea-pigs. In these studies Paul Brickell has been the leader.

Organisation of biochemists

Rejuvenation of The Biochemical Society

When I joined UCL for my Ph.D. it seemed natural for me to attend the meetings of The Biochemical Society, encouraged by Professor Young who had been Chairman of the Editorial Board of the *Biochemical Journal* from 1942–1946. At the National Institute for Medical Research I gained a reputation as an innovative organiser and so I came to join the Committee of the Society in 1957. In truth, I found a largely moribund situation and I recall that after about the third meeting of the Committee virtually all the matters on the agenda were those raised by myself. I had a row with the senior Honorary Secretary who was due to retire so in order to keep me quiet I was elected Meetings Secretary (junior of the two Secretaries) in 1958. There were three broad areas of the activities of the Society to which I concentrated my attention – meetings, publications and international affairs. I shall deal with each in turn.

At that time, apart from the one annual symposium, which was published, the meetings consisted of the presentation of 10-min papers of original work by the members. A 400-word abstract of the papers was published in the *Biochemical Journal*. The Secretary was expected to listen to all the papers. The meetings rotated around the biochemistry departments in the various universities, and occasionally industry. In 1959 I proposed that a new form of meeting should be tried at which a topic chosen by the host department should be discussed. The proceedings would not be published and we proposed the name 'Colloquium'. This was eventually allowed but only on an experimental basis so the consequence was that, although it was hailed as a great success at the meeting in Liverpool in 1959, there had to be a pause for two years before another could be held. I and others also encouraged the foundation of nominated lectures at the meetings such as The Hopkins and Jubilee Lectures, to be followed by others later.

The Society only published the *Biochemical Journal* and the annual symposium. The journal was published and printed by the Cambridge University Press, as it had been since its foundation in 1911. The Press charged for publication at a standard rate that applied to all their journals. I recall meeting the Manager and asking what the Press did for this charge. The Treasurer apologised on behalf of the Society for such a rude question. I was not surprised when the Manager answered 'nothing'. There was another snag in that the printing capacity of the Cambridge University Press was quite limited so that the printing of accepted papers was delayed. By this time Bill Whelan had joined me as Meetings Secretary and Cuthbertson had become Treasurer. We set about a scheme for the Society itself to publish the journal and to use The Chemical Society for distribution purposes. We found an alternative printer since the Cambridge University Press had a rule that they would only print journals which they also published. The Editors of the Journal refused to accept the new arrangements on the grounds that the standard of the Journal would be jeopardised. As a result the already difficult relations between those responsible for the Journal and the officers of the Society were exacerbated so that eventually the editorial board resigned and a very difficult situation in the affairs of the Society was reached. Henry Arnstein replaced Whelan as Secretary in 1962. I recall the two of us wondering how we could edit the Journal single handed. Fortunately the new Chairman of the Editorial Board, Howard Rogers, cooled the atmosphere, the transfer was delayed a year and all has gone well ever since. Indeed it is now acknowledged that we were right and that it was a turning point in the financial fortunes of the

Society, but for some time eminent members of the society who had not been involved spread the news that I was a most difficult customer.

Another venture in which I was involved was the publication of *Essays in Biochemistry*. I was struck at the time that the *Annual Reviews of Biochemistry*, while excellent in many respects, were not very suitable for students and others who wanted to get a quick overview of a subject. I therefore suggested 'a series of small volumes of essays which might be read with pleasure and profit by the senior student and would be of a size and price suitable to his pocket'. This proposal was met with considerable criticism by those who claimed that the role of the Society should be confined to the publication of original research with the exception of the annual symposia. It was even suggested that such reviews would endanger the dignity of the Society. In the event the proposal was passed by the narrowest of majorities at the Annual General Meeting in 1964 and the first volume was published in 1965. One condition I had to concede was that the Society should lose no money over the venture. Hence, although the Society held the copyright, Academic Press took all the financial risk and the Society merely gained the royalties. A pity about this, for the first volume sold 10,000 copies. I was joined by Guy Greville as co-editor and there could not have been a finer choice. I am glad to say that I enjoyed editing the first 20 volumes.

The foundation of FEBS

I turn now to international matters and especially the foundation of the Federation of European Biochemical Societies (FEBS). The Biochemical Society was for long conscious of its international responsibilities and in 1949 organised the first International Congress of Biochemistry in Cambridge during which an International Committee for Biochemistry was set up with a view to the formation of an international union which was eventually founded in 1955. There followed a number of joint meetings between the Biochemical Society and other societies, such as that with the Belgian Society in 1953 and with the Scandinavians in 1956. When I became Secretary I noted that when these meetings were held overseas the delegates from the UK consisted mainly of the Officers of the Society, for there were no funds to support the travel of younger people. I was very much encouraged when the Director of The Wellcome Trust agreed to provide funds for young biochemists.

In 1959 I was invited to a meeting in Stockholm and I recall being at the house of the Director of the Wenner Gren's Institute, Professor Olav Lindberg. He complained that they had recently had a meeting in Italy and that the Italians spoke their own language so that the Swedes could not follow the proceedings. He asked whether the British could not organise European meetings in such a way that English was the common language. I thought about this together with the fact that many of us were worried that the centre of biochemical research had moved to the USA. We decided that all european biochemists should be especially invited to the meeting of the Society in the summer of 1960 which was to be held in Cambridge. This was repeated in 1962 by which time Bill Whelan had returned as International Secretary. It was at this meeting that the idea of FEBS arose. It was agreed that there should be an annual congress, the location of which should rotate among the countries of FEBS, care being taken to see that English was the common language.

FEBS was formally constituted by 17 delegates at the Society's meeting at Oxford in July 1963 with its foundation in January 1964. The first meeting of FEBS was in London in March 1964. I and Henry Arnstein were the two Secretaries of the Society with Bill Whelan as the first Secretary of FEBS. Prakash Datta was elected Treasurer. We had 1050 registered members with 450 from overseas. There was a splendid dinner as the guests of Arthur Guinness Son & Co Ltd at the Goldsmiths' Hall in the City of London with everyone in evening dress. The occasion is still remembered by those who attended [43].

Those who started FEBS were determined that their objectives should not be frustrated by politics but I recall that on at least two occasions politics did protrude. The first was when the West German society refused to recognise East Germany as a country. It was eventually agreed that since FEBS was a Federation of Societies and not of countries that a country like Germany could have two societies, one of which could be in the West and the other in the East. This was an important principle when later the problem of China and Taiwan arose in connection with the International Union of Biochemistry. The other problem was when Israel applied to join. Was Israel to be regarded as situated in Europe? We decided Yes. Later when I was in Israel I met the Director of the Weizmann Institute who, although an atomic physicist, spoke highly of the biochemists. When I asked him why, he told me that it was FEBS that decided that Israel was in Europe after which all the other disciplines had followed the example.

It will be seen that the foundation of FEBS was not without its problems, but we were compensated by its success. I recall sitting on the steps of the University at the FEBS meeting in Vienna in 1965 and meeting biochemists who had been allowed to travel from East Germany for the first time. As a result we all thought that the efforts had been amply justified.

I retired as Secretary of The Biochemical Society in 1964 but returned to FEBS as the second Chairman of the Summer Schools Committee which Henry Arnstein had started. The Council of FEBS had from the outset aimed to organise short courses at various sites in Europe to introduce new techniques and fields of research mainly for the benefit of younger scientists. Arnstein had succeeded in getting funds from the Volkswagen Foundation and I continued to negotiate with them for what today would be regarded as rather modest but was, at that time nevertheless, crucial support. I recall with pleasure various events concerned with the Summer Schools such as a course at the Weizmann Institute in Israel during which, while bathing in the Dead Sea, I was able to recruit Fred Sanger to hold a practical course. The only possible time proved to be the last days of December 1968. The USSR invaded Czechoslovakia in August of that year. We were due to receive two Czechs for the course and had grave doubts as to their ability to be able to attend. When they did turn up on time we asked how it was that they had passed the frontier guards. They said that when they were stopped they pleaded that since FEBS had invited them as guests it would be rude to let down their hosts. The guards were much amused that the British should hold a Summer School in mid winter so let them through. After that we changed the name to 'Advanced Courses'. Another innovation was that of the Wintersbergers who combined skiing instruction with their FEBS courses.

In those early days of FEBS we all endeavoured to stride the 'iron curtain' by holding congresses and courses on both sides. I was glad to welcome in Leeds Jo Stahl from Berlin Buch as the first Unilever Fellow of The Biochemical Society to come from East Germany. (Some 22 years later he kindly sent me a piece of the Berlin Wall when eventually it was demolished. I recall traversing the Wall many times and reflecting on man's inhumanity to man.) On one occasion I was asked on a visit to East Berlin if I would mind being interviewed on their radio. I rapidly agreed, which brought looks of amazement to my hosts who could not believe that I did not have to seek permission first. In the event the first question from Professor Sam Rapoport was 'What do you think of East Berlin?' At that point I thought that our Foreign Office might be listening so I replied 'better than I had expected'.

It was these contacts with Eastern Europe that prompted me in 1989 to suggest to FEBS that they should set up a 'Scientific Apparatus Recycling Scheme' (SARS). The object of SARS is to enable those in Eastern Europe to make use of apparatus, books and journals that are surplus to the current needs of those in Western Europe. It seemed to me bizarre that a place like the Institut Pasteur should have about 150 copies of *Nature* each week while an institute in, say, Poland has none. Similarly, while we throw out bench centrifuges because of some over-enthusiastic safety precaution there are plenty of laboratories in Eastern Europe without one. FEBS has been generous with its support of SARS but could not have been so if it had not first launched its journals.

Work for the International Union of Biochemistry

In 1968 my old professor Frank Young, who was then Treasurer of the International Union of Biochemistry (IUB), invited me to become Chairman of the Committee on Symposia which had become rather moribund. I agreed and set up an international advisory committee and tried to bring some democracy into the affairs. I then joined the Council of IUB in 1970. Harland Wood became General Secretary and thus began a long and happy association with IUB which I served in many capacities.

In 1971 FEBS held their annual congress in Varna in Bulgaria under the Chairmanship of Todor Nikolov. He invited me to conduct a round-table discussion on biochemical education. Apart from a Summer School which I had previously organised in London on the role of practicals in the teaching of biochemistry to medical students, FEBS had never been concerned with education. I organised a panel of speakers but since we were in competition with many scientific sessions I hardly expected an audience. In the event the room was overfull and the venture was deemed a success. I suggested to IUB that a Committee on Education should be created. This was the start of the IUB newsletter *Biochemical Education*. It was fortunate that Bernard Kilby, a Reader at Leeds, agreed to become the first editor. After the first 8 years it was decided to convert the news letter into a journal. Robert Maxwell of Pergamon Press made a generous offer and they have published it ever since on behalf of IUB. Ed Wood, also of the department in Leeds, took over from Bernard Kilby and has been a great inspiration. I believe that the journal is still unique among the various scientific disciplines at least in being a commercial venture.

During my chairmanship of the Committee on Education various other initiatives were launched. We organised film and video sessions at congresses and when invited also held discussions on various topics. However, we really wanted to influence the teachers on their own ground especially in developing countries. IUB seemed to be in a unique position to fulfill this role and so we started the workshops.

The first was in Karachi in 1977 and this set a pattern for nearly 30 similar events. The format was for the Workshop to be led by about four experienced teachers from overseas together with some from the host country. The hosts covered the internal expenses while IUB or other organisations, such as the British Council, paid for the overseas travel. If possible, the visitors came from a mixture of countries and had no preset message. The object was to give our hosts the benefit of our various experiences in the hope that at least some of them would prove applicable. This was important for I recall that one speaker told us at Karachi that they did not want to be told what they should do by some old imperialists. Not only did we speak about the methods of teaching including practicals but we also talked about our special fields of interest so that it was clear that we were biochemists and not mere educationists. If possible we liked to start by holding the workshop in one centre and then see as many different universities and medical schools as possible. At the conclusion the visitors together with their hosts wrote a report which was usually seen by the government and/or those who had provided financial support. I recall participating in workshops in not only Karachi but Taiwan, Shanghai, Hong Kong, Japan, Egypt and Venezuela. After I retired as chairman and Frank Vella took over he con-

tinued with the work with even increased enthusiasm. At the
very least IUB became better known. There were many memorable incidents. Thus in Egypt we were handed postcards by
the most attractive girl medical students with messages encouraging us to continue our work, for we were the first visitors to give them the opportunity to express their views concerning their teachers.

IUB has done much to assist university teachers by the provision of text books under their textbook fund and often, but
not sufficient, copies of *Biochemical Education* have been provided. The Biochemical Society in London was the first to sub-
scribe and they still sell T shirts in support. They were followed by the biochemical societies in the USA, Japan and the
Netherlands. Frank Vella has in this way distributed thousands of textbooks on biochemistry to developing countries.

Another venture of the Education committee has been to
establish a Fellows conference before each of the triennial congresses. IUB provides support in the form of Travel Fellowships for young biochemists, mainly from developing countries, to attend the congresses but we observed that the
Fellows often felt lost on arrival especially since they were
probably visiting a foreign country for the first time. We resolved that the Fellows should be invited to spend a few days
together before the congresses under the guidance of two experienced hosts. The Fellows have a chance to show their posters
and also to hear a general review on some field of work from
the hosts. The first such conference was at Perth in 1982 and
they have since been held regularly with considerable success.

I was also concerned with the Federation of Asian and
Oceanian Biochemists (FAOB), a regional organisation along
the lines of FEBS founded on the initiative of Bill Whelan after
he succeeded Harland Wood as General Secretary of IUB, and
enjoyed working with the Japanese who were keen supporters.
One major problem concerned the membership of mainland
China, already referred to, since the society in Taipei was already a member of IUB and FAOB. Thanks to the fine arduous

work of Bill Whelan and Bill Slater the problems were eventually solved not only for the benefit of the biochemists but also for the other unions attached to the International Council of Scientific Unions (ICSU). I visited Beijing in 1980 to explore the ways in which IUB could help in the development of biochemistry. Later in 1982 I attended a meeting in Changchun, in Manchuria, at which I was the only foreign visitor among some 150 Chinese from medical schools and universities. After travelling in a sleeper for 16 hours I was greeted at 7 a.m. on the station platform by the entire organising committee and then informed that I was to give three one-hour lectures. I was certainly relieved when the projector broke down after $1\frac{1}{2}$ hours. The problem of the two Chinas reminds me that sometimes the union had to act to unify rival groups within a country. Two societies in Greece sought to adhere to IUB. My task was to interview the representatives of the societies who would not speak to each other and suggest the formation of a national committee. This was successful. The same formula proved successful in Nigeria. A rather different problem arose in South Africa where I went in 1973 to advise on the foundation of a multiracial biochemical society. This was done, but we could not put it to the test for the lack of black biochemists. South Africa subsequently joined the union and we look forward to the society gaining in activity in the now improved situation in that country. Indeed perhaps they will be influential in setting up a regional organisation for biochemists in Africa.

I came to be the Editor-in-Chief of one of the IUB journals, the *Journal of Applied Biochemistry*, now renamed *Biotechnology and Applied Biochemistry*, by accident. Unfortunately, the first editor could not continue soon after the start of the journal and since I was then the Chairman of the Committee on Publications I took over the editorship. It has proved to be a hard task to revive the journal but we have great hopes for it under its new publishers Portland Press, the publishing subsidiary of The Biochemical Society.

432   P.N. CAMPBELL

The creation of British Council Links

Another aspect of my work in developing countries has been
the creation of Links under the auspices of The British Council. My activities started in 1967 when I went to Leeds. Frank
Happold, my predecessor, had had relationships with the Departments of Food Science and Biochemistry at the University
of Ghana. The main campus is at Legon in the suburbs of
Accra, the Medical School being in the centre of town at Korle
Bu. The concept of the links is that they should set up a relationship between a UK university and a similar institution
overseas preferably at the level of departments. The links
allow for the exchange of academic and technical staff and also
the provision of some equipment and running expenses. The
arrangements are usually reviewed at 3-yearly intervals, Ideally the Link should provide mutual benefits for the two departments, but this is seldom the case for biochemistry, an
exception being molecular parasitology. A visit overseas for
even 2 weeks certainly makes one question the rationale of
one's teaching at home and I have found such arrangements to
have a very beneficial effect on the academic staff. It also helps
a lot when receiving overseas students at one's home university.

The experience of the link with Ghana has been mixed. As
usual the students were excellent but rapid changes of government and consequent changes in the university coupled with
food shortages has complicated matters. If it had not been for
the devotion of one person, namely Norman Woolhouse, who
taught the entire course of biochemistry to medical students,
we would have been very depressed.

After I returned to London in 1976, I was approached by the
British Council, who wished to know how we could help with
the education of the medical students from Ghana. At the
Middlesex Hospital Medical School, as is usual in the UK, we
had a scheme whereby, on a voluntary basis, the students
could at the conclusion of their preclinical course spend a year

studying for a B.Sc. in a science subject. We agreed to take a few of the best students from Ghana to attend our courses in London and let them sit the same exams as our students. We then advised the medical school in Ghana concerning our assessment and they were awarded a degree of the University of Ghana. In this way no fees were paid in London. Over 6 years some 26 students came to us. The scheme was a great success both academically and on a personal basis. I am still in touch with several of the students and have acted as a referee for them. Unfortunately I believe few if any of them now are working in Ghana.

Another British Council Link was with the University of Khartoum. The staff there are medically qualified and were usually excellent academics so it was a pleasure to help them. Unfortunately many of the best academics are now scattered around the Middle East and are doing stirling work for many Arab medical schools. Going to Khartoum was a tough assignment, for it was usually very hot and the town was never very clean. I recall the pleasure of lugging a large spectrophotometer through customs at the airport in the middle of the night. The fine personality of the Sudanese made it all worthwhile.

As a result of my visit to China for IUB we established a British Council Link between our medical school and hospital in London and the Shanghai Medical University as it is now called. Professor Gu Tian-jue the Chairman of the Department of Biochemistry was a prime mover in this link. I well remember arriving in Shanghai for the first time in 1980 and being met by Professor Tsao from the Institute of Biochemistry. He had worked with Dr Kenneth Bailey in Cambridge. I explained to him that I wanted to visit a medical school. He said he would try to fix that but it would mean my missing a trip to Suzhou. When I reached the medical school they showed me that they were using the textbook that I and Bernard Kilby had edited with our colleagues in Leeds, *Basic Biochemistry for Medical Students*. Under the Link we received several visitors in London and several of our staff went to Shanghai. Even better was our success in sending several pairs of medical students on their electives. Unfortunately, enthusiasm for the Link subsequently wained with the change in the political scene in China.

Perhaps the most fruitful link has been with the Department of Biochemistry at the University of Nairobi. We have been able to provide them with a life-line for biochemicals and spare parts since the Link started in 1984.

Activity as an External Examiner

Associated with the Links has been my work as an External Examiner. In the UK all final exams for degrees are monitored by an expert who is not a member of the staff of the university awarding the degree. This tradition has continued for many years in the universities of the British Commonwealth and elsewhere. Thus I have been to Malaysia, Kuwait, Singapore, Sudan and Kenya and have had a close concern with the Chinese University of Hong Kong. These visits have provided an excellent and pleasurable opportunity to meet the youth of these countries even though they would not regard the circumstances as being ideal. I am a great believer in the system although sometimes there were difficult decisions concerning standards. The students in general are excellent. The defects arise as a result of poor teaching and facilities [44]. Having spent time among many Chinese universities in mainland China, Taiwan, Hong Kong and Singapore I have come to the view that they are the worst offenders when it comes to over-teaching. They seem not to understand that our role as teachers is to enable the students to develop to the best of their ability. (Our medical students sometimes make the same comments about us.) Over-teaching is not, however, confined to the Chinese. It seems difficult for teachers to encourage such activities as journal clubs or to understand that students can teach one another. Too often journals are available but they are not read by the staff.

Some concluding thoughts

I am not one who has carefully planned his life; it has been more a case of taking one's opportunities in the hope that something interesting and worthwhile will emerge. Medical students on entry to university judge the teaching in terms of its relevance for their future career. I tell them that even a train journey can be very relevant if you keep your eyes open. From a young age I developed the habit of asking questions, those questions that everyone else wanted to ask but had not got the nerve to do so. As a result I have sometimes been accused, quite correctly, of tactlessness. It is because of this habit of questioning others that I chose the title for this contribution to the series on the history of biochemistry.

It is obvious that an essential characteristic for a research scientist is the ability to ask questions* and to be excited when the answer turns up something novel. In the first section on research I have tried to describe the good fortune I have had in this respect. As to activities overseas I have seen my role as being one to stir the pot; to ask people to think why they are undertaking their jobs and to introduce them to one another. Too often I have found scientists working in close proximity but knowing little about the work going on nearby.

But I would like to end by emphasizing the importance of having at least a reasonable reputation as a research worker,

---

* That a useful response is not always forthcoming by a question was brought home to me by a lady from Armenia who came to Leeds around 1970 under the British Council. During a journey by car to a meeting in Birmingham she suddenly pronounced 'Do you know, Professor, that Stalin's mother was an Armenian?'. I replied with another question 'Bela, are you happy or saddened by this?'. After a deathly pause she replied 'Professor, you should not ask me awkward questions.'

for I believe that any success I may have had as an organiser depended in large measure on the respect due to me as an expert in at least one field of science. Indeed, the joy of being a scientist derives from being a member of an international club that knows no boundaries, so that the members may discuss matters in such a way that they rapidly get to the heart of a problem. I have been very fortunate in having been a member of the club and to have benefitted from it.

REFERENCES

1 P.N. Campbell and F.G. Young, Biochem. J., 52 (1952) 439–444.
2 P.N. Campbell, Biochem. J., 52 (1952) 444–447.
3 P.N. Campbell and H. Davson, Biochem. J., 43 (1948) 426–429.
4 T.Z. Csaky, Ztschr. Physiol. Chem., 277 (1942) 47–50.
5 H.R. Bentley, E.E. McDermott, J. Pace, J.K. Whitehead and T. Moran, Nature Lond., 165 (1950) 150–151.
6 P.N. Campbell, T.S. Work and E. Mellanby, Nature Lond., 165 (1950), 345–346.
7 P.N. Campbell, T.S. Work and E. Mellanby, Biochem. J., 48 (1951) 106–113.
8 P.N. Campbell and T.S. Work, Biochem. J., 50 (1952) 449–454.
9 P.N. Campbell and T.S. Work, Biochem. J., 52 (1952) 217–228.
10 P.N. Campbell and T.S. Work, Nature Lond., 171 (1953) 997–1001.
11 B.A. Askonas, P.N. Campbell and T.S. Work, Biochem. J., 58 (1954) 326–331.
12 B.A. Askonas, P.N. Campbell and T.S. Work, Biochem. J., 61 (1955) 105–115.
13 B.A. Askonas, P.N. Campbell, J.H. Humphrey and T.S. Work, Biochem. J., 56 (1954) 597–601.
14 P.N. Campbell, Biochem. J., 61 (1955) 496–503.
15 P.N. Campbell, Adv. Cancer Res., 5 (1958) 97–155.
16 P.N. Campbell and N.E. Stone, Biochem. J., 66 (1957) 19–31.
17 P.N. Campbell and B.A. Kernot, Biochem. J., 82 (1962) 262–266.
18 J.R. Sargent and P.N. Campbell, Biochem. J., 96 (1965) 134–146.
19 H.M. Dintzis, Proc. Nat. Acad. Sci., Wash., 47 (1961) 247–261.
20 W.W. Richardson, Brit. J. Cancer, 10 (1956) 415–423.
21 N.R. Rose and E. Witebsky, J. Immunol., 76 (1956) 417–427
22 I.M. Roitt, D. Doniach, P.N. Campbell and R. Vaughan Hudson, Lancet, ii (1956) 820–821.

23 I.M. Roitt, P.N. Campbell and D. Doniach, Biochem. J., 69 (1958) 248–256.
24 J. Vallance-Owen, E. Dennes and P.N. Campbell, Lancet, ii (1958) 336–338.
25 J. Vallance-Owen, E. Dennes and P.N. Campbell, Lancet, ii (1958) 696.
26 J.S. Hothersall, R.P. Muirhead and S. Wimalawansa, Biochem. Biophys. Res. Commun., 169 (1990) 451–454.
27 N.F. Gonzalez-Cadavid and P.N. Campbell, Biochem. J., 105 (1967) 427–442.
28 N.F. Gonzalez-Cadavid and P.N. Campbell, Biochem. J., 105 (1967) 443–450.
29 N.F. Gonzalez-Cadavid, M. Bravo and P.N. Campbell, Biochem. J., 107 (1968) 523–529.
30 J. Stahl, G.R. Lawford, B. Williams and P.N. Campbell, Biochem. J., 109 (1968) 155–157.
31 A. von der Decken and P.N. Campbell, Biochem. J., 84 (1962) 449–445.
32 K. Brew and P.N. Campbell, Biochem. J., 102 (1967) 265–274.
33 K. Brew and P.N. Campbell, Biochem. J., 102 (1967) 258–264.
34 H.A. McKenzie and F.H. White, Jr., Adv. Protein Chem., 41 (1990) 173–315.
35 U. Brodbeck, W.L. Denton, N. Tanahashi and K.E. Ebner, J. Biol. Chem., 242 (1966) 1391–1397.
36 P.N. Campbell, D. McIlreavy and D. Tarin, Biochem. J., 134 (1973) 345–347.
37 R.K. Craig, P.A. Brown, O.S. Harrison, D. McIlreavy and P.N. Campbell, Biochem. J., 160 (1976) 57–74.
38 I.C. Bathurst, R.K. Craig and P.N. Campbell, Biochem. J., 181 (1979) 501–504.
39 L.J. Burditt, D. Parker, R.K. Craig, T. Getova and P.N. Campbell, Biochem. J., 194 (1981) 999–1006.
40 L. Hall, R.K. Craig, M.S. Davies, D.N.L. Ralphs and P.N. Campbell, Nature Lond., 290 (1981) 602–604.
41 L. Hall, R.K. Craig, M.R. Edbrooke and P.N. Campbell, Nucleic Acids Res., 10 (1982) 3503–3515.
42 H. Lubon and L. Hennighausen, Biochem. J., 256 (1988) 391–396.
43 P.N. Campbell, Chem. Ind. (1964) 961–965.
44 P.N. Campbell, Biochem. Education, 19 (1991) 13–14.

E.C. Slater, R. Jaenicke and G. Semenza (Eds.)
Selected Topics in the History of Biochemistry: Personal Recollections, IV
(Comprehensive Biochemistry Vol. 38) © 1995 Elsevier Science B.V.

Chapter 10

The Paths of my Life

A.A. BAYEV

Engelhardt Institute of Molecular Biology, Moscow (Russia)

Charles Darwin calles his autobiography 'Reflections on the development of my mind and character'. My biography deals with the same subjects. It is, however, rather difficult to follow the quiet style of the great biologist who, in his words,

'...tried to write as if I were no longer alive and were looking back at my life from another world'.

For someone of my generation, who lived during a period of cruel wars, bloody revolutions, social reform and unrest, it is difficult to step back from events in which one was involved, and at the same time extinguish the sparks of emotion. Taking up a pen forces you once again to relive the past.

My early life showed no signs of future misfortunes. I was born on January 10th, 1904 (December 28th 1903, according to the old Russian calendar) in the far-away Siberian town of Chita. The childhood and education that I consciously remember, however, are connected with Kazan, a town on the great Russian river, the Volga. It was here that I completed my secondary and higher education and began my professional life.

My father died early and we lived with my grandfather's family (on my mother's side). He owned a small enterprise, building and repairing ships on the Volga. My grandfather was born a serf and was poorly educated, but I remember him

as a naturally talented person. Most of his numerous children received a higher education. My grandfather's family used to live according to the Russian patriarchal mode of life, respecting authorities. They were moderately religious and strictly observed Russian folk traditions.

When the February and October revolutions took place, I was 14 years old and my moral and intellectual principles were almost completely formed, although they had not quite fully matured. It was only under the conditions of social chaos that this maturity was to reach its completion.

The years of revolution, civil war, and poverty, brought to each generation its own advantages and disadvantages. These events caught me at the threshold of my spiritual evolution. Older people were involved in the civil war, and sometimes brothers found themselves on opposite sides of the barricades. Even some of my classmates, in spite of their youth, took part in the internecine struggle and were killed before they reached maturity. For some, post-revolution Russia became a totally alien country. My grandfather's family lost its material wealth and in a short time all but disappeared. This was the first turning point in my life.

The normal course of my secondary education was interrupted by the closure of the 2nd Kazan gymnasium, where I had been studying. I devoted a year to educating myself with text-books, then spent another year studying on evening courses. I finally received my certificate from Soviet school No. 2 (formerly my gymnasium).

From the beginning of autumn 1918, I was the bread-winner for my family. I started by selling cigarettes, but was then forced by circumstances into a rather unsuitable organisation, the Kazan detective service (criminal police), where I worked on statistics of numerous crimes. From that day onwards, I have been constantly employed. It was only during my studies in the third, fourth, and fifth years of the Medical Department, that I had to make do with casual earnings.

My time at school, and my first university years, were very

difficult for me. During the day, I had to work at the office, and in the evening I would sit at my school desk, and then later I went to lectures at the University. Once I got home I would sleep for several hours, get up during the night, work at a book, then, towards morning, fall asleep again for a short time. This went on for several years.

You could say that I picked up my education en route, when I had time off from work. As a result, I spent the rest of my life feeling (perhaps, wrongly) that I was a half-educated person. But in any case, I completed my secondary education.

When you are young and all your perceptions are fresh, your mind is open to new ideas, and extremes of emotion are natural. For my generation, this was a time of revolutions, wars, ideological struggle, and poverty. It was a cruel time which left our youthful spirits devastated!

D.V. Grigorovich in his 'Literary Reminiscences' wrote about the same period of his youth:

> '...one could see in everything a tide of fresh strengths, the lively courage of youth, the appearance of new ideas; suddenly amid this passionate enthusiasm, could be heard the words: 'down with oppression and injustice'.

We were deprived of any such possibilities. We did have similar experiences, which resembled some sort of psychological summer lightning, but the struggle for survival took first place and any outbursts of abstract thoughts were restricted to bland and cautious political discussions. Some of my classmates at the gymnasium did emerge in opposite political camps and took up arms.

Getting into university was complicated for me. In 1921 I was refused admission to the Medical Department of Kazan University because of my so-called non-proletarian origin – my father was a lawyer rather than a worker or a peasant. Neither my father's nor my mother's family belonged to the aristocracy. As I have said, my grandfather was actually born a serf, but by that time, an official version had already taken root. Only a year later, in 1922, did I manage to obtain permission to be transferred to the Medical Department. In 1923, however, I was even expelled from there (and later re-admitted) for the same reason – having an inappropriate background. In general, following the 1917 Revolution, the authorities would act above all according to the current ideology. They would avoid any kind of deliberation, preferring most of all to act swiftly and decisively.

During the 1920's at Kazan University, there was a power struggle between the so-called proletarian students and the faculty. The proletarian student body was actually only a small group of communists (almost all of whom, incidentally, were imprisoned in 1937 – and the majority were annihilated). This struggle and the reasons behind it was not of much interest to me. I think that the faculty was only fighting to preserve the moderate freedom of the universities, which had been established as far back as the Charter of the Tsar.

Kazan University is one of the oldest higher education establishments in Russia. When founded on 5th November 1804, it had four departments and, among other things, was famous for its rector, N.I. Lobachevskii, a brilliant mathematician who discovered non-Euclidean geometry.

In spite of all the difficulties, the teaching standards at Kazan University were kept at a high level. Among our professors there were many outstanding figures: A.N. Mislavskii, the physiologist, B.I. Lavrentiev, the histologist, S.S. Zimnitskii, M.N. Cheboksarov and N.K. Goriaev, the clinical professors etc. We were well-instructed and were expected to acquire a sound knowledge. As a result, I would often recall my university teachers with gratitude. After graduating from the department, I spent three years working as a general practitioner in the countryside, about 100 km from Kazan. Then in 1930, I returned to Kazan and was admitted as a post-graduate student in the biochemistry department of the Kazan Medical Institute.

One day, therefore, during the Indian summer of 1930, I

made my way up the worn staircase of the third block at Kazan University (by that time the administration of the Medical Institute and the University had already been separated) and with some trepidation, I entered the domain of the Biochemistry Department, where I was to spend the next four years of my life. At that time, V.A. Engelhardt, the professor of the department, was still young (36 years old), but already had noticeably grey hair. He was a tall, lean man with pleasant, courteous manners and he met me without any particular enthusiasm. As we got to know each other, however, our friendship developed. Indeed, V.A. Engelhardt was to become my benefactor during the more difficult times of my life.

As professor of the department, which had been founded back in 1862, V.A. Engelhardt was giving his lectures at the very turning point of the formation of dynamic biochemistry, and he completely changed the content of the former lectures and practical courses that had previously existed.

It was in Kazan that V.A. Engelhardt made his historic discovery: the resynthesis of ATP during respiration in avian nuclear erythrocytes. At that time, the idea of the energetic role of ATP, and its energy-rich (macroergic, according to V.A. Engelhardt) pyrophosphate bond had not yet been conceived. V.A. Engelhardt perceived the breakdown and resynthesis of the ATP phosphate group as the manifestation of the cyclic recurrence of chemical processes in a living cell.

The research into the supposed cyclic behaviour of the ATP amino group fell to me. I established, however, that the NH group of ATP, unlike the pyrophosphate group, was split off under anaerobic conditions, and was not resynthesized when oxygen was supplied, *i.e.* this process was irreversible. A more valuable part of this work was a direct enzymatic analysis of the transformation of ATP, using the Schmidt's adenylate deaminase. A more detailed analysis of the breakdown of ATP and the resynthesized products was carried out much later in Moscow, at the Institute of Biochemistry of the USSR Academy of Sciences, where I once again found myself in V.A.

Engelhardt's laboratory (he left Kazan in 1933, and moved first to Leningrad and then to Moscow).

The experimental material thus obtained, enabled me to write a dissertation and compete for my Ph.D. in biological science. The completed dissertation was already on V.A. Engelhardt's desk in 1937, but I was not destined to present it at that time: I was arrested, and began a different life that was to last for 17 years. During that period of my life, I had only two goals: to survive and, with luck, to return to science. The year 1937 in our country saw the culmination of Stalin's reprisals on his enemies, a few of whom were real but most were fictitious. Political trials and mass repressions went hand in hand with this despot until his death in 1953.

I was charged with having become a member, during the 1930s in Kazan, of an underground organisation, 'the Young Supporters of Bukharin', who planned to kill Stalin and restore capitalism to the country. According to the investigators, the organisation consisted of biologists and medics from Kazan University and other higher education establishments in Kazan. The KGB named V.N. Slepkov, a former genetics professor at Kazan University, as head of this organisation, which numbered about 150 members.

In fact, there was no real underground organisation, and the whole business was fabricated by the KGB during their preparations for a court case against N.I. Bukharin. The latter took place in 1938, and resulted in Bukharin's execution. V.N. Slepkov had attracted the KGB's attention because his elder brother, A.N. Slepkov, was close to N.I. Bukharin and was one of the Party activists, who were considered to be in opposition. But the imaginary members of this terrorist organisation were simply the audience of V.N. Slepkov's lectures on the general problems of biology, which he delivered at Kazan University and at other higher education institutions. The KGB aimed to prove that during his politically active period, N.I. Bukharin had established active underground terrorist organisations. It seemed that I was only one of the numerous pawns in this (for me, very strange) political chess game, or theatrical production – call it what you will.

The inquiry into my case seemed relatively simple by current standards. There were several reasons for this. Firstly, between my arrest and my departure for Kazan, I spent about two months in Butyrskaya prison. In 1937, this institution gathered together quite a lot of people of different nationalities, party membership, and background. For people such as myself, Butyrskaya prison was a real university. It was there that I received a complete picture of what was happening in the country, the aims of the KGB, and how the investigation was being carried out. Like all prisoners, I assumed that my detention was only an unfortunate mistake, and that I would be freed in the very near future. But more experienced people explained that this institution never relinquished its victims, and I should not, therefore, entertain any vain hopes. I remember, as I was leaving Butyrskaya prison for Kazan, one of my fellow prisoners said in parting, 'Don't sign any false confessions: a person who doesn't feel sorry for himself, would be even less likely to feel sorry for others' (he was talking about false evidence against other people).

However, my training at this institution was not the only important factor; there was also another, very significant one involved. Each inquest which took place within KGB walls was, above all, a theatrical production, a scenario for which even the choice of cast was made in advance. I was late arriving in Kazan, and came in at the end. Perhaps they had intended to keep an important role for me, but I arrived there too late and all the main participants had already been chosen. Only the role of an 'extra', so to speak, was left for me, and because of this, I was not particularly interesting to the investigator.

I made a firm decision to resist the investigator's tricks, and indeed, I did not sign any false evidence, nor slander anybody either. However, I am not entirely sure that I can attribute this to any special kind of determination on my part. I know for sure that after July of 1937, the use of torture was officially sanctioned. I myself never experienced this, but was able to endure insults, threats, and being deprived of sleep during long interrogations (or so-called 'conveyor belts'). My examination ended with ten days' imprisonment in the punishment cell (or 'stone sack', as it is known), where, according to legend, Emelyan Pougachev, a leader of the rebellious peasants, was imprisoned during the reign of Catherine II (18th Century). I don't know what would have happened, if I had been tortured. Perhaps I would not have been able to bear it.

I was sentenced to ten years' imprisonment by the Military College of the Supreme Court of the USSR in the infamous Lefortovskaya prison. In order to reach this verdict, three soldiers spent all of several minutes on me. When the sentence was passed, I froze in terror for the first time, because only then did I realise that these three could have sentenced me to death with just the same ease. According to some sources, in September 1937 the Military College sentenced to death by execution seven out of every eight prisoners.

Following the Lefortovo trial, I went to Orlovskaya prison, which had been recently painted and was cold. However, some time later, we, the inmates of one cell, were put into railway cars and taken off to an unknown destination.

It was not a long journey, and late one cold autumn evening we were transferred onto a small ship, whose hold had been divided into sections by hanging up blankets (in order to separate the individual groups of prisoners from each other). One of my fellow prisoners happened to see the name of the ship – 'SLON'. Suddenly everything became clear. This was the abbreviation of 'Solovetskii Special Purpose Camp'. We arrived at the White Sea, and were evidently on our way to the Solovetskie Islands. And so it turned out.

I remember how, during our short sea journey, I was unable to resist the temptation, and turning back the corner of a blanket, I looked at our neighbours through the peep-hole I had created. To my surprise, in among the unremarkable figures,

I spotted the Englishman, whom I had met in Butyrskaya prison. This young man had been a radio operator at the Soviet Embassy in Madrid. Under some pretence or other, he had been lured onto a Soviet ship, arrested, and secretly sent to Moscow. And now here he was at Solovki. I do not know what happened to him; I expect he was killed.

Our landing on the main Solovetskii island was very impressive. It was night, the air was still, the moon bright and full, and the coarse pebbles crunched under our feet. We walked along, escorted by a large convoy and angry dogs. In front of us, out of the darkness, emerged the ancient walls of the Solovetskii Kremlin, built of huge, glacial boulders, and stocky, peaked towers. The iron gate opened with a moan...

A memorable incident took place once we were dressed in our prison clothes. One of my companions, quite a young man, had a new felt hat, which we evidently valued very highly. It goes without saying that it was confiscated, but for some reason it was not mentioned on the receipt, which listed all the things taken away. And there, crouching like a bird of prey, the head of Solovetskii prison sat in silence. The owner of the felt hat appealed to him, and the following dialogue took place:

'My felt hat is not mentioned on the receipt'.
'What do you need it for?'
'What do you mean, one day I shall be free – and I shall need something to put on my head'.
'Nobody ever leaves this place'.

These were the words which greeted us at the Solovetskaya cloister, which was to become my 'home' until 1939.

So the question arose: how was I to live? I don't mean in the biological sense of the word, because the physical living conditions in Solovetskaya prison were reasonable. I am talking about spiritual life. Prisoners have neither a past, nor a future. They don't have any future because the existence of such depends entirely on the powers that be and their whims. As I said above, the KGB is reluctant to part with people once it has got them in its clutches. As far as political prisoners were concerned, the USSR penitentiary system was obviously intent on their destruction. Indeed, it was the totally random policies of most USSR prisons and camps that were the largest contributing factor in this. For this reason, it was pointless holding out any hope for the future, or seeking strength from such hopes. Neither the past itself, nor our reminiscing about it, can help our existence; on the contrary, they make it worse.

In Solovetskaya prison, my recollections of freedom were still fresh and bright, but they only caused me intense pain, and I had to curtail them, and force them out of my memory. They used to appear suddenly, as if escaping from a cage. I remember that they often occurred during my meals; I would be lifting a spoon of prison soup to my lips, and my psychological control would weaken for a moment, and just then, a image of the past, alive and bright, would flash into my mind like a streak of lightning. My heart would stop beating, a spasm of hysteria constrict my throat, and a lump of food would stop short, trapped by cramp. It would take an enormous effort to overcome this condition.

And the present is depressing. First of all, it is senseless. I am neither a politician nor a conspirator by nature, but I was artificially turned into both. In searching for some logical justification for my situation, I concluded that my imprisonment was retribution for my indifference towards the lie that surrounded me. On his death, my father left behind a large legal library, and so in my childhood, I became familiar with the speeches of our famous national lawyers, even before I read Dostoevsky's 'Crime and Punishment'. It was clear to me that the trials, organised according to Stalin's instructions, were coarse dramatisations. And, in spite of understanding this, I remained silent; I felt no internal reaction. So the Solovetskaya prison was, therefore, a punishment not for my supposed involvement in an underground political organisation, but rather it was retribution for my sleeping conscience. Thus, my stay there was to a certain extent justifiable.

Prison life, with its strict routine, numerous petty restrictions, and monotony was a killer. Contact with other inmates was limited, as there were people of different characters, intellectual levels, and unknown moral standards all together in the same cell. The prison day itself lacked any substance, so it became necessary to fill it somehow. In the end, my reflections did prove to be useful. First of all, life in the monastery cells, which had been transformed into prison cells, reminded one of solitary monks: they were not merely vegetating, but their lives must have had some higher, spiritual meaning. After that I remembered the experiences of people who were imprisoned for long periods of time. Many of them noticed an increase in intellectual activity. As one example I could cite Nikolai Morozov, a member of the 'Narodnaya volya' party, who spent twenty years in the casemates of the Shlisselburg fortress. He was protected from spiritual degradation by his intense intellectual work.

Briefly then, my reflections led me to the conclusion that the empty prison day should be filled with intellectual activity. I should make an effort to create my own internal world, a life without any past or future, which would act as a counterbalance to the unbearable prison routine, and the external world, which no longer existed for me. This plan actually turned out to be feasible: I managed to create such a world.

Solovetskaya prison had been preceded by Solovetskii camp, where numerous intellectuals, such as the mensheviks, social-revolutionaries, religious activists, etc., had been imprisoned. They created an excellent library, which the Solovetskaya prison had inherited. This library contained fiction, historical literature, textbooks, and books on natural sciences, including many books in foreign languages. The prisoners had the right to choose two books a week from the catalogue and take them out. While in prison, I studied a complete course of higher mathematics (it had not been included in the syllabus of the Natural Science Department of the Physical and Mathematical Faculty, and much less the Medical Faculty). There was a mathematics textbook by Poussen, which contained many exercises, and I worked out the answers to all of them, sometimes spending a whole day on a difficult one. I also occupied myself there in reading books in French, German, and English. I had no dictionaries, so I had to work out the meanings of unknown words from the contexts. I was kept busy for a whole day, and even that was not enough.

My time was filled with intellectual pursuits; I created my own internal world, destroyed the intellectual vacuum, and put up barriers against memories of the past and hopes for the future. My emotional requirements were satisfied to some extent by a somewhat random selection of fiction.

My cell-mate was A.Y. Veber, the People's Commissary for Education of the German Republic on the Volga. He was a nice young man, and I used to practise conversational German on him.

Sometimes I was even happy. The moral factor plays an enormous role in a prisoner's life, and often the road to death begins with the destruction of one's personality.

After two years, however, the lifestyle I had settled into, came to an end. Suddenly, one day in July, 1939, the bolts of the cell door began to rattle. Everybody pricked up their ears – what was this? General execution? But something quite different happened. The inmates were allowed to go out into the huge courtyard of the monastery, and for the first time we all saw the Solovetskaya prison inmates, who until then had been kept carefully hidden away by the prison administration.

As a consequence of this, after a short time, we were taken by the North Arctic Sea route to the Norilsk camp. This remote place, beyond the Artic Circle, was a mining centre, primarily for nickel, but also for platinum, non-ferrous metals and coal. It was here that I had to spend the next eight years of my life, up until 1947.

Following my arrival in this inhospitable region, I spent some time chipping away at the permafrost, but later my medical education was put to use and I spent my first arctic winter living in an improvised dispensary, which was housed in a tarpaulin tent, through snowstorms and temperatures of −50°C. After this I was made a doctor in a hospital for the free citizens of Norilsk. I was in charge of the therapeutic, children's and infectious diseases departments.

I lived in Norilsk until 1947, by which time I got used to its arctic days and nights, its severe cold and heavy snowfalls, when the strong winds would carry the smallest dry snowflakes and set them into a granite-like mass.

It was during this time that I became grateful to my university teachers for instilling in me a sound medical knowledge. This helped me to survive unscathed the severe climatic conditions of my camp imprisonment.

In 1944, two and a half years before the end of my ten year sentence, I was released as a result of the work I had done during the War. However, I still had no right to leave Norilsk and so continued to work there as a doctor.

In the same year of 1944, I married Ekaterina Vladimirovna Kosyakina (nee Yankovskaya). It had been the war that had brought her to Norilsk. In June 1941, she and her husband, together with their young son, were on their way home along the Yenisei, after three years' work with a geological party at Tixi Bay on the North Arctic Sea. War was declared. Right there on the steam boat, her husband was mobilised and within a short time he had been killed at the front. She arrived in Norilsk and began to work at the Norilsk metallurgical plant.

We were brought together by tragic circumstances. In the spring of 1943, in a moment of carelessness on the part of his nanny, her son fell into a fast-flowing stream. He was rescued, but later died of pneumonia in my department of the hospital.

Two of our children were born in Norilsk. Since 1944 we have never been apart from each other and together we have shared all our sorrows and all our joys. For me she always was and always will be a moral support and a dear friend.

In 1944, fate brought me to the threshold of a new beginning in life – I started to try to get back into science. I believe that nature itself instilled the instincts of a researcher in me.

Wherever I was and whatever I did, there were always subjects which I would start to research, acting on my own initiative. This I would do for the love of research itself and not for any other reasons. During my medical practice in the countryside, I wrote some articles for publication on the health of young conscripts and on tuberculosis among the Tartar and Chuvash people (Pirquet reactions for tuberculosis were carried out on 4160 people). As a reminder of my time in Norilsk, I kept the following manuscripts: 'The Radiation Resources of Norilsk', 'The Principles of Personnel Selection for Work in the North', 'Reference Book on the Nutrition of Children of Breast-feeding Age', and a few others. From this it can be seen that it was only natural for me to try to return to biochemistry.

I decided to defend my aforementioned thesis on biological sciences and with this in mind I applied to V.A. Engelhardt. As it turned out, he still had my dissertation. I was absolutely amazed by this: eight years had passed, he had been through a war, evacuation and similar events and yet still my manuscript had survived. Only the review of my thesis was out of date, mainly because of the new understanding of the role of ATF in cellular metabolism, which had been discovered by F. Lipman and colleagues.

Thanks to the efforts of V.A. Engelhardt, I was allowed to come to Moscow for a month. It was here that I worked on my thesis, sitting at his writing table, in his flat behind the Bolshoi theatre (while V.A. and his family were away at a resort). It was approved by V.A. Engelhardt and one year later, on June 6th 1947, I defended it at the Institute of Physiology of the USSR Academy of Sciences, in Leningrad. This was made possible by the kind permission of Academician L.A. Orbeli.

Even before this event in 1945 and 1946, V.A. Engelhardt and L.A. Orbeli tried to help me return to Moscow. First they applied to the administration of the Norilsk metallurgical plant and then to the People's Commissariate of Internal Affairs to the dreaded L.P. Beria. Both applications were categorically refused.

The times of repressions, morbid suspicions and unrestrained cruelty were still continuing throughout the country and it was therefore dangerous to defend me. Nevertheless, V.A. Engelhardt and L.A. Orbeli did just that. I would probably not be wrong to interpret their actions as a silent protest against the country's political despotism and the breakdown (if not total destruction) of social morals. Their behaviour was a good reflection of their individual moral standards.

In 1947 my family and I managed to leave Norilsk together. I came to Syktyvkar in the Komi Republic and was made head of the biochemical laboratory, which was affiliated to the USSR Academy of Sciences. But I had only been there for one and a half years, before I was rearrested (for no other reason than that of having been imprisoned in 1937) and sent into permanent exile in a small village on the bank of the Yenisei river in Siberia.

Following my arrest, my wife tried hard to change the path that fate had chosen for me, but without success. In the autumn of 1949, she, together with our small children, made the long train journey from the Komi Republic to Krasnoyarsk, then on by steamboat along the Yenesei all the way to my place of exile. It was a late autumn evening when I met her as she came ashore. In total darkness, we made our way across the turbulent and dangerous river and eventually found ourselves in the place which was to be our home for the next five years.

While in exile, I was the head of a small, but comfortable hospital, where my wife worked as a technical assistant in the hospital laboratory we had organised. The winter of 1949 on the Yenisei was extremely severe. For two whole months, temperatures did not rise above −50°C. The Taiga looks incredibly beautiful at such a time: a short day, a low sun and a pink sky. The trees are covered with white hoarfrost and partially frozen white partridges sit on their branches; everything stands still, numbed by the cold. In general, however, the climate was not quite so severe – we even managed to have a small garden where we grew corn and sunflowers.

This whole region was a place of exile, with Kulaks and their descendents as well as exiled Lithuanians. My hospital treated them all.

On the eve of 1953 we received the news that all the exiles from the region were to be sent north to the Tundra, near the river Podkamennaya Tunguska, and that our present region was to be occupied by new exiles – Jews. I was extremely anxious when I shared this news with my wife; she said that she would always share my fate, regardless of what it might be. We ended our conversation embracing in tears.

But people are not immortal. In 1953 Stalin died and within a year I was rehabilitated 'for lack of corpus delicti' and became a free man.

In accordance with the law at that time, I was accepted back into the Academy of Sciences as if I had never left and no time had passed. It was officially recorded that I had spent 17 years working peacefully and continually at the Academy of Sciences and not in prison or in exile. I even received compensation for all these years, equal to two months' salary in 1937.

Of course, my life in prison and the time I spent in hard labour camps and in exile, left its mark both on my character and on my behaviour. The amount of forced effort required to live with my own internal world brought with it a degree of autism. But I did not become a misanthrope, I felt neither revenge nor hatred towards my surroundings. I was aware that the majority of people in our country had all suffered deprivations, injustice, misfortunes and grief. Nobody had been able to choose his lot in this huge collection of misfortunes and I had simply received my share, that was all. For much of my time, I gave medical assistance to prisoners, exiles and those who had been injured by fate, who were living a deprived existence in remote places. This gave me a sense of moral satisfaction and helped me to live my own life. Being engaged in science, especially in fundamental science, for the most part makes it impossible to do any good directly and be really useful to people. Instead, the motives for deeds become more abstract 'to serve society'. Having come back into science, I was inevitably condemned to a more egotistical type of activity than before.

I became extremely hard working, even to my own detriment. But, having got used to being satisfied with a small income, I did not try to exploit my love of hard work to my advantage or to further my career – it was all that I wanted, just to be actively occupied. On returning to Moscow, I intended first of all to continue my experimental work and not to spend time defending my doctor's thesis as the first step in my academic career.

To some extent I became a fatalist: both misfortune and success had come to me of their own accord, and not through any efforts or wishes of my own. I believe it was with these characteristics that I began, practically from scratch, the second stage of my life.

The year 1953 was critical for me – the real instigator of the country's suffering and of my suffering – Josef Stalin, had died. At the same time, J. Watson and F. Crick had discovered the double helix of DNA, and this was the beginning of molecular biology. It was this science that became the focus of my scientific activity during the second half of my life.

Coming back into science was not easy for me. I was fifty years old, and nature had not left me with much time for scientific creativity. In order to reach even a medium standard of knowledge in biochemistry, I read a great deal and was fortunate to be able to combine my biochemical education with earning a living. To do this, I borrowed books from the publishing house for 'New Books from Abroad' and wrote critical reviews, which were then printed in the bulletin of the same publishing house.

My scientific activity can be divided into five periods which overlap with each other: *1*. cyclic transformations of ATP during cell respiration (1930–1939, when I was a post-graduate student at the Kazan Medical Institute); 2. primary structure of transport RNAs and 'cut molecules' (1960–1969); 3. recombinant DNAs (from 1969); 4. biotechnology (from 1972); 5. the human genome (from 1987).

An object of research, however, is not sufficient in itself to exhaust the entire process of scientific activity. Scientific activity is made up of a combination of several components: reading scientific literature – to feed the mind; intense intellectual work – to stimulate creativity; and practical 'hands-on' activity by the researcher – to provide physical verification of his intellect. All these things contribute towards a harmonious blend of scientific activity.

These factors have their own optima which are relatively independent and do not necessarily coincide in time. They depend very much on circumstances. Unfortunately, in my case, they were never able to develop harmoniously. I began my experimental work when I was 26 years old – this was a good age to be starting, but then I was forced to stop for 17 years, and was only to begin again when I was 50. This was very late. It was natural that my work at the laboratory table was impaired. My mental activity, however, continued and my sense of initiative was just as keen as before; it was just that the hands of my students and colleagues did the work, instead of mine. As the experimental process itself gradually became less attractive, the conceptual aspect of my research began to take precedent. I lost interest in tactical problems and became more involved with theories of strategy, i.e. the research which forms a part of a whole scientific plan. And now, as I approach my 90th year, it is evident that further evolution will take place, if it is possible to use such a word to describe the process of decay and passing away.

The first thing I did when I came back to V.A. Engelhardt's laboratory, was to finish the experiments on ATP resynthesis in pigeon erythrocytes, which had been so abruptly interrupted in 1937. I identified the products of ATP breakdown by switching on the respiration and its resynthesis under aerobic conditions. To be honest, the current ideas available at the time would have enabled me to do the same without analysis, but out of stubbornness, I wanted to finish what I had started in May–June 1937. This was the end of my first period of scientific activity.

I carried out several investigations on the myosine ATPase which was a subject of interest to V.A. Engelhardt. But then I moved on towards problems of a different kind – the study of cellular nucleotides. I worked together with V.A. Engelhardt until the very end of his life. He never preached at his students. His influence on them came from the example he set and from an atmosphere of creativity, intellectuality, kindness and devotion, which naturally seemed to emanate from him. He was not interested in active politics and avoided coming into contact with it. His actions (as they effected my life) were indicative of his independent ideas which he never openly proclaimed, but which were always evident in his behaviour.

Today, as our country is making its way forward towards democracy, we should in theory be seeing more scientists with these positive individual qualities, but as yet, in my opinion, no figure has appeared on the scientific scene, who could even come close to emulating the nature and merits of V.A. Engelhardt.

Towards the end of the 50s, molecular biology had begun to take shape in our country, and for me it became a passion. There were many chemical elements in biochemistry that had not attracted me, especially in the 20s, mainly because biological principles had been so uppermost in my education and thoughts.

During the 30s, the cyclic transformation of substances in a living cell, propagated by V.A. Engelhardt, was an essential step towards the combination of chemical and biological ideas. Biopolymers, however, especially DNA as a carrier of genetic information, the primary and spatial structures of which have been understood since 1953, aroused real enthusiasm among Russian biologists. This had already taken place when I arrived at the Institute of Molecular Biology in 1959. At that time it was known as the Institute of Radiation and Physico-chemical Biology. This somewhat strange name was supposed to be a cover. When the institute was being set up in 1959, T.D. Lysenko continued to dictate in biology and was not prepared to recognise molecular biology at all. Lysenko's powerful pa-tron – Stalin – had already died, but to everyone's surprise, the ambitiously determined founder of so-called Michurin bi-ology, found himself another sponsor, N.S. Krushchev. I re-member that Lysenko once visited our institute, probably as an inspector. He insisted that we show him a preparation of DNA, something which was particularly distasteful to him. We presented it to him, but just exactly what his thoughts were as he looked at the tube of white powder, remains a mys-tery to this day – in all likelihood, he was expecting to see something more significant.

This took place at the beginning of the 60's, the time when the main idea of protein synthesis was beginning to take shape. The study of its code had begun, creating what was to be an outstanding era in the history of molecular biology. The period of cryptographic research, which was carried out mainly by theoreticians (D. Gamov, M. Ychas, F. Crick and A. Orgel) had come to an end, but it had left behind it conclusions that would later prove to be of use. In place of the theoreticians came the experimenters, M.W. Nierenberg, S. Ochoa and oth-ers. Like so many others, I was captivated by this magnificent cascade of discoveries and although, like the majority of soviet molecular biologists, I stayed on the sidelines of this powerful movement, I still followed the publications with great interest and gave many reports on the subject.

My attention was drawn to transport RNAs. At that time I was already moving towards structural research and my dream was to understand the functional properties of bioplym-ers through their structures. tRNA seemed to be a particularly

suitable object for this purpose. I admit that our comprehen-
sion of a connection between structure and function was
rather primitive in reality and it did not bring us the results
we expected. We put together a group of predominately young
scientists in my laboratory, which included T.V. Venkstern,
A.D. Mirzabekov, R.I. Tatarskaya, V.D. Axelrod, N.M. Abrosi-
mova-Amelyanchik, A.I. Krutilina and L. Lee.

Valine transport RNA of baker's yeasts was chosen for the
determination of primary structure.

The group's work was based on a strict division of labour. In
the beginning A.D. Mirzabekov was responsible for the isolation of tRNA Val, T.V. Venkstern for that of paper chromatography and spectrophotometric determination, R.I. Tatarskaya
for the isolation of the necessary enzymes and I myself worked
with column chromatography etc.

We either had to develop some analytical methods (for example the isolation and application of guanyl-RNase from
Actinomyces) or we had to improve on those already existing
(such as the spectrophotometry of nucleotides). The entire
combination of the methods we used for the tRNA analysis,
was completely new for our group (with perhaps the exception
of chromatography).

Finally, in 1967, this work resulted in the complete determination of Valine tRNA 1 primary structure. It proved to be
among the first five tRNAs with known primary structures. In
1969 this work was awarded the State Premium of the USSR.

Following this the primary structures of Valine tRNA 2 and
3 were also determined. But after that, the work on the tRNA
primary structure came to a halt. Our expectation that the
tRNA primary structure would enable us to read with ease
something about its function (in particular the enzymatic aminoacylation) was of course not entirely correct. Neither the
structures of the valine tRNA, nor of the other tRNAs themselves, brought any functional answers. Next came the creation of the method of 'cut molecules' in our laboratory i.e. the
study of the functional peculiarities of halves, quarters and

tRNA fragments in general. This idea had come to me in 1966, if not earlier.

Our experiments have shown that the self-assembly of halves and quarters of a valine tRNA 1 molecule takes place in solution, and to some extent this is accompanied by the restoration of its ability for aminoacylation i.e., its interaction with aminoacyl-tRNA synthetase. The self-assembly of molecules was achieved by I. Fodor in 1966 and one year later, the first publication on this phenomenon appeared.

As time passed, our experiments became broader. We studied the self-assembly of molecule quarters and its various fragments, from which 1 to 5 nucleotides were removed. We did this to determine the points of the valine tRNA interaction with the specific aminoacyl synthetase. At one time, we were particularly interested that the splitting off of two nucleotides of anticodon A_{36}-C_{37} in valine tRNA completely inactivated the aminoacylation. We also undertook other research, such as the interaction with ribosomes, methylation and also the secondary structure of aggregated molecules. The research into this cycle was published in a review (*Methods in Enzymology, v. XXIX, part E, pp. 643–661*, Academic Press, N.Y.-London, 1974), in which the results of the work carried out mainly by A.D. Mirzabekov were described. This work was stopped for the same reasons as was the analysis of primary structure of the tRNA, in spite of all the successes achieved in this field.

By the middle of the 70s our friendly group gradually dispersed and the study of tRNA came to a halt. Personally, I was busy organising a laboratory in the Institute of Biochemistry and Physiology of Microorganisms of the USSR Academy of Sciences, in Puschino. Here it was possible to organise research into molecular genetics of microorganisms. A.D. Mirzabekov had other interesting ideas. R.I. Tatarskaya and V.D. Axelrod decided to emigrate; A.I. Krutinlina and L. Lee transferred to work in Puschino at the Institute of Biochemistry and Physiology of Microorganisms, the USSR Academy of Sciences. I very much regretted all this, since investigation into the method of cut molecules was far from exhausted and the same was even more true for the primary structures of tRNA. But circumstances proved stronger than my personal feelings. Of course, if it had not been for the fact that I was busy with my new projects, I would not have allowed this research in the laboratory to stop. As it was, in the end, the research took a new direction.

At this point I must regress slightly and return to the chronology of events.

In 1964 I became a member of the CPSU. This decision of mine requires some explanation on my part, as events in my life up until this point hardly fell in line with this move. I knew of course, how much the country had suffered under the Stalin regime and as a result, how much damage had been done to the consciousness and morals of society. Even though at the same time, the country did become literate, better educated and as a result of great sacrifices, built up a powerful industry.

Following Krushchev's exposure of Stalin in 1964, it was hoped by many that the errors of the past would be corrected and the crimes would be punished. The history of socialism in Russia, extending from the last century, shows that it was originally characterised by qualities of humanity and idealism. It was only later that the warped ideology and inhumanity were cultivated by Lenin and in particular by Stalin.

N.A. Berdyaev, who could hardly be suspected of sympathising with communism, maintained that this happened as a result of an evolution of the earlier flow of Russian social ideas. He even thought that

'... at the moment this is the only authority which can at least to some extent protect Russia from the dangers threatening her. A sudden collapse of the Soviet regime would be even more dangerous for Russia and would threaten anarchy'

(N.A. Berdyaev, Sources and Interpretation of Russian Communism. YMCA Press, Paris, p. 120)

I had not taken into account either the conservative mentality of the party members and statesmen within the Communist Party or the rigidity of the contrived methods of state government. There was no evolution and my hopes were dashed.

It is possible, however, that this was not all that important. My generation was already in the first stages of developing its awareness, when it became familiar with the former and current socialist movements in Russia, as portrayed by bright, intellectual and morally-upright individuals. I, together with many of my contemporaries, felt that socialist values were an integral part of nature and were absorbed together with the impressions of childhood and youth, which were current in the cultural environment of the time. It would be impossible to attribute these same values at a later date to Stalin, his accomplices and the dark events of our history.

I have to confess that some traces of these values still remain in me today, in spite of my age and all my experiences. It is obviously too late for me to part with them. And indeed, is it necessary to do so?

It was not only these considerations that influenced my behaviour. The Party and power apparatus of the USSR were characterised by extreme bureaucracy and formality. This allowed the Party and state officials the freedom to avoid making decisions and taking action on their own initiative. It eliminated the individual factor and made it easier to control performance.

I had to endure so much hardship, due to the simple fact that my father was not a worker or a peasant, but a lawyer (during the 20s). In 1954, the charge of terrorism against me was withdrawn, but this was only court rehabilitation. One could never be quite sure that party or social circles would not one day decide that court rehabilitation did not completely prove the innocence of people with a past like mine. It would be quite simple to think of several examples that could justify this. I had neither ideological, nor career motives when I received the red book which confirmed my loyalty. I only wanted the smallest bit of freedom to be able to live as I chose. In our country, the despotic regime had deprived the people of their right to live. A person who wanted to live and act would always have to pay the price. Now is not the time to look into what this price is, or to discuss whether or not it should be paid. In 1964, the price I paid was joining the Party. My behaviour can be judged in different ways, but I certainly paid my price. In saying this, it would be easy to accuse me of self-justification, but there is nothing I can do about this.

Scientific events were continuing to develop, however, and my period of genetic engineering began.

The works of P. Berg, S.N. Cohen and H.W. Boyer (1972–1973) heralded the beginning of the era of recombinant DNAs. Even before this, however, my attention was drawn to J. Beckwith's publication in 'Nature' (vol. *224, p. 768, November 22nd, 1969)* on the isolation of lactose operon. I was similarly affected by the news that the Congress of the USA had granted 10 million dollars from the 1971 budget to support genetic scientists, represented by J. Lederberg. At that time I had already sensed that there were important events on the horizon in biology, and I began to prepare my research into molecular genetics, starting with prokaryotes.

My own Institute of Molecular Biology, at the USSR Academy of Sciences, had neither the equipment nor the personnel for such work. I, therefore, had to seek some kind of solution. Around 1970, my position changed. In 1968, I was elected as a correspondent member, and a year later was made an academician. Also in 1970, I was elected academician-secretary of the Department of Biochemistry, Biophysics, and Chemistry of physiologically active compounds, and a member of the Presidium of the Academy of Sciences of the USSR.

I only accepted this appointment after serious consideration. This election meant my moving away from the laboratory table, but at the same time it gave me the opportunity to influence the direction of the research into physico-chemical biology. It was difficult for me to come to terms with the fact that my age (I was 65) was proving to be a hindrance in experimental work. In the end, I decided to take on the duties of academician-secretary, hoping that I could be of some use in the scientific and organisational activities.

A few words about my work as academician secretary. At first glance, my duties seemed to be purely bureaucratic or administrative. In reality, however, they were neither one nor the other. This Department of the USSR Academy of Sciences combined at that time, just as it does today, 17 strong biological institutes, which focussed on physical and chemical approaches to the understanding of life. It was in such an environment, that molecular biology, molecular genetics, genetic engineering and the modern branches of biotechnology, were born and developed in our country. All of them developed between 1970–1988, in an exceptionally dynamic way. New institutes were established, many young people came into science, talented new individuals appeared and a lively flow of new investigations sprang up. This was an interesting time of creativity. And I was among the pioneers and initiators of this powerful movement.

The Council of Ministers of the USSR and The Central Committee of the CPSU adopted three decrees on the development of the above-mentioned sciences. These were as follows:

– 'On the measures of acceleration in the development of molecular biology and molecular genetics and the use of their achievements in the national economy and industry' (April 19th, 1974, No. 304)

– 'On the further development of physico-chemical biology and biotechnology and the application of their achievements in medicine, agriculture and industry.' (July 24th, 1981, No. 652)

– 'On the further development of new branches in biology and biotechnology' (August 26th, 1985, No. 607)

I was quite actively involved in the preparation of all these decrees, and my task was mainly to substantiate the scientific conceptions. All these decrees provided essential practical support for the new directions in biology, and had a great influence on their development in our country; this is still apparent today, even in such difficult times.

All the above-mentioned decrees testified to the fact that the Soviet Government and the Central Committee of the CPSU were giving their unreserved and unrestricted support to the new branches of biological science.

The stringent ideological control exercised by the party organisations, which in the past had been unrestrained, began to diminish, and from the beginning of the 70s, all but disappeared.

Generally speaking, it was mainly the humanities and the sociological sciences that had suffered the most from this control. In the field of biological science, it is well known that genetics suffered but this was a direct result of the influence of an overwhelming power – that of Stalin himself, who had decided that he should reform biology. During my time as the academician-secretary, the Department of Science of the Central Committee of the CPSU, officially had absolute power, it was not, however, actually involved in the subjects or the direction of scientific research into physico-chemical biology. Research projects were designed by the heads of the laboratories at the department institutes, and I cannot recall a single project that was altered by the Department or the Presidium and certainly never by the Science Department of the Central Committee of the CPSU. Because of this, the development of new directions in physico-chemical biology went ahead without any evident interference from State or Party circles. This does not mean, however, that the science itself was without shortcomings, errors, conservatism or deviations, but these were all mainly its own doing.

Quite a considerable part of the Department's work involved the coordination of biological research within the Academies of

Science of the Soviet Republics, and international cooperation. First and foremost, the latter concerned all countries within the so-called socialist camp, but at the same time it covered broader links with western countries. Particularly valuable was the cooperation in molecular biology with scientists from France (M. Gruenberg-Manago, J.P. Ebel), Germany (where Prof. H.G. Zachau was and still is the permanent organiser), and Italy (Prof. P. Volpe).

I worked as an academician secretary until 1988, when I became an advisor at the Presidium of the Russian Academy of Sciences, where I remain to this day.

Meanwhile, events in experimental science were developing according to plan. The year 1967 brought with it another event which was to prove beneficial to my projects. I accepted an invitation given by G.K. Skryabin, Director of the Institute of Biochemistry and Physiology of Microorganisms (IBPM) in Puschino (125 km from Moscow), to organise a laboratory of molecular biology and the genetics of microorganisms in his institute. This institute had everything necessary for working with microorganisms, and could, therefore, attract the appropriate researchers.

In the beginning, as the direction of research in the laboratory was not quite clear, we opted for genetic enzymology as our main task (we had considered the study of DNA polymerases, DNA ligases, and restriction endonucleases). The first research assistants arrived as early as 1969: they were V.I. Tanyashin, I. Fodor, A.I. Krutilina and L. Lee. We had barely even started to reach a firm concensus on the theme of our research before the year 1972 arrived, and with it the works of P. Berg, S.N. Cohen, and H.W. Boyer. Our laboratory at IBPM became the first centre of genetic engineering in the country, and before long, it was a Department, consisting of seven laboratories. We obtained the first non-intricate recombinant DNA molecule in 1975, and the first vector in 1977.

The systematic research into phages lambda, T2, T4, T5, endonucleases and ligases, DNA methylation in microorganisms, and the genetics of nitrogen fixation was first started in our laboratory. We spent a long time studying the primary structures of microbial tRNAs, before moving on to the structure of prokaryotic DNA.

It is not within the framework of my biography to review all the research carried out by the laboratory (later, Department). Gradually the young scientists became doctors of science, and heads of laboratories. With time, my role also changed, as I became less personally involved. It was the death of G.K. Skryabin, the permanent director of the Institute, that marked the end of my continuous, intensive work at the Institute and its gradual reduction. My connections with the laboratory, however, still remain to this day.

I expended a great deal of time and effort as genetic engineering was beginning to develop in our country. I presented several reports to a variety of audiences, wrote articles for popular journals and the general press, and on various occasions, letters to help popularise this new branch of biology. I organised three workshops in Puschino on May 4th 1972, June 4th 1973, and February 4th 1974, as well as symposia on chromosome structure, nucleic acids, and the molecular genetics of plasmids. These were of vital significance to the development of genetic engineering and molecular biology. I also wrote and published the first rules in our country on safety in working with recombinant DNAs.

Genetic engineering, which was originally perceived as an exotic explosion of ideas and laboratory technology, gradually moved onto firmer ground within many academic institutes (not only the Academy) and became the natural form of their research practices. My part in the first stages of this was to create the laboratory at IBPM of the USSR Academy of Sciences and to publicise it, implementing the above-mentioned organisational measures, which made my position as academician-secretary of the USSR Academy of Sciences easier.

Now for a few words about biotechnology. As soon as genetic engineering had achieved recognition, it was immediately followed by ideas for its application. At first this was confined to logic, and then our original suppositions were confirmed in practice. 'The DNA industry' became a household term. In my reports, I stressed that it would become possible to increase the production of amino acids, vitamins, antibiotics, nutrient protein, etc., in our country, based on the discoveries of cell and molecular biology. Whereas the channels of public information, such as public lectures and articles in the press, were used in order to advance the case of genetic engineering, in the case of biotechnology it took countless official and often confidential letters.

Biotechnology developed slowly, buried in the depths of the academic collectives (I am referring to those aspects which concerned me and were within my grasp) and only appeared officially in the Government decrees in 1981 (Decree of the Council of Ministers of the USSR and the Central Committee of the CPSU, July 24th, 1981).

In 1982, a Scientific Council on the problems of biotechnology was established at the USSR Academy of Sciences. I became its chairman, and left only recently – on February 12th, 1993. The Council was a platonic organisation: it had no finances of its own, and was dependent on IBPM. There was money available for biotechnology through the usual channels, but the practical activity of the Scientific Council was limited to its representation at various institutions, and the organisation of conferences and symposia. The research into biotechnology was strongly supported by conferences on new developments in biotechnology; the Fifth, and latest of these being in 1992. In 1993, the first All-Russian Congress on biotechnology took place. I believe that useful decisions were made, which were later to have practical significance. In 1994 special conferences on plant biotechnology began to be held.

Many institutes of the Russian Academy of Sciences carry out research into the fundamental problems of biotechnology. On the whole, however, this research has not produced the results we expected. As a result of the general economic collapse and partly because of mistakes made in the USSR (now Russian) biotechnological industry, I believe that the industry itself has fallen into a state of complete decay. This has in turn resulted in a breakdown in the natural process of transferring biotechnological developments from their laboratory stage to the industrial sectors.

But the general aspects of this problem would extend beyond the framework of my biography. They do, however, relate to it and I shall go on to mention those events which effected me personally.

In my laboratory in the Institute of Molecular Biology at the Russian Academy of Sciences, it was suggested that a study should be carried out into tissue plasminogen activator and calcitonin. This did not, however, arouse any enthusiasm among my colleagues, even though several articles had already been published on this subject. The initial research into the human somatotropin (growth hormone) was more successful. In 1980 cDNA was obtained from a hypophysis tumour of an acromegalic, and a recombinant system for hormone expression was created. In 1981 I gave a report on this at the Presidium at the USSR Academy of Sciences, but the results of the research were only published in 1984. The work on this hormone is now being continued in IMB RAS by P.M. Rubtsov, who was involved in this from the very beginning.

At the beginning of the 80's, the first signs of plant genetic engineering appeared and the first experiments began. Some new ideas emerged, but things were already developing without my direct participation.

In the end, my role in biotechnology was restricted to its promotion, overcoming various official obstacles, and initiating certain research. I was probably the only one ever emotionally involved at the beginning of the research into human somatotropin.

The fifth stage, which I mentioned earlier, is that of the human genome, and this belongs not to the past, but to the present. I devoted a great deal of energy – perhaps the last I had – to this period, and experienced many problems as a result of this. In the beginning, I was the only person actively involved in this enterprise, and I was unable to share either the work or the responsibility with anyone else. The transfer from genetic engineering to the human genome was not unnatural. It followed on naturally from previous research carried out by myself and my colleagues. In this I am referring not so much to its methodical aspects, but rather to its conceptual basis.

Even in the 60's, research into tRNA in my laboratory was concerned with the primary structure of nucleic acids, and it included an attempt to reveal their functional properties on the basis of structure. We only achieved partial success in this, but the idea itself stayed alive.

In the 1970s, we decided to move on in the laboratory from a simple object tRNA to one more complex and palatable from a biological point of view, such as DNA. This was carried out by K.G. Skryabin, a young member of my laboratory, who used a new technique to study the primary structures of yeast ribonucleic acid genes. This research led to an extensive series of works and an influx of young researchers, who to this day are still studying DNA structure.

During the 70s, the study of DNA developed at an increasing rate in many countries. At the end of this period, came the news that the Americans intended to sequence the human genome, and so be able to read all the instructions for a person's life, as imprinted in his DNA.

One of the first, and still fairly remote indications of this, was the news that the USA Congress of 1988 was to grant special funds for the study of the human genome. Our scientists were attracted to the works of R. Sinsheimer, Di Lisi [1985, 1987], to articles by R. Dulbecco (*Science, 231, p. 1055, 1986*), and the conferences held in Santa Cruz (1985), Cold Spring Harbor (1986), Los Alamos (1986), and other similar events. Everything testified to the serious intentions of the American scientists, and this prompted us to get to work. Our

science was prepared to take on this challenge, both theoretically and technologically.

In December 1987, I wrote a letter about the research into
the human genome, which I sent, via the academician I.T.
Frolov, directly to M.S. Gorbachev, asking for his support in
this study. Some time later, I received a positive reply. In
1988, we were invited to give a presentation to the Government on the materials of the programme for human genome
research. On August 31st 1987, the Council of Ministers of the
USSR (which still existed at that time) announced its decision
regarding this problem.

In 1989, 'The Human Genome' Programme became one of
the fourteen USSR State Programmes; all possible means
were employed to make the public aware of the content of the
programme; these included public lectures and articles in the
press. I myself gave a report to the Presidium of the Russian
Academy of Sciences, the Ministry of Science and Scientific
and Technical Politics, and the Parliament Information Centre of the Russian Federation.

A Scientific Council was established to lead the programme,
with A.A. Baev as Chairman, A.D. Mirzabekov as Vice-Chair-
man, and N.N. Belyaeva as Scientific Secretary. The Institute
of Molecular Biology became the administrative centre of the
organisation.

The research into the human genome extends beyond the
boundaries of my biography. I should perhaps mention, however, the three principles that formed the basis of 'The Human
Genome' Programme from the very beginning. These are as
follows:

1. Publicity – information concerning all the activities of the
Council was made available to the public in regularly issued
information bulletins and other sources.

2. Grants – here, for the first time in our country this system
was used as a single method for distributing money between

researchers. The financing was carried out directly through the Ministry of Science and came as an additional budget, which was independent from that of the Academy of Sciences of the USSR.

3. Democracy – meaning that all decisions were collective (through the Council). Grants were distributed according to results of competition and the decisions of specialist commissions, finally approved by the Council.

Research into the human genome was very well funded in the beginning (32,000,000 roubles in 1990), and this was only for operation expenses, since the researchers' salaries and general expenses came out of the Academy budget. The Programme received $ 10,000,000 to purchase foreign reagents and equipment. The present-day situation is critical. Due to inflation, roubles obtained under the programme have been devalued and there is no hard currency at all, but still the work continues.

In a few months I shall be ninety years old, and it is likely that the study of the human genome will be my last. The time has come to move on to my conclusions.

Looking back at my whole life, which has been long and complicated, I am left with the impression that much of it has been dictated by the events and impressions of my childhood and youth, in spite of all the grim events throughout our Russian history and my private life.

I tried to become a doctor, and finally achieved this. Such was the wish of my late father who died early, and as a lawyer had the highest respect for the doctor's profession.

My last two years at school played an important part in my life, when philosophical and sociological problems were the subjects of lively discussions among my schoolmates. When I was in the last grade, we were taught a subject called 'An introductory course to philosophy', because of which I became familiar with philosophical problems for the rest of my life, or at least for a significant part of it. The teacher, V. Panova, was a firm follower of Indian Yoga. Thanks to her, while still at school, I became acquainted with some books by Yoga Ramacharaki, but I did not become a follower – in fact quite the reverse. I was strongly impressed by Ernest Heckel's 'World Puzzles'. Thanks to them (I can say this quite sincerely) an interest in the general problems of biology grew and the basis of my materialistic world outlook was consolidated for the rest of my life.

When I became a student of the Natural Science Branch of the Physico-Mathematical Department of Kazan University, I already had a lively and active interest in the general problems of biology and philosophy. This also stayed with me while I was a postgraduate student of the Kazan Medical Institute. I was an avid listener and active member of a course on dialectical materialism which was oriented towards biological problems. It was delivered to the postgraduate biologists of Kazan higher educational institutions by V.N. Slepkov, the instigator of my future misfortunes in life.

Moreover, in 1933 and 1934 I ran an optional course on the general problems of medicine for an audience of the Advanced Courses for medical doctors in Kazan, and it was quite a success. Unfortunately, not one of my notes remains in my archives; everything was confiscated at the time of my arrest in Moscow.

The way my life worked out was not favourable for the development and maintenance of my interests in the general problems of biology, science and philosophy. Of course, I was still interested in these problems and everything relating to them, but to my regret, they could not be properly developed later, as there was never any physical opportunity for this to happen in my life. This was only one of the alternative routes of spiritual evolution that my life could have taken, but did not. Nobody knows where it could have brought me, but this route, so dangerous in our country, was never taken. I have mentioned above that the events in my life seemed stronger than my wishes, and my fate was a manifestation of this over-whelming power. My life was such that it gave me specific opportunities for practical and scientific activity.

Trying to look at myself critically from the outside is not always successful and often meets with disapproval. Some-times signs of self-congratulation are evident in the text, or the reverse, my estimates are lower than in reality, or there is evidence of a desire to justify some biographical episodes.

In the declining years of life, one's enthusiasm and emotions become calmer; appraisals become unnecessary and reproofs are almost meaningless, because you cannot change the past and the future may disappear any moment. This is why I feel justified in saying a few conclusive words.

I do not feel any particular satisfaction in summing up my life, although I do not blame myself or renounce what I did. Everything, or almost everything, in my actions had sufficient and evident grounds. Periods of sadness from my past, regrets concerning unfulfilled opportunities and unfinished actions, come to my memory more often than do images of the opposite kind. Although, on the other hand, cold reason questions whether it would have been possible to choose something es-sentially better. And one's memory brings back to life the events which brought spiritual satisfaction and joy.

The most attractive and memorable aspect of my scientific activity was the research into the primary structure of valine tRNA and 'cut molecules'; especially the latter. The ideas and methodical approaches used here, anticipated the molecular engineering (of hereditary molecules, proteins) which is now being developed in so many different and interesting ways. And in actual fact, the idea of 'cut molecules' included not only the assembly of tRNA fragments, but also their shortening. Further to this, we planned the splicing and substitution of individual nucleotide links. This approach would have created a whole direction in itself, but (at least for me) it remained a single episode and nothing more.

When I became academician-secretary of the Academy Department, I was enthusiastic about the possibilities for the development of genetic engineering in our country and quite consciously chose this route. My age gave me cause to hurry and limited my strength, which was insufficient to cope with all that attracted me. Now, with everything behind me, I regret that I sacrificed all my projects to the 'common weal', which as is well-known, rarely reaps its reward. From this example, it can be understood that in my declining years I evaluate events of the past differently than I did at the time.

One more remark concerning my scientific activity. I have already mentioned that I preferred and enjoyed working with my own hands. It was only the experiments with ATP of nuclear avian erythrocytes, however, that I actually carried out with my own hands. In the experiments with valine tRNAs, I did my share of manual work and all the rest was done by my colleagues. My input was only that of intellect and pen. Such was the ratio of desire to the outcome in reality. Instead of things working out according to my ideas and requirements, my life was carried away by a turbulent stream and, like my whole generation, was a toy in the midst of powerful events such as wars, revolutions, social quakes and ideological extremes. External events dictated the necessary behaviour and influenced one's psychology, but did not completely suppress a person's spiritual individuality. In spite of all the dramatic events, the possibility always remained – albeit not to any large extent – of putting forward the word 'I' and acting in accordance with this. This was possible not only because of my persistence, but also because, despite all the cunning of the despotic power in its attempts to build individual lives according to strictly determined patterns of behaviour, there were always some ecological niches for dissidents. Inevitably we had to search for such a niche, or sometimes even create it.

I think that in my case, my compromises never went beyond the bounds of what was morally permissable, although my joining the CPSU may be interpreted in different ways. I cannot accuse myself of doing anything for the sake of my career or material welfare. Usually I was driven first and foremost by my scientific interests or a longing to be useful to society and mankind. The latter, i.e. good directed towards concrete individuals, was a beneficial stimulus to my medical work and it provided me with moral support during my life's difficult moments.

Once, in conversation, someone asked me why I did not become a dissident. I told him that I believed that only a special kind of people became dissidents. By nature I was neither a denouncer, nor a searcher for truth, nor a preacher, nor a conspirator, even though it had been announced that I was a member of a terrorist underground organisation. The only kind of activity acceptable to me was the creation of something positive and real, something that was useful to my native country, to society and to man. I cannot and do not want to act in any other way.

My scientific activity was characterized by its own logic, although this did not always totally correspond with my ideals. In my last years, a feeling of haste arose, a desire to cover everything that seemed interesting: the years were passing and I was in a rush to do things while still alive.

I devoted almost two decades of my life to scientific and organisational activity. But was this a substitute for scientific activity, or was it something that was really valuable? It was not done for honour or for money; it was rather my showing a tendency towards doing something useful which was within my power.

I have already spoken about my 17-year separation from science. I was no more than an ordinary victim of political struggle, passions, vice and crime. People living in the Soviet Union resemble crocodile birds that have got used to visiting the crocodile's mouth, but occasionaly this mouth slams shut. This is what happened to me. In our country, justice and morality were determined by one's duty to the State and the CPSU. According to the official terminology, these duties were to be useful towards the construction of socialism. Nobody took

pity on the people. On the subject of the idea of the common
weal, Lev Tolstoy said: 'for your own sake you would not decide
to kill a human being – it is easier for the sake of your neighbour; and for the sake of the 'common good' thousands are
killed'.

It would be only natural to conclude my reminiscences with
an answer to the question: was the author happy? Only very
rarely (or perhaps never) does the whole of my life appear in
my memories as one continuous happy song. For the most
part, it is a question of separate situations which differ in
length of time, and which compensate in a man's consciousness for all the grief, losses and bitter mistakes. They are only
episodes in his life.

As I see it, happiness is not found in simple satisfaction with
the material aspects of life: prosperity, comfort, and an attractive official and social position. I had all of these in my life –
but they are not important.

Above, all, happiness is a state of mind, the harmony of your
own internal world with that around you. It requires an intui-
tive feeling and perception of the value and uniqueness of your
life and its expressions. It can be brought on by short-lived,
accidental, and even what appear at first to be insignificant
events. Sometimes, I experienced moments or hours of happiness under the most unlikely conditions.

Happiness is a state of mind that requires neither logic nor
analysis: it is something that is spontaneous. Elements of irrationality can be found in this interpretation, but I am convinced that neither the irrationality nor the spirituality of the
world does not contradict its materiality.

In answer to the above question, I should conclude by saying: fortune did not entirely pass me by – it even gave me
moments of happiness. In most cases it seems, the course of
life is stronger than the individual. I seldom used to pick up a
Bible, but now, in moments of contemplation, I sometimes do
so. In one of the passages I have read, the prophet Ecclesiastes
says, 'as the fishes that are taken in an evil net, and as the

birds that are caught in the snare; so are the sons of men snared in an evil time, when it falleth suddenly upon them.'

Seventy years of CPSU supremacy in Russia did not create a prosperous and morally-healthy society. This is not only true of the past, but even after the events of 1985, it is still the case. Eliminating the repressive structure of the Stalin era immediately gave rise to a terrible growth in crime, profiteering, corruption, dirty politics, lies, etc. For a short time after the revolution of 1917, some traces of the social ideals from the end of the last century and the beginning of this, could still be found. Now I do not think anybody believes in them.

I must end my reminiscences on this sad note: it cannot, unfortunately, be helped. I have never once departed from reality, right up to this point, as I finish my manuscript and lay my pen on the desk.

Name Index

Datta, P., 425
Davies, R.E., 311
Davis, B.D., 271
Davson, H., 408
Dean, A., 391
De Duve, C., 52
De Matteis, F., 389, 390
de Vries, J., 66
de Vries, S., 251
Deboben, A., 84
Desnuelle, C., 90
Desnuelle, P., 32
Determann, H., 75
Devenish, R.J., 342
Dickens, F.D., 411, 415
Dixon, M., 355
Doniach, D., 415
Dose, K., 62
Dostoievsky, F.M., 361
Doy, C., 272
Dresel, E.I.B., 376, 387, 388, 389
Dreyer, J.-L., 234
Du Toit, P.J., 368
du Vigneaud, V., 150
DuBuy, M.G., 313
Dunham, D., 231

Eales, L., 390, 393, 394
Ebel, J.P., 467
Eckermann, G., 326
Eckstein, F., 180
Edelman, G., 151
Edsall, J.T., 356
Ehrenberg, A., 211
Ehrlich, H., 122
Ehrlich, P., 57, 135, 155
Eigen, M., 173
Einstein, A., 157
Eisen, H.N., 169
Eisenhower, D.D., 128
Ek, A., 142
Ellinger, P., 136

Elson, E.L., 183
Embden, G., 8
Emptage, M., 234
Engelhardt, V.A., 444, 445, 453, 454, 457, 458
Ennor, Sir Hugh, 278
Ernster, E., 329

Fahrenholz, F., 82
Fairley, N.H., 382
Falk, J.E., 376, 385, 387, 388, 389
Farrant, J.L., 346
Farrer-Brown, H., 387
Faulstich, H., 60, 70, 80
Feder, D., 181
Feldmann, K., 175
Fieser, L.F., 57, 130
Fischer, E.H., 43, 50, 138, 171, 174, 360
Fischer, H., 30, 130, 167, 365, 366, 375, 376, 377, 379, 380
Fisher, E., 41
Fiume, L., 62, 64
Florey, H., 266
Fodor, I., 467
Fourie, P.J.J., 377
Frank, J., 197
Freise, V., 125
Frieden, C., 169
Frimmer, U., 65, 68
Froncisz, W., 213
Fukui, T., 177

Gaffron, H., 197, 198
Gale, E., 295
Garland, P., 345
Gattermann, L., 23
Gebert, U., 70
Gerlach, W., 121
Gibson, F., 342
Gibson, J.F., 222
Gibson, M.I., 274

Woods, D.D., 266, 267
Woods, M.W., 313
Woodward, R.B., 55, 88, 130, 133
Work, T.S., 409, 410
Wright, B.J., 339
Wünsch, E., 52

Yanagida, T., 83
Yanovsky, C., 295
Yeats, W.B., 112
Yonemitsu, O., 145
Yonetani, T., 211

Young, F.G. 407, 422, 428
Young, I.G., 277
Young, S.E., 339

Zachau, H.G., 74, 467
Zalkin, H., 241
Zamecnik, P.Z., 413
Zanotti, G., 80
Zheng, L., 242
Zumft, W.G., 214
Zweig, S., 113